Lecture Notes in Statistics 141

Edited by P. Bickel, P. Diggle, S. Fienberg, K. Krickeberg,
I. Olkin, N. Wermuth, S. Zeger

Springer

New York
Berlin
Heidelberg
Barcelona
Hong Kong
London
Milan
Paris
Singapore
Tokyo

Peter Müller
Brani Vidakovic (Editors)

Bayesian Inference in Wavelet-Based Models

 Springer

Peter Müller
Institute of Statistics and Decision Sciences
Duke University
Box 90251
Durham, NC 27708-0251

Brani Vidakovic
Institute of Statistics and Decision Sciences
Duke University
Box 90251
Durham, NC 27708-0251

CIP data available.
Printed on acid-free paper.

Camera ready copy provided by the editors.
Printed and bound by Braun-Brumfield, Ann Arbor, MI.
Printed in the United States of America.

9 8 7 6 5 4 3 2 1

ISBN 0-387-98885-8 Springer-Verlag New York Berlin Heidelberg   SPIN 10731344

Preface

This volume presents an overview of Bayesian methods for inference in the wavelet domain. The papers in this volume are divided into six parts: The first two papers introduce basic concepts. Chapters in Part II explore different approaches to prior modeling, using independent priors. Papers in the Part III discuss decision theoretic aspects of such prior models. In Part IV, some aspects of prior modeling using priors that account for dependence are explored. Part V considers the use of 2-dimensional wavelet decomposition in spatial modeling. Chapters in Part VI discuss the use of empirical Bayes estimation in wavelet based models. Part VII concludes the volume with a discussion of case studies using wavelet based Bayesian approaches.

The cooperation of all contributors in the timely preparation of their manuscripts is greatly recognized. We decided early on that it was important to referee and critically evaluate the papers which were submitted for inclusion in this volume. For this substantial task, we relied on the service of numerous referees to whom we are most indebted.

We are also grateful to John Kimmel and the Springer-Verlag referees for considering our proposal in a very timely manner. Our special thanks go to our spouses, Gautami and Draga, for their support.

Durham, NC   *Peter Müller and Brani Vidakovic*

April 1999

Contents

IV   PRIOR MODELS - DEPENDENT CASE

V   SPATIAL MODELS

VI EMPIRICAL BAYES

VII CASE STUDIES

Contributors

Felix Abramovich, *felix@math.tau.ac.il*, Department of Statistics and Operations Research, School of Mathematical Sciences, Tel Aviv University, Israel.

Omar Aguilar, *omar@stat.duke.edu*, Global Investments, CDC Investments Co., New York, NY.

John D. Albertson, *dalbertson@virginia.edu*, University of Virginia, Charlottesville, VA.

Mark Berliner, *mb@stat.ohio-state.edu*, Department of Statistics, Ohio State University, Columbus, OH.

Adhemar Bultheel, *Adhemar.Bultheel@cs.kuleuven.ac.be*, Department of Computer Science, Katholieke Universiteit Leuven, Heverlee, Belgium.

Hugh A. Chipman, *hachipma@uwaterloo.ca*, Department of Statistics and Actuarial Science, University of Waterloo, Canada.

Merlise Clyde, *clyde@stat.duke.edu*, Institute of Statistics and Decision Sciences, Duke University, Durham, NC.

Fabio Corradi, *corradi@ds.unifi.it*, Dipartimento di Statistica "G. Parenti", Universitá di Firenze, Italy.

Noel Cressie, *ncressie@stat.ohio-state.edu*, Department of Statistics, Ohio State University, Columbus, OH.

Edward I. George, *egeorge@mail.utexas.edu*, Department of MSIS, University of Texas, Austin, TX.

Hsin-Cheng Huang, *hchuang@stat.sinica.edu.tw*, Institute of Statistical Science, Academia Sinica, Taipei, Taiwan.

Maarten Jansen, *Maarten.Jansen@cs.kuleuven.ac.be*, Department of Computer Science, Katholieke Universiteit Leuven, Heverlee, Belgium.

Jerome Kalifa, *kalifa@cmapx.polytechnique.fr*, Centre de Mathematiques Appliquees, Ecole Polytechnique, 91128 Palaiseau Cedex, France.

Gabriel Katul, *gaby@duke.edu*, School of the Environment, Duke University, Durham, NC.

Robert Kohn, *robertk@agsm.unsw.edu.au*, Australian Graduate School of Management, University of New South Wales, Sydney, N.S.W. Australia.

Eric Kolaczyk, *kolaczyk@math.bu.edu*, Department of Mathematics and Statistics, Boston University, MA.

Hamid Krim, *ahk@eos.ncsu.edu*, Electrical and Computer Engineering Department, North Carolina State University, Raleigh, NC.

David Leporini, *David.Leporini@lss.supelec.fr*, CNRS, Supélec, associé à l'Université Paris Sud, Paris, France.

Jim Lynch, *lynch@stat.sc.edu*, Department of Statistics, University of South Carolina, Columbia, SC.

Stephane Mallat, *mallat@cmapx.polytechnique.fr*, Centre de Mathematiques Appliquees, Ecole Polytechnique, 91128 Palaiseau Cedex, France.

J. Stephen Marron, *marron@stat.unc.edu*, Department of Statistics, University of North Carolina, Chapel Hill, NC.

Ralph Milliff, *milliff@ncar.ucar.edu*, National Center for Atmospheric Research, Boulder, CO.

Peter Müller, *pm@stat.duke.edu*, Institute of Statistics and Decision Sciences, Duke University, Durham, NC.

Robert D. Nowak, *nowak@egr.msu.edu*, Electrical Engineering, Michigan State University, East Lansing, MI.

Todd Ogden, *ogden@stat.sc.edu*, Department of Statistics, University of South Carolina, Columbia, SC.

Luis Pastor, *lpastor@escet.urjc.es*, ESCET, Universidad Rey Juan Carlos, Madrid, Spain.

Marianna Y. Pensky, *mpensky@pegasus.cc.ucf.edu*, Department of Mathematics, University of Central Florida, Orlando, FL.

Jean-Christophe Pesquet, *Jean-Christophe.Pesquet@lss.supelec.fr*, CNRS, Supélec, associé à l'Université Paris Sud, Paris, France.

David Ríos Insua, *d.rios@escet.urjc.es*, ESCET, Universidad Rey Juan Carlos, Madrid, Spain.

Angel Rodríguez, *arodri@dtf.fi.upm.es*, Dpto. Tecnologia Fotonica, Facultad de Informatica, Universidad Politecnica de Madrid, Spain.

Fabrizio Ruggeri, *fabrizio@iami.mi.cnr.it*, Consiglio Nazionale delle Ricerche, Istituto per le Applicazioni della Matematica e dell'Informatica, Milano, Italy.

Theofanis Sapatinas, *T.Sapatinas@ukc.ac.uk*, Institute of Mathematics and Statistics at the University of Kent at Canterbury, U.K.

Eero Simoncelli, *eero.simoncelli@nyu.edu*, Center for Neural Sciences and Courant Institute for Mathematical Sciences, New York University, NY.

Marina Vannucci, *mvannucci@stat.tamu.edu*, Department of Statistics, Texas A&M University, College Station, TX.

Brani Vidakovic, *brani@stat.duke.edu*, Institute of Statistics and Decision Sciences, Duke University, Durham, NC.

Yazhen Wang, *yzwang@stat.uconn.edu*, Department of Statistics, University of Conneticut, Storrs, CT.

Chris Wikle, *wikle@stat.missouri.edu*, Department of Statistics, University of Missouri, Columbia, MO.

Lara J. Wolfson, *ljwolfso@uwaterloo.ca*, Department of Statistics and Actuarial Science University of Waterloo, Canada.

Paul Yau, *pauly@agsm.unsw.edu.au*, Australian Graduate School of Management, University of New South Wales, Sydney, Australia.

1

An Introduction to Wavelets

Brani Vidakovic and Peter Müller

1.1 What are Wavelets

Wavelets are functions that satisfy certain requirements. The very name *wavelet* comes from the requirement that they should integrate to zero, "waving" above and below the x-axis. The diminutive connotation of *wavelet* suggest the function has to be well localized. Other requirements are technical and needed mostly to ensure quick and easy calculation of the direct and inverse wavelet transform.

There are many kinds of wavelets. One can choose between smooth wavelets, compactly supported wavelets, wavelets with simple mathematical expressions, wavelets with simple associated filters, etc. The simplest is the *Haar wavelet,* and we discuss it as an introductory example in the next section. Examples of some wavelets (from the family of Daubechies' wavelets) are given in Figure 1. Like sines and cosines in Fourier analysis, wavelets are used as basis functions in representing other functions. Once the wavelet function (sometimes informally called *the mother wavelet*) $\psi(x)$ is fixed, one can form translations and dilations of the mother wavelet $\{\psi(\frac{x-b}{a}), (a, b) \in R^+ \times R\}$. It is convenient to take special values for a and b in defining the wavelet basis: $a = 2^{-j}$ and $b = k \cdot 2^{-j}$, where k and j are integers. This choice of a and b is called *critical sampling* and generates a sparse basis. In addition, this choice naturally connects multiresolution analysis in discrete signal processing with the mathematics of wavelets.

Wavelet novices often ask, why not use the traditional Fourier methods? There are some important differences between Fourier analysis and wavelets. Fourier basis functions are localized in frequency but not in time. Small frequency changes in the Fourier transform will produce changes everywhere in the time domain. Wavelets are local in both frequency/scale (via dilations) and in time (via translations). This localization is an advantage in many cases.

Second, many classes of functions can be represented by wavelets in a more compact way. For example, functions with discontinuities and functions with sharp spikes usually take substantially fewer wavelet basis functions than sine-cosine basis functions to achieve a comparable approximation.

This parsimonious coding makes wavelets excellent tools in data compres-

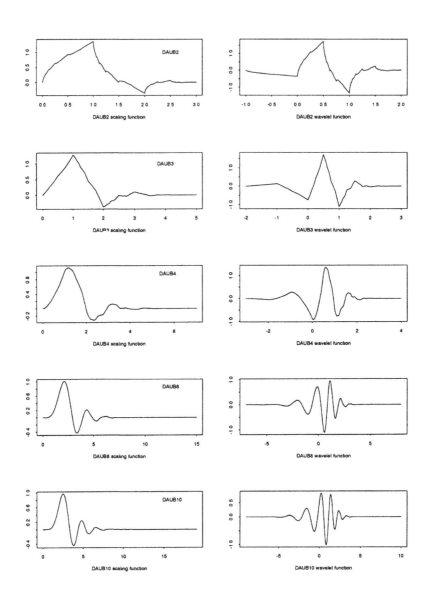

FIGURE 1. Wavelets from the Daubechies family (Daubechies, 1988)

sion. For example, the FBI has standardized the use of wavelets in digital fingerprint image compression. The compression ratios are on the order of 20:1, and the difference between the original image and the decompressed one can be told only by an expert.

This already hints at how statisticians can benefit from wavelets. Large and noisy data sets can be easily and quickly transformed by the discrete wavelet transform (the counterpart of the discrete Fourier transform). The data are coded by the wavelet coefficients. In addition, the epithet "fast" for Fourier transform can, in most cases, be replaced by "faster" for the wavelets. It is well known that the computational complexity of the fast Fourier transformation is $O(n \cdot \log_2(n))$. For the fast wavelet transform the computational complexity goes down to $O(n)$.

Many data operations can now be done by processing the corresponding wavelet coefficients. For instance, one can do data smoothing by shrinking the wavelet coefficients and then returning the shrunk coefficients to the "time domain" (Figure 2). The definition of a simple shrinkage method, thresholding, and some thresholding policies are given in Section 3. A Bayesian approach to shrinkage will be discussed later in the volume.

Figure 2: Data Analysis by Wavelets: Donoho and Johnstine's Shrinkage Paradigm.

1.2   How Do the Wavelets Work

1.2.1   The Haar Wavelet

To explain how wavelets work, we start with an example. We choose the simplest and the oldest of all wavelets (we are tempted to say: mother of all wavelets!), the Haar wavelet, $\psi(x)$. It is a step function taking values 1 and -1, on $[0, \frac{1}{2})$ and $[\frac{1}{2}, 1)$, respectively. The graph of the Haar wavelet is given in Figure 3.

The Haar wavelet has been known for more than eighty years and has been used in various mathematical fields. Any continuous function can be approximated uniformly by Haar functions.   Dilations and translations of the function ψ, $\psi_{jk}(x) = \text{const} \cdot \psi(2^j x - k)$, define an orthogonal basis of $L^2(R)$ (the space of all square integrable functions). This means that any element in $L^2(R)$ may be represented as a linear combination (possibly infinite) of these basis functions. The orthogonality of ψ_{jk} is easy to check. The constant that makes this orthogonal basis orthonormal is $2^{j/2}$. The functions $\psi_{10}, \psi_{11}, \psi_{20}, \psi_{21}, \psi_{22}, \psi_{23}$ are depicted in Figure 3. The set $\{\psi_{jk}, j \in Z, k \in Z\}$ defines an orthonormal basis for L^2. Alternatively we will consider orthonormal bases of the form $\{\phi_{j_0,k}, \psi_{jk}, j \geq j_0, k \in Z\}$,

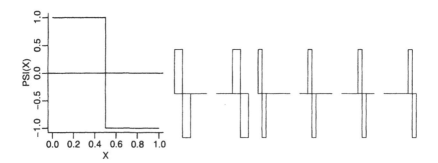

FIGURE 3. Haar wavelet $\psi(x)$ (left panel) and dilations and translations $\psi_{jk}(x)$, $j = 1, 2$, $k = 0, \ldots, 2^j - 1$ (rescaled)

where ϕ is called the *scaling function* associated with the wavelet basis ψ_{jk}, and $\phi_{jk}(x) = 2^{j/2}\phi(2^j x - k)$. The set $\{\phi_{j_0 k}, k \in Z\}$ spans the same subspace as $\{\psi_{jk}, j < j_0, k \in Z\}$. We will later make this statement more formal and define ϕ_{jk}. For the Haar wavelet basis the scaling function is very simple. It is unity on the interval $[0,1)$, i.e. $\phi(x) = \mathbf{1}(0 \leq x < 1)$.

The statistician may be interested in wavelet representations of functions generated by data sets. Let $y = (y_0, y_1, \ldots, y_{2^n - 1})$ be the data vector of size 2^n. The data vector can be associated with a piecewise constant function f on $[0,1)$ generated by y as follows, $f(x) = \Sigma_{k=0}^{2^n - 1} y_k \cdot \mathbf{1}(k2^{-n} \leq x < (k + 1)2^{-n})$. The (data) function f is obviously in the $L^2[0, 1)$ space, and the wavelet decomposition of f has the form

$$f(x) = c_{00}\phi(x) + \Sigma_{j=0}^{n-1}\Sigma_{k=0}^{2^j - 1}d_{jk}\psi_{jk}(x). \tag{1}$$

The sum with respect to j is finite because f is a step function, and everything can be exactly described by resolutions up to the $(n - 1)$-st level. For each level the sum with respect to k is also finite because the domain of f is finite. In particular, no translations of the scaling function ϕ_{00} are required.

We fix the data vector y and find the wavelet decomposition (1) explicitly. Let $y = (1, 0, -3, 2, 1, 0, 1, 2)$. The corresponding function f is given in Figure 4. The following matrix equation gives the connection between y and the wavelet coefficients. Note the constants 2^j $(1, \sqrt{2}$ and $2)$ with Haar wavelets on the corresponding resolution levels $(j=0, 1,$ and $2)$.

$$
\begin{bmatrix} 1 \\ 0 \\ -3 \\ 2 \\ 1 \\ 0 \\ 1 \\ 2 \end{bmatrix}
=
\begin{bmatrix}
1 & 1 & \sqrt{2} & 0 & 2 & 0 & 0 & 0 \\
1 & 1 & \sqrt{2} & 0 & -2 & 0 & 0 & 0 \\
1 & 1 & -\sqrt{2} & 0 & 0 & 2 & 0 & 0 \\
1 & 1 & -\sqrt{2} & 0 & 0 & -2 & 0 & 0 \\
1 & -1 & 0 & \sqrt{2} & 0 & 0 & 2 & 0 \\
1 & -1 & 0 & \sqrt{2} & 0 & 0 & -2 & 0 \\
1 & -1 & 0 & -\sqrt{2} & 0 & 0 & 0 & 2 \\
1 & -1 & 0 & -\sqrt{2} & 0 & 0 & 0 & -2
\end{bmatrix}
\cdot
\begin{bmatrix} c_{00} \\ d_{00} \\ d_{10} \\ d_{11} \\ d_{20} \\ d_{21} \\ d_{22} \\ d_{23} \end{bmatrix}
$$

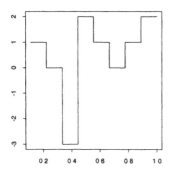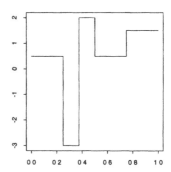

FIGURE 4. "Data function" on $[0,1)$ (left panel) and smoothed sequence (right panel, see discussion in Section 1.3).

The solution is

$$(c_{00}, d_{00}, d_{10}, d_{11}, d_{20}, d_{21}, d_{22}, d_{23})' = (\frac{1}{2}, -\frac{1}{2}, \frac{1}{2\sqrt{2}}, -\frac{1}{2\sqrt{2}}, \frac{1}{4}, -\frac{5}{4}, \frac{1}{4}, -\frac{1}{4})'.$$

Thus,

$$f = \frac{1}{2}\phi - \frac{1}{2}\psi_{00} + \frac{1}{2\sqrt{2}}\psi_{10} - \frac{1}{2\sqrt{2}}\psi_{11} + \frac{1}{4}\psi_{20} - \frac{5}{4}\psi_{21} + \frac{1}{4}\psi_{22} - \frac{1}{4}\psi_{23} \quad (2)$$

The solution is easy to check. For example, when $x \in [0, \frac{1}{8})$,

$$f(x) = \frac{1}{2} - \frac{1}{2} \cdot 1 + \frac{1}{2\sqrt{2}} \cdot \sqrt{2} + \frac{1}{4} \cdot 2 = 1.$$

An obvious disadvantage of the Haar wavelet is that it is not continuous, and therefore choice of the Haar basis for representing smooth functions, for example, is not natural and economic.

The reader may already have the following question ready: "What will we do for vectors y of much bigger length?" Obviously, solving the matrix equations becomes impossible.

1.2.2 Mallat's Multiresolution Analysis, Filters, and Direct and Inverse Wavelet Transformation

The Multiresolution Analysis (MRA) is a tool for a constructive description of different wavelet bases (Mallat, 1989). We start with the space $L^2(R)$ of all square integrable functions.[1] The MRA is an increasing sequence of closed subspaces $\{V_j\}_{j \in Z}$ which approximate $L^2(R)$.

---

[1] A function f is in $L^2(S)$ if $\int_S f^2$ is finite.

Everything starts with a clever choice of the *scaling function* ϕ. Except for the Haar wavelet basis for which ϕ is the characteristic function of the interval $[0, 1)$, the scaling function is chosen to satisfy some continuity, smoothness and tail requirements. But, most importantly, the family $\{\phi_{0k}(x) = \phi(x - k), \ k \in Z\}$ constitutes an orthonormal basis for the *reference* space V_0. The following relations describe the analysis.

MRA 1 $\cdots \subset V_{-1} \subset V_0 \subset V_1 \subset \cdots$
The spaces V_j are nested. The space $L^2(R)$ is the closure of the union of all V_j. In other words, $\cup_{j \in Z} V_j$ is dense in $L^2(R)$. The intersection of all V_j is empty.

MRA 2 $f(x) \in V_j \Leftrightarrow f(2x) \in V_{j+1}, \ j \in Z.$
The spaces V_j and V_{j+1} are "similar." If the space V_j is spanned by $\phi_{jk}(x), k \in Z$ then the space V_{j+1} is spanned by $\phi_{j+1,k}(x), k \in Z$. The space V_{j+1} is generated by the functions $\phi_{j+1,k}(x) = \sqrt{2}\phi_{jk}(2x)$.

We now explain how the wavelets enter the picture. Because $V_0 \subset V_1$, any function in V_0 can be written as a linear combination of the basis functions $\phi_{1k}(x) = \sqrt{2}\phi(2x - k)$ of V_1. In particular:

$$\phi(x) = \Sigma_k h(k)\phi_{1k}(x). \tag{3}$$

The coefficients $h(k)$ are defined as $\langle \phi(x), \sqrt{2}\phi(2x - k)\rangle$. Consider now the orthogonal complement W_j of V_j to V_{j+1} (i.e. $V_{j+1} = V_j \oplus W_j$). Define

$$\psi(x) = \sqrt{2}\Sigma_k(-1)^k h(-k + 1)\phi(2x - k). \tag{4}$$

It can be shown that $\{\sqrt{2}\psi(2x - k), k \in Z\}$ is an orthonormal basis for W_1.[2]

Again, the similarity property of MRA gives that $\psi_{jk}(x) = \{2^{j/2}\psi(2^j x - k), k \in Z\}$ is a basis for W_j. Since $\cup_{j \in Z} V_j = \cup_{j \in Z} W_j$ is dense in $L_2(R)$, the family $\{\psi_{jk}(x) = 2^{j/2}\psi(2^j x - k), j \in Z, k \in Z\}$ is a basis for $L^2(R)$.

For a given function $f \in L^2(R)$ one can find N such that $f_N \in V_N$ approximates f up to preassigned precision (in terms of L_2 closeness). If $g_i \in W_i$ and $f_i \in V_i$, then

$$f_N = f_{N-1} + g_{N-1} = \Sigma_{i=1}^M g_{N-i} + f_{N-M}. \tag{5}$$

Equation (5) is the MRA decomposition of f. For example, the data function in Figure 4 is in V_n, if we use the MRA corresponding to the Haar wavelet. Note that $f \equiv f_n$ and $f_0 = 0$.

---

[2]This can also be expressed in terms of Fourier transformations as follows: Let $m_0(\omega)$ be the Fourier transformation of the sequence $1/\sqrt{2} \ h(n), n \in Z$, i.e. $m_0(\omega) = 1/\sqrt{2} \ \Sigma_n h(n)e^{in\omega}$. In the "frequency domain" the relation (3) is $\hat{\phi}(\omega) = m_0(\frac{\omega}{2})\hat{\phi}(\frac{\omega}{2})$. If we define $m_1(\omega) = e^{-i\omega}m_0(\omega + \pi)$ and $\hat{\psi}(\omega) = m_1(\frac{\omega}{2})\hat{\phi}(\frac{\omega}{2})$, then the function ψ corresponding to $\hat{\psi}$ is a *wavelet associated with the* MRA.

We repeat the multiresolution analysis story in the language of signal processing theory. Mallat's multiresolution analysis is connected with so called "pyramidal" algorithms in signal processing. Also, "quadrature mirror filters" are hidden in the MRA.

Recall from the previous section that

$$\phi(x) = \Sigma_{k \in Z} h(k)\sqrt{2}\phi(2x - k) \text{ and } \psi(x) = \Sigma_{k \in Z} g(k)\sqrt{2}\phi(2x - k). \quad (6)$$

The l^2 sequences[3] $\{h(k), k \in Z\}$ and $\{g(k), k \in Z\}$ are *quadrature mirror filters* in the terminology of signal analysis. The connection between h and g is given by: $g(n) = (-1)^n h(1 - n)$ (compare with equation (4)).

The sequence $h(k)$ is known as a *low pass* or *low band* filter while $g(k)$ is known as the *high pass* or *high band* filter. The following properties of $h(n), g(n)$ can be proven by using Fourier transforms and orthogonality: $\Sigma h(k) = \sqrt{2}, \ \Sigma g(k) = 0$.

The most compact way to describe the Mallat's MRA as well to give effective procedures of determining the wavelet coefficients is the *operator representation of filters*. For a sequence $a = \{a_n\}$ the operators H and G are defined by the following coordinatewise relations:

$$(Ha)_n = \Sigma_k h(k - 2n)a_k \text{ and } (Ga)_n = \Sigma_k g(k - 2n)a_k.$$

The operators H and G correspond to one step in the wavelet decomposition. The only difference is that the above definitions do not include the $\sqrt{2}$ factor as in equations (6).

Denote the original signal by $\underline{c}^{(n)}$. If the signal is of length 2^n, then $\underline{c}^{(n)}$ corresponds (in Haar's MRA) to the function $f(x) = \Sigma \underline{c}_k^{(n)} \phi_{nk}, \ f \in V_n$. At each step of the wavelet transformation we move to a coarser approximation $\underline{c}^{(j-1)}$ with $\underline{c}^{(j-1)} = H\underline{c}^{(j)}$ and $\underline{d}^{(j-1)} = G\underline{c}^{(j)}$. Here, $\underline{d}^{(j-1)}$ is the "detail" lost by approximating $\underline{c}^{(j)}$ by the averaged $\underline{c}^{(j-1)}$. The discrete wavelet transformation of a sequence $y = \underline{c}^{(n)}$ of length 2^n can then be represented as another sequence of length 2^n (notice that the sequence $\underline{c}^{(j-1)}$ has half the length of $\underline{c}^{(j)}$):

$$(\underline{d}^{(n-1)}, \underline{d}^{(n-2)}, \ldots, \underline{d}^{(1)}, \underline{d}^{(0)}, \underline{c}^{(0)}). \quad (7)$$

Thus the discrete wavelet transformation can be summarized as:

$$y \longrightarrow (Gy, GHy, GH^2y, \ldots, GH^{n-1}y, H^ny).$$

The reconstruction formula is also simple in terms of H and G; we first define adjoint operators $H^\star$ and $G^\star$ as follows:

$$(H^\star a)_k = \Sigma_n h(k - 2n)a_n \text{ and } (G^\star a)_k = \Sigma_n g(k - 2n)a_n.$$

---

[3]A sequence $\{a_n\}$ is in the Hilbert space l^2 if $\Sigma_{k \in Z} a_k^2$ is finite.

Recursive application leads to:

$$(\underline{d}^{(n-1)}, \underline{d}^{(n-2)}, \ldots, \underline{d}^{(1)}, \underline{d}^{(0)}, \underline{c}^{(0)}) \longrightarrow \underline{y} = \Sigma_{j=0}^{n-1} (H^\star)^j G^\star \underline{d}^{(j)} + (H^\star)^n \underline{c}^{(0)}.$$

Equations (6) which generate filter coefficients (sometimes called *dilation equations*) look very simple for the Haar wavelet:

$$\phi(x) = \phi(2x) + \phi(2x-1) = \frac{1}{\sqrt{2}}\sqrt{2}\phi(2x) + \frac{1}{\sqrt{2}}\sqrt{2}\phi(2x-1),$$

$$\psi(x) = \phi(2x) - \phi(2x-1) = \frac{1}{\sqrt{2}}\sqrt{2}\phi(2x) - \frac{1}{\sqrt{2}}\sqrt{2}\phi(2x-1).$$

The filter coefficients in the above equations are $h(0) = h(1) = 1/\sqrt{2}$ and $g(0) = -g(1) = 1/\sqrt{2}$.

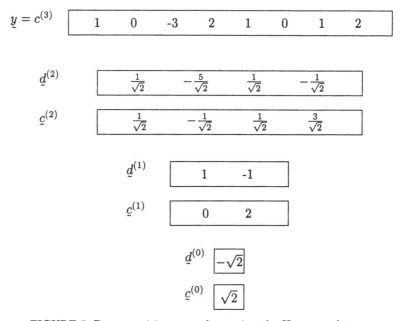

FIGURE 5. Decomposition procedure using the Haar wavelet.

Figure 5 schematically gives the decomposition algorithm applied to our data set. To get the wavelet coefficients as in (2) we multiply components of $\underline{d}^{(j)}$, $j = 0, 1, 2$ and $c^{(0)}$ with the factor $2^{-N/2}$. This multiplication would not be necessary if we defined our "data function" on the interval $[0,8]$ instead of $[0,1]$, $d_{jk} = 2^{-N/2}d_k^{(j)}$, $0 \le j < N \ (= 3)$.

It is interesting that in the Haar wavelet case $2^{-3/2}c_0^{(0)} = c_{00} = \frac{1}{2}$ is the mean of the sample $\underline{y}$. Figure 6 schematically gives the reconstruction algorithm for our example.

The careful reader might have already noticed that when the length of the filter exceeds 2, boundary problems occur. (There are no boundary problems with the Haar wavelet!) There are several ways to handle the boundaries. The two main approaches are: *periodic* and *symmetric*.

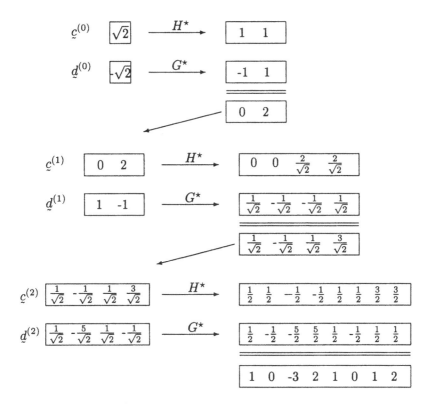

FIGURE 6. Reconstruction procedure.

1.3 Classical Thresholding Methods

In wavelet decomposition the filter H is an "averaging" filter while its mirror counterpart G produces details. The wavelet coefficients correspond to details. When details are small, they might be omitted without substantially affecting the "general picture." Thus the idea of thresholding wavelet coefficients is a way of cleaning out "unimportant" details considered to be noise. We illustrate the idea on our old friend, the data vector $(1, 0, -3, 2, 1, 0, 1, 2)$.

Example 1. The data vector $(1, 0, -3, 2, 1, 0, 1, 2)$ is transformed into the vector $(1/\sqrt{2}, -5/\sqrt{2}, 1/\sqrt{2}, -1/\sqrt{2}, 1, -1, -\sqrt{2}, \sqrt{2})$. If all coefficients less than 0.9 (well, our choice) are replaced by zeroes, then the resulting ("thresholded") vector is $(0, -\frac{5}{\sqrt{2}}, 0, 0, 1, -1, -\sqrt{2}, \sqrt{2})$. The graph of "smoothed data", after reconstruction, is given in Figure 4 (right panel).

An important feature of wavelets is that they provide unconditional bases[4] for not only L^2, but variety of smoothness spaces such as Sobolev, Besov, and Hölder spaces. As a consequence, wavelet shrinkage acts as a smoothing operator. The same can not be said about Fourier basis. By shrinking Fourier coefficients one can get bad results in terms of mean square error. Also, some unwanted visual artifacts can be obtained, see Donoho (1993).

Why is thresholding good? The parsimony of wavelet transformations ensures that the signal of interest can be well described by a relatively small number of wavelet coefficients. A simple Taylor series argument shows that if the mother wavelet has L vanishing moments and the unknown "signal" is in C^{L-1}, then $|d_{jk}| \leq \text{const} \cdot 2^{-j(L-1/2)} \int |y|^L |\psi(y)| dy$. For j large (fine scales) this will be negligible. For a nice discussion on a compromise between regularity (number of vanishing moments) and the mother wavelet support see Daubechies (1992), page 244.

The process of thresholding wavelet coefficients can be divided into two steps. The first step is the policy choice, i.e., the choice of the threshold function T. Let λ denote the threshold, and let d generically denote a wavelet coefficient. Two standard choices are: **hard** and **soft** thresholding with corresponding transformations given by:

$$T^{hard}(d, \lambda) = d \; \mathbf{1}(|d| > \lambda), \tag{8}$$

$$T^{soft}(d, \lambda) = (d - sgn(d)\lambda) \; \mathbf{1}(|d| > \lambda). \tag{9}$$

The second step is the choice of a threshold. In the following subsections we briefly discuss some of the standard methods of selecting a threshold.

1.3.1 Universal Threshold

Donoho and Johnstone (1994) and Donoho et al. (1995) proposed a threshold λ based on the following result.

Result: If the random variables $X_1, \ldots, X_n, \ldots$ are independent standard normal, then $P\left(|X_{(n)}| > \sqrt{c \log n}\right) \sim \frac{\sqrt{2}}{n^{c/2-1}\sqrt{c\pi \log n}}$, when n is large. Thus, for $c = 2$,

$$\lambda = \sqrt{2 \log n} \; \hat{\sigma}, \tag{10}$$

---

[4]Informally, a family $\{\psi_i\}$ is an unconditional basis for a space S if one can decide if the element $f = \Sigma_i a_i \psi_i$ belongs to S by inspecting $|a_i|$, only.

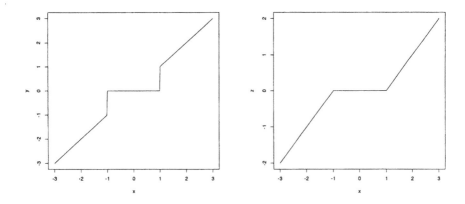

FIGURE 7. Hard and soft thresholding with $\lambda = 1$.

ensures, with high probability, that no noise (of size σ) is present in the data after thresholding. The threshold λ in (10) is one of the first proposed and provides an efficient thresholding.

There are several possibilities for the estimator $\hat{\sigma}$ in (10). Almost all methods involve the wavelet coefficients in the finest scale since signal-to-noise ratio is smallest at high resolutions in the wavelet domain for almost all reasonably behaved signals.

Some standard estimators are:

(i) $\hat{\sigma}^2 = 1/(N/2 - 1) \sum_{k=1}^{N/2}(d_{n-1,k} - \bar{d})^2$, where $\bar{d} = 1/(N/2) \sum d_{n-1,k}$;

(ii) $\hat{\sigma}^2 = 1/0.6745 \, MAD(\{d_{n-1,k}, k = 1, \ldots, N/2\})$, where $n - 1$ is the level of finest detail.

In some problems, especially with large data sets, and when the σ is over-estimated, the universal thresholding gives under-fitted models.

1.3.2 A Few Other Methods

Donoho and Johnstone (1995) propose an adaptive method, *SUREShrink*, based on Stein's unbiased estimator of risk. Nason (1996) proposed a cross-validatory threshold selection procedure. Wang (1996) generalizes Nason's crossvalidation technique by removing more than half of the data each time. The motivation is to robustify the threshold selection procedure against the effect of a correlated noise (with a long range dependence). Saito (1994) incorporates the hard thresholding into a minimum description length paradigm.

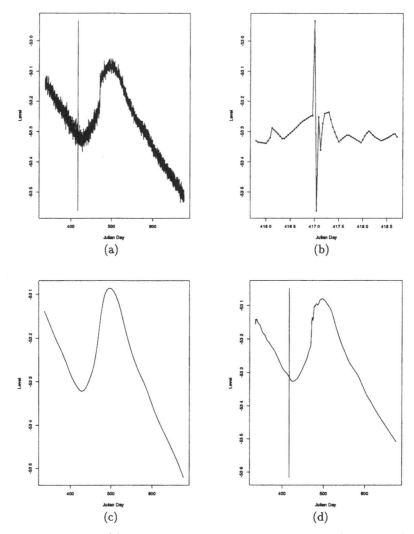

FIGURE 8. Panel (a) shows $n = 8192$ hourly measurements of the water level for a well in an earthquake zone. Notice the wide range of water levels at the time of an earthquake around $t = 417$. Panel (b) focuses on the data around the earthquake time. Panel (c) demonstrates action of a standard smoother supsmo, and (d) gives a wavelet based reconstruction.

1.4 An Application: California Earthquakes

A researcher in geology was interested in predicting earthquakes by the level of water in nearby wells. She had a large ($8192 = 2^{13}$ measurements) data set of water levels taken every hour in a period of time of about one year in a California well. Here is the description of the problem.

The ability of water wells to act as strain meters has been observed for centuries. The Chinese, for example, have records of water flowing from wells prior to earthquakes. Lab studies indicate that a seismic slip occurs along a fault prior to rupture. Recent work has attempted to quantify this response, in an effort to use water wells as sensitive indicators of volumetric strain. If this is possible, water wells could aid in earthquake prediction by sensing precursory earthquake strain.

We have water level records from six wells in southern California, collected over a six year time span. At least 13 moderate size earthquakes (magnitude 4.0 - 6.0) occurred in close proximity to the wells during this time interval. There is a a significant amount of noise in the water level record which must first be filtered out. Environmental factors such as earth tides and atmospheric pressure create noise with frequencies ranging from seasonal to semidiurnal. The amount of rainfall also affects the water level, as do surface loading, pumping, recharge (such as an increase in water level due to irrigation), and sonic booms, to name a few. Once the noise is subtracted from the signal, the record can be analyzed for changes in water level, either an increase or a decrease depending upon whether the aquifer is experiencing a tensile or compressional volume strain, just prior to an earthquake.

A plot of the raw data for hourly measurements over one year ($8192 = 2^{13}$ observations) is given in Figure 8a. After applying the DAUB2 (Figure 1.) wavelet transformation and thresholding by the Donoho-Johnstone "universal" method, we got a very clear signal with big jumps at the earthquake time. The cleaned data are given in Figure 8b. The magnitude of the water level change at the earthquake time did not get distorted in contrast to usual smoothing techniques. This is a desirable feature of wavelet methods. Yet, a couple of things should be addressed with more care.

(i) Possible fluctuations important for the earthquake prediction are cleaned as noise. In post-analyzing the data, having information about the earthquake time, one might do time-sensitive thresholding.

(ii) Small spikes on the smoothed signal (Figure 8b) as well as 'boundary distortions" indicate that the DAUB2 wavelet is not the most fortunate choice. Compromising between smoothness and the support shortness of the mother wavelet with help of wavelet banks, one can develop ad-hoc rules for better mother wavelet (wavelet model) choice.

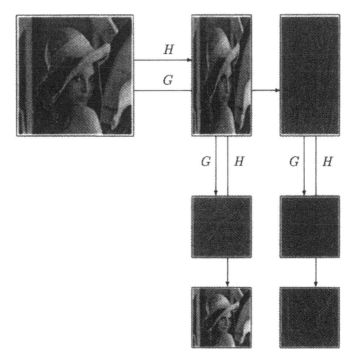

FIGURE 9: Wavelet decomposition of an image

1.5   Image Processing Using Wavelets

We will briefly explain how wavelets may be useful in matrix data processing. The most remarkable application is, without any doubt, image processing. Any black and white image can be approximated by a matrix A in which the entries a_{ij} correspond to gray intensities for the pixel (i, j). For reasons that will be obvious later, it is assumed that A is a square matrix of dimension $2^n \times 2^n$, n integer.

The process of the image wavelet decomposition goes as follows. On the rows of the matrix A the filters H and G are applied. Two resulting matrices are obtained: $H_r A$ and $G_r A$, both of dimension $2^n \times 2^{n-1}$ (Subscript r suggest that the filters are applied on rows of the matrix A). Now on the columns of matrices $H_r A$ and $G_r A$, filters H and G are applied again and the four resulting matrices $H_c H_r A, G_c H_r A, H_c G_r A$ and $G_c G_r A$ of dimension $2^{n-1} \times 2^{n-1}$ are obtained. The matrix $H_c H_r A$ is the average, while the matrices $G_c H_r A, H_c G_r A$ and $G_c G_r A$ are details (Figure 9).

The process can be continued with the *average* matrix $H_c H_r A$ until a single number ("an average" of the whole original matrix A) is obtained. Two examples are given below.

An interesting generalization of wavelets are wavelet packets. The wavelet packets are a result of applications of operators H and G, discussed on page 7, in *any* order. That generates a library of bases from which the best basis can be selected for the specific data set. For more details on wavelet packets we direct the reader to Wickerhauser (1995).

Example 2. This example shows that the best wavelet packet basis compresses with smaller mean square error than a comparable discrete wavelet transformation. Figure 10 (a) gives the original image of Reverend Thomas Bayes (1702-1761) with superimposed tilling determined by the best wavelet packet basis. The vertical stripes visible in the original image are artifacts of the scanning process.

Panel (b) in Figure 10 shows relative MSE (energy losses) as a function of compression ratio. The vertical line is positioned at 20:1 compression ratio.

Panels (c) and (d) give reconstructed images from the 5%-best coefficients of the best wavelet packet basis and discrete wavelet transformation, respectively.

The best wavelet packet has slightly better MSE performance [the panel (b)]. Unfortunately it also better preserves the vertical artifacts in the original image, exemplifying that minimizing MSE may not always be desirable when enhancing and smoothing images.

1.6 Can You Do Wavelets?

Yes, you can! There are several free packages (and many commercial ones, that we selected not to advertise here) which support wavelet calculations.

The most comprehensive is WAVELAB. Wavelab is a library of MATLAB programs for wavelet and general time/frequency analysis. It is maintained at Department of Statistics at Stanford University, and can be downloaded from [http://www-stat.stanford.edu/~wavelab/] The latest version 0.800, compatible with Matlab 5, is developed by David Donoho, Xiaoming Huo and Thomas P.-Y. Yu. Some contributors to the earlier versions of the software include Jeffrey Scargle, Iain Johnstone, Shaobing Chen, Jonathan Buckheit and Eric Kolaczyk.

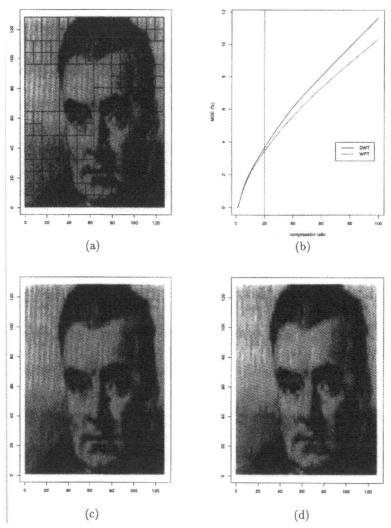

(a)

(b)

(c)

(d)

FIGURE 10. (a) Figure and tilling induced by the best two-dimensional wavelet packet. (b) The percent of the mean square error compared with the compression ratio. The solid curve corresponds to the DWT, the dotted curve corresponds to the wavelet packet transformation. (c) and (d) Reconstruction from 5% coefficients in the best basis and standard wavelet transformation, respectively.

Some other free packages include WAVETHRESH and RICE WAVELET TOOLS. WaveThresh is an add-on package for S-Plus (or its free clone R). The author of the software is Guy Nason, University of Bristol. WaveThresh contains S-Plus functions for performing one- and two-dimensional wavelet transforms and their inverses for basic wavelet families. It also contains the rudiments of a thresholding scheme for doing wavelet based curve-estimation. The current version of WaveThresh available on the web is 2.2.

Rice Wavelet Tools is a collection of Matlab M-files and MEX-files implementing wavelet and filter bank design and analysis. It can be downloaded from [http://www-dsp.rice.edu/software/RWT/].

Some resourceful web pages are

(i) The Wavelet Digest

[http://www.wavelet.org/wavelet/index.html] is a free monthly newsletter edited by Wim Sweldens which contains information concerning wavelets: announcement of conferences, preprints, software, questions and answers, etc., and

(ii) Mathsoft Wavelet resources

[http://www.mathsoft.com/wavelets.html] is a web page for "one-stop shopping" for wavelet manuscripts. This page contains a list of many preprints on the subject of wavelets and their various applications in the worlds of mathematics, statistics, engineering and physics.

References

Daubechies, I. (1988), Orthonormal bases of compactly supported wavelets. Commun. Pure Appl. Math., 41 (7), 909-996.

Daubechies, I. (1992), *Ten Lectures on Wavelets*, Society for Industrial and Applied Mathematics.

Donoho, D. (1993). Nonlinear Wavelet Methods for Recovery of Signals, Densities, and Spectra from Indirect and Noisy Data. *Proceedings of Symposia in Applied Mathematics,* American Mathematical Society. Vol. 47, 173–205.

Donoho, D. and Johnstone, I. (1994). Ideal spatial adaptation by wavelet shrinkage. *Biometrika,* 81, 3, 425–455.

Donoho D., and Johnstone, I. (1995). Adapting to unknown smoothness via wavelet shrinkage. *J. Amer. Statist. Assoc.,* 90, 1200–1224.

Donoho, D., Johnstone, I., Kerkyacharian, G, and Picard, D. (1995). Wavelet shrinkage: Asymptopia? *J. Roy. Statist. Soc. Ser. B*, 57, 2, 301–369.

Mallat, S. G. (1989). A theory for multiresolution signal decomposition: the wavelet representation. IEEE Transactions on Pattern Analysis and Machine Intelligence, 11 (7), 674-693.

Nason, G. (1996). Wavelet regression by cross-validation. *J. Roy. Statist. Soc. Ser. B*, 58, 463–479.

Saito N. (1994). Simultaneous noise suppression and signal compression using a library of orthonormal bases and the minimum description length criterion. In: *Wavelets in Geophysics*, Foufoula-Georgiou and Kumar (eds.), Academic Press.

Wang, Y. (1996). Function estimation via wavelet shrinkage for long-memory data, *Ann. Statist.*, 24, 2, 466–484.

Wickerhauser, M. V. (1994). *Adapted Wavelet Analysis from Theory to Software*, A K Peters, Ltd.

2

Spectral View of Wavelets and Nonlinear Regression

J. S. Marron

ABSTRACT   This chapter reviews one of the most widespread uses of wavelets in statistics: nonlinear nonparametric regression. Simple insight into the utility and power of the wavelet approach comes from a discrete spectral analysis point of view. Some of the ideas here are the same as some introduced in Chapter 1, but the different viewpoint is intended to give additional insights.

2.1   Background

2.1.1   Spectral Analysis

Spectral methods for curve estimation are based on "spectral decompositions" of vectors in $\Re^n$. These involve a "transformation", i.e. a change to a different basis of $\Re^n$. Such transformations can provide very powerful and useful new ways to work with given sets of vectors. For example, vectors that are a sum of a few sinusoids, are effectively summarized (and thus studied in many ways) by changing to the Fourier orthonormal basis. Wavelets are simply viewed as another such orthonormal basis, whose good properties are shown in this chapter.

The mathematics of discrete spectral analysis starts with $\{\psi_1, ..., \psi_n\}$, an arbitrary orthonormal basis of $\Re^n$, i.e.

$$\psi_i^t \psi_{i'} = \left\{ \begin{array}{ll} 0 & \text{if } i \neq i' \\ 1 & \text{if } i = i', \end{array} \right. \tag{1}$$

The spectral representation of a vector $\mu \in \Re^n$ is defined as

$$\mu_i = \sum_{i'=1}^{n} \theta_{i'} \psi_{i',i}, \ i = 1, ..., n, \tag{2}$$

where $\psi_{i',i}$ is the i-th entry of $\psi_{i'}$, and where the coefficients $\theta_{i'}$ are given by the inner products

$$\theta_{i'} = \mu^t \psi_{i'}, \tag{3}$$

which give the lengths of the projection of μ in the direction of $\psi_{i'}$. The coefficients can be collected into a vector $\theta = (\theta_1, ..., \theta_n)^t$ which is called the "transform" of μ. Useful intuition comes from thinking of θ as a "rotation in $\Re^n$" (only an approximation, since other transformations such as mirror images are also orthonormal).

The particular rotation of this type called the "discrete Fourier Transform" has been a workhorse tool, especially in fields where the study of periodicities and frequencies are important, perhaps most noticeably in electrical engineering. This is because the Fourier rotation of vectors is particularly adept at revealing their frequency structure. In statistics, the best known use of this rotation of data vectors is the Fourier analysis of time series. See Bloomfield (1976) for a very readable account.

A potentially confusing aspect of spectral analysis is that it can often be viewed both "discretely", in terms of ordinary vectors in $\Re^n$, and "continuously" where the "vectors" are functions in a suitable Hilbert space. The two are typically related to each other because summation (the operation underlying the discrete inner product) is related to integration (the operation underlying the continuous inner product) through Riemann integration. However, orthonormality and the properties following from it are exact in both contexts. For example, in the Fourier basis, the familiar sin and cos waves are orthogonal as functions (with respect to the integral inner product), but also as vectors when evaluated at appropriate equally spaced grids (with respect to summation). The distinctions between these parallel theories is often blurred, because of the approximation. Only the discrete theory is studied in this chapter, with occasional reference made to the continuous theory where it adds insight.

2.1.2   Nonlinear nonparametric regression

An important statistical application of spectral analysis is to nonparametric regression. Here one models a data vector as

$$\mathbf{Y} = \mu + \varepsilon,$$

where μ is some "smooth" underlying mean (i.e. "signal") vector, and ε is a random "noise" vector (assumed to have mean $\mathbf{0}$, which results in $E(\mathbf{Y}) = \mu$).

Orthogonal transformation allows recasting the problem of using $\mathbf{Y}$ to estimate μ, into that of using the transform of $\mathbf{Y}$ to estimate the transform of μ, which can often be much simpler. Let $\widetilde{\theta} = \left(\widetilde{\theta}_1, ..., \widetilde{\theta}_n\right)^t$ denote the transformation of the data, where the "empirical coefficients" are

$$\widetilde{\theta}_{i'} = \mathbf{Y}^t \psi_{i'}, \tag{4}$$

which are unbiased estimates of the $\theta_{i'}$, the transform coefficients of μ from (2). A further important property of the rotation aspect of an orthonormal transformation is that when the errors are independent and homoscedastic the "spherical" covariance structure of $\tilde{\theta}$ is the same as that of $\mathbf{Y}$.

For a good choice of basis, most of the "power of μ" (a useful concept from signal processing), which is conveniently quantified as a sum of squares,

$$P_\mu = \sum_{i=1}^n \mu_i{}^2 = \sum_{i=1}^n \theta_i{}^2$$

(where the last equality follows by the Parseval identity), will be "contained in a few" of the θ_i. In this case, a reasonable reconstruction of the "signal" μ can be obtained from the data $\mathbf{Y}$ by inverting the transform, but using only the "important" coefficients:

$$\hat{\mu}_i = \sum_{i' \in S} \tilde{\theta}_{i'} \psi_{i',i}, \ i = 1, ..., n, \tag{5}$$

(sometimes grouped into the vector $\hat{\mu} = (\hat{\mu}_1, ..., \hat{\mu}_n)^t$) where S is some set of "high power coefficients". If the set S is small, but at the same time the restricted power $\sum_{i' \in S} \theta_{i'}^2$ is a large fraction of P_μ, the estimator will be very effective, since most of the power (i.e. component of the sum of squares) of the noise will be contained in terms that do not appear in $\hat{\mu}$, which will thus be eliminated.

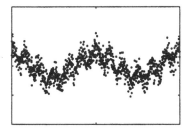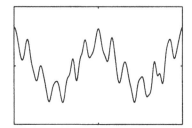

FIGURE 1. Left panel: $n = 1024$ simulated data points, using Gaussian white noise. Right Panel: Fourier basis nonparametric regression estimate, using hard thresholding.

This principle is demonstrated in Figure 1. The left panel shows a simulated data vector $\mathbf{Y}$, which was generated by adding homoscedastic Gaussian noise to a smooth underlying mean vector μ. The right panel shows a signal recovery (i.e. a nonparametric regression estimate), $\hat{\mu}$, using the Fourier basis, where the set S of coefficients was chosen using the "Universal Hard Thresholding" method discussed in Section 3 of Chapter 1. While the data vector $\mathbf{Y}$ only suggests a single periodic component, note that $\hat{\mu}$ suggests the underlying mean curve has two periodic components, which is in fact the actual structure of this underlying μ. This illustrates the potential power of the idea of rotating the data vector in a carefully chosen direction, and then using just a few coefficients to dampen noise in signal recovery, i.e. nonparametric regression.

This noise damping effect has a simple, useful quantification when the errors are uncorrelated and homoscedastic. In that case the average variance is:

$$n^{-1} \sum_{i=1}^{n} var\left(\widehat{\mu}_i\right) = \frac{\sigma^2}{n} \#\left(S\right), \qquad (6)$$

which is smaller when S has fewer members. On the other hand smaller S means more average squared bias:

$$n^{-1} \sum_{i=1}^{n} \left(E\widehat{\mu}_i - \mu_i\right)^2 = n^{-1} \sum_{i' \notin S} \theta_{i'}^2 = n^{-1} \left(P_\mu - \sum_{i' \in S} \theta_{i'}^2\right). \qquad (7)$$

This shows that the size of S works as a "smoothing parameter" in terms of controlling the usual trade off between variance (which quantitates "wiggliness") and squared bias (which quantitates "goodness of fit"). Also, the spectral type estimator $\widehat{\mu}$ will be most effective when there is a small set S which contains most of the power of the signal, and when that set can be approximately identified. I.e. the potential good performance of $\widehat{\mu}$ is linked to concepts of "signal compression". An example of this is when the Fourier Transform is applied with a smooth and periodic underlying signal μ, and S is a set of low frequency terms, as shown in Figure 1. However, when the underlying regression function μ and the basis are such that "power of the signal is spread across the spectrum" (e.g. the Fourier transform of a rough, aperiodic signal), meaning most of P_μ is shared by many of the θ_i, there will not be a good choice of S, and this type of estimator will not perform well. Figure 2 shows an extreme example of this.

 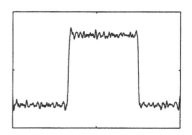

FIGURE 2. Left panel: $n = 1024$ simulated data points, using Gaussian white noise. Right Panel: Fourier basis nonparametric regression estimate, using hard thresholding.

The data in the left panel of Figure 2 are again simulated with independent Gaussian noise. It is visually clear that the underlying signal μ has two large jumps. The Fourier basis does a poor job of compressing such a signal (i.e. the power of the signal is spread all across the spectrum), which has a serious negative impact on the nonparametric regression estimate $\widehat{\mu}$, shown in the right panel. Note that at many locations the wiggliness of $\widehat{\mu}$ is visually almost as large as the range of the data. This is especially true near the jump points, where "ringing" effects, i.e. Gibb's phenomena, are quite noticeable. Furthermore because the Fourier basis is smooth, the crisp jumps become smoothed. Other choices of the set S, result in either increased ringing, or else increased rounding of the jumps. The choice of S used here is (as in Figure 1) the Universal Hard Thresholding method discussed in Section 3 of Chapter 1.

One may wonder why the estimator $\widehat{\mu}$ is called "nonlinear", since the representations (5) and (4) suggest that it is a linear function of the data vector $\mathbf{Y}$. The distinction is in the choice of the set S. When S is chosen independently of the data, e.g. when using the Fourier basis with just a set of low frequency terms, then $\widehat{\mu}$ is a linear estimator. However, when S is chosen as some function of the data $\mathbf{Y}$, e.g. according to various types of thresholding rules as in Section 3 of Chapter 1, then $\widehat{\mu}$ is (perhaps mildly) nonlinear.

2.2   The Wavelet Bases

The excitement that has surrounded the development of wavelet bases comes from their surprisingly good signal compression of a wide new range of signal types.

2.2.1   Wavelet Fundamentals

The Fourier basis is effective when signals are "spatially homogeneous", meaning the amount of smoothness is roughly the same in different locations. A set of bases which can be effective when the smoothness is not homogenous are the wavelet bases. Some wavelet bases are good at compressing smooth signals, as well as those that are "somewhat unsmooth in some locations", such as the step function underlying Figure 2. This is accomplished by having basis vectors which are smooth, but allow "localization in time", as well as in frequency.

As with Fourier theory, wavelets have closely parallel discrete and continuous theories, which are connected by Riemann summation. Again, the discrete case is treated here, although sometimes it is useful to think in terms of continuous functions evaluated at $x_1, ..., x_n$, equally spaced on the unit interval, e. g. $x_i = i/n$. The presentation is simplest when n is a power of 2, so that will be assumed throughout this section.

A good starting point for understanding the special structure of wavelet bases is the Haar basis. Some of these basis functions are shown in Figure 4 of Chapter 1. A way of organizing these is shown here in Figure 3. This basis provides coefficients θ_i, conveniently reindexed as $\theta_{j,k}$, which give both a "scale" (this is wavelet terminology that should be viewed as a synonym for "frequency" by those familiar with Fourier analysis) and a "location" decomposition of μ. A convenient index for scale is $j = 0, ..., \log_2(n/2)$. At scale j, basis vectors essentially consist of disjoint step functions (thinking continuously for the moment), indexed by $k = 0, ..., 2^j - 1$, whose intervals of support are the dyadic intervals $(k2^{-j}, (k+1)2^{-j}]$. Orthogonality with coarser scale basis vectors comes from making the step function assume positive and negative values that are equal in magnitude, on the two halves of the support. More precisely (and now thinking discretely), using the vector notation

$$1(n) = (1, ..., 1)^t, \ \mathbf{0}(n) = (0, ..., 0)^t \text{ and } \mathbf{0}(0) = \{\},$$

define

$$\psi_{j,k} = \left(\frac{2^j}{n}\right)^{1/2} \left(\begin{array}{c} \mathbf{0}\left(\frac{kn}{2^j}\right) \\ \mathbf{1}\left(\frac{n}{2^{j+1}}\right) \\ -\mathbf{1}\left(\frac{n}{2^{j+1}}\right) \\ \mathbf{0}\left(n - \frac{(k+1)n}{2^j}\right) \end{array} \right), \tag{8}$$

for $j = 0, ..., \log_2(n) - 1$ and $k = 0, ..., 2^j - 1$. The factor of $\left(\frac{2^j}{n}\right)^{1/2}$ makes the lengths one. Basis vectors with the same scale j are orthogonal because the nonzero entries are disjoint. Orthogonality across frequencies follows either from this disjointness, or from inner products resulting in a subvector of the form $\left(\begin{array}{c} 1 \\ -1 \end{array} \right)$ being multiplied by a constant vector. These $n - 1$ vectors, together with $\varphi_{0,0} = n^{-1/2}\mathbf{1}(n)$ form an orthonormal basis of $\Re^n$, which is visually represented in Figure 3.

The Fast Fourier Transform was considered to be a major computational breakthrough, since it allows calculation of the n Fourier coefficients with only $O(n \log n)$ computations, instead of the $O(n^2)$ operations needed for the obvious naive computation using (4). The key is a very clever reorganization of the computation, which makes heavy use of special properties of trigonometric functions. The Haar coefficients $\theta_{j,k}$ can also be calculated very efficiently by clever reorganization, which turns out to be both simpler and even faster than the FFT.

Insight into the fast algorithm comes from the relationship (Section 2.2 of Chapter 1 gives another way of looking at this) between the "mother

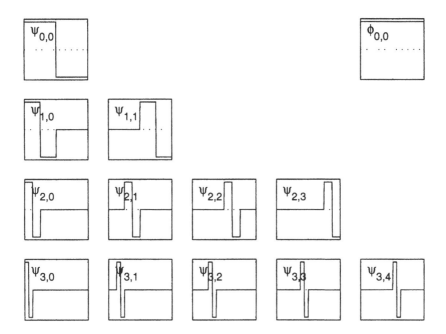

FIGURE 3. The Haar basis, with "father vector" $\varphi_{0,0}$ and "mother vectors" $\psi_{j,k}$. The row for scale $j = 3$ is to the right. The pattern continues below the bottom for scales $j = 4, 5, ...$

basis vectors" $\psi_{j,k}$ from (8), and the "father basis vectors"

$$\varphi_{j,k} = \left(\frac{2^j}{n}\right)^{1/2}\left(\begin{array}{c} \mathbf{0}\left(\frac{kn}{2^j}\right) \\ \mathbf{1}\left(\frac{n}{2^j}\right) \\ \mathbf{0}\left(n - \frac{(k+1)n}{2^j}\right) \end{array}\right),$$

for $j = 0, ..., \log_2(n)$ and $k = 0, ..., 2^j - 1$. Note that for any j_0, $\varphi_{j_0,0}, ...,$ $\varphi_{j_0,2^{j_0}-1}$ and $\varphi_{0,0}, \psi_{0,0}, \psi_{1,0}, ..., \psi_{j_0-1,2^{j_0}-1}$ span the same subspace (essentially step functions over intervals of length $\frac{1}{2^{j_0}}$) of $\Re^n$. In Section 2.2 of Chapter 1, this subspace is denoted as V_{j_0}. Hence the "tip of the pyramid" shown in Figure 3 (down to any given row) can be replaced by a row of father basis vectors, to yield another orthogonal basis. The wavelets also have a "magnification property" in that (modulo location) higher frequency basis vectors $\varphi_{j,k+1}$ ($\psi_{j,k+1}$) are simple horizontal rescalings of $\varphi_{j,k}$ ($\psi_{j,k}$ respectively), by a factor of 2. This is why the index j is said to index "scale", even though "frequency" is also a useful interpretation.

The crucial relationship between these is that the mother and father basis vectors at each frequency are a simple linear function of father vectors at the next higher frequency:

$$\psi_{j,k} = \frac{1}{\sqrt{2}} \left(\varphi_{j+1,2k} - \varphi_{j+1,2k+1} \right)$$
$$\varphi_{j,k} = \frac{1}{\sqrt{2}} \left(\varphi_{j+1,2k} + \varphi_{j+1,2k+1} \right). \tag{9}$$

Since most of the wavelet coefficients are of the inner product form $\theta_{j,k} = \mu^t \psi_{j,k}$, they can be quickly obtained through an analogous recursion. For this define the father coefficients $f_{j,k} = \mu^t \varphi_{j,k}$. Applying inner products to both sides of both equations in (9) gives the same relation between coefficients

$$\theta_{j,k} = \frac{1}{\sqrt{2}} \left(f_{j+1,2k} - f_{j+1,2k+1} \right)$$
$$f_{j,k} = \frac{1}{\sqrt{2}} \left(f_{j+1,2k} + f_{j+1,2k+1} \right). \tag{10}$$

Using these equations iteratively over scales, starting from the values $f_{\log_2(n),k} = \mu_k$, for $k = 1, ..., n$ results in a fast and simple algorithm. As the FFT provides a big improvement over the naive implementation, this approach to Haar wavelet decomposition also reduces the $O(n^2)$ matrix multiplication illustrated in Section 2.1 of Chapter 1, to an $O(n)$ calculation, that is both slightly faster and also simpler than the FFT.

An important feature of this linear transformation is that it preserves power at each step in the sense that $(\theta_{j,k}^2 + f_{j,k}^2) = (f_{j+1,2k}^2 + f_{j+1,2k+1}^2)$. Allocation of signal power is the key to understanding the usefulness of wavelet transforms in general. In this Haar case, the transformation handles the power of "constant vectors" by putting it entirely into the father coefficient at the next lower frequency (this is an "averaging operation"), with a 0 contribution to the wavelet coefficient $\theta_{j,k}$ (this is a "differencing operation"). Hence the mother basis provides very effective compression of step functions (since there are many zero coefficients), and usually will be effective for estimating them.

The Haar basis is often not good at compressing signals that are smooth, as shown in the left panel of Figure 4, which uses the data from Figure 1. The result does not look much like the sum of two sinusoids, although the essential structure is recovered. Since the Haar basis is a set of step functions, signal compression is much better for the step function data from Figure 2, as shown in the left panel of Figure 4. Here the signal recovery is excellent (again the Universal Hard Threshold method was used to choose the set S of coefficients). Note that the behavior of the nonparametric regression estimate is rather different at the two jumps. The jump on the right is at the dyadic point $x = \frac{3}{4}$, and their is a very clean jump (because this feature is well described by a single Haar basis element). The jump on the left is at the very non-dyadic point $x = \frac{1}{3}$, which involves a non-zero Haar coefficient at each scale. The error entailed by the estimation of all of these gives the slightly rougher effect near this jump.

Wavelet bases with "smoother" basis functions are much more useful for recovering smooth signals. These have a structure very similar to the Haar

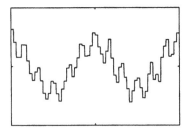
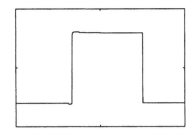

FIGURE 4. Haar basis nonparametric regression estimates, using hard thresholding, for the data in, Left panel: the left panel of Figure 1, Right Panel: the left panel of Figure 2.

basis, in particular sharing the same indexing system, the "mother - father" relationships, the magnification property and analogous fast iterative algorithms for computation (with the same "pyramid structure"). Development of such bases has been fairly recent, perhaps because (in the continuous domain) orthonormal bases of compactly supported functions exist, but unfortunately have no closed form, and indeed seem to be smoothed versions of fractals, see Figure 1 of Chapter 1.

Generalization of the fast algorithm for the Haar basis to smoother basis functions follows from writing the relationships between coefficients of different scales in terms of operators:

$$\theta_j = (\downarrow 2_o)\,(\mathbf{hi} *_{F,F} \mathbf{f}_{j+1})$$
$$\mathbf{f}_j = (\downarrow 2_e)\,(\mathbf{lo} *_{F,B} \mathbf{f}_{j+1})\ ,$$

where the coefficients of scale j have been combined into vectors as:

$$\theta_j = \begin{pmatrix} \theta_{j,0} \\ \vdots \\ \theta_{j,2^j-1} \end{pmatrix}, \ \mathbf{f}_j = \begin{pmatrix} f_{j,0} \\ \vdots \\ f_{j,2^j-1} \end{pmatrix},$$

where $(\downarrow 2_e)$ and $(\downarrow 2_o)$ denotes the even and odd "decimation operator of order 2", which take "every other entry", i. e. for $m = 1, 2, \ldots$

$$(\downarrow 2_e)\begin{pmatrix} x_0 \\ x_1 \\ x_2 \\ \vdots \\ x_{2m-2} \\ x_{2m-1} \end{pmatrix} = \begin{pmatrix} x_0 \\ \vdots \\ x_{2m-2} \end{pmatrix}, \ (\downarrow 2_o)\begin{pmatrix} x_0 \\ x_1 \\ x_2 \\ \vdots \\ x_{2m-2} \\ x_{2m-1} \end{pmatrix} = \begin{pmatrix} x_1 \\ \vdots \\ x_{2m-1} \end{pmatrix},$$

where $*_{F,B}$ denotes a "back padded" circular filter operation

$$
\begin{pmatrix} a_0 \\ \vdots \\ a_{\ell-1} \end{pmatrix} *_{F,B} \begin{pmatrix} b_0 \\ \vdots \\ b_{m-1} \end{pmatrix} = \begin{pmatrix} \sum_{i=0}^{\ell-1} a_i b_i \\ \sum_{i=0}^{\ell-1} a_i b_{1+i} \\ \vdots \\ a_0 b_n + \sum_{i=1}^{\ell-1} a_i b_{-1+i} \end{pmatrix},
$$

and $*_{F,F}$ denotes a "front padded" circular filter operation

$$
\begin{pmatrix} a_0 \\ \vdots \\ a_{\ell-1} \end{pmatrix} *_{F,F} \begin{pmatrix} b_0 \\ \vdots \\ b_{m-1} \end{pmatrix} = \begin{pmatrix} \sum_{i=0}^{\ell-2} a_i b_{m-\ell+1+i} + a_{\ell-1} b_0 \\ \sum_{i=0}^{\ell-3} a_i b_{m-\ell+2+i} + a_{\ell-2} b_0 + a_{\ell-1} b_1 \\ \vdots \\ \sum_{i=0}^{\ell-1} a_i b_{m-\ell+i} \end{pmatrix},
$$

and where the "lo pass filter vector", which averages coefficients, thus passing the power of constants (i.e. the low frequency component of the sum of squares) into the $f_{j,k}$, is

$$
\text{lo} = \frac{1}{\sqrt{2}} \begin{pmatrix} 1 \\ 1 \end{pmatrix}
$$

and the "hi pass filter vector", which differences coefficients, thus passing the power of the residuals about the mean (i.e. the high frequency component of the sum of squares) into the $\theta_{j,k}$, is

$$
\text{hi} = \frac{1}{\sqrt{2}} \begin{pmatrix} -1 \\ 1 \end{pmatrix}.
$$

There are other ways to formulate these operations, but this particular form is chosen to make the coefficients correspond well to the location suggested by the indices.

Smoother wavelet bases are obtained by clever choices of the filter vectors lo and hi. The key to more smoothness is modification of the low pass filter to pass polynomials of a certain degree (the Haar filters are the degree 0 case). This entails filter vectors of length more than two, which entails more complicated orthogonality conditions. The shortest filter vectors which meet these requirements give the "Daubechies" family, see Table 6.1 of Daubechies (1992) , for numerical values of the coefficients of $\text{lo} = (h_0, ..., h_{\ell-1})^t$, where the symbol h is used for consistency with Daubechies (1992). These basis functions are illustrated in Figure 1 of Chapter 1. Longer filter vectors allow bases that can satisfy additional requirements. For example, slightly longer filter vectors allow one degree of freedom that can be used to make the resulting basis functions "as symmetric as possible". This results in the family of filter vectors whose coefficients are in Table 6.3 of Daubechies (1992) , but the table values must be divided by $\sqrt{2}$ to give an orthonormal basis. The "Symmlet 8" ($N = 8$ in Daubechies' notation) basis is shown in Figure 5, and its usefulness for nonparametric regression is demonstrated in Figure 6.

Only the low pass coefficients are given in Daubechies' tables, because the high pass coefficients, $\mathbf{hi} = (g_0, ..., g_{\ell-1})^t$ are simply related as:

$$g_i = (-1)^{\ell-1} h_{\ell-1-i}, \ i = 0, ..., \ell - 1$$

Insight into the inversion of the wavelet transform comes from realizing that it is just a matrix multiplication, by an orthonormal matrix (a Haar version of this matrix is shown in Section 2.1 of Chapter 1). Hence the inverse is multiplication by the transpose matrix. Plots of basis functions are easy to construct by calculating the inverse transform of unit vectors (i. e. vectors with a single 1, and all other entries 0, which are the transforms of the basis vectors).

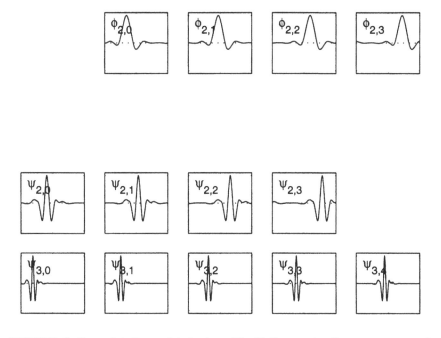

FIGURE 5. Symmlet 8 wavelet basis, with "father vectors" $\varphi_{2,0}, ..., \varphi_{2,3}$ and "mother vectors" $\psi_{j,k}$, $j \geq 2$. As for the Haar basis, the row for scale $j = 3$ is incomplete, with panels for $k = 5, 6, 7$ truncated on the right side. The pattern continues below the bottom for scales $j = 4, 5, ...$

An organization of the Symmlet 8 basis is given in Figure 5. The layout is the same as Figure 3, except now instead of a single top level father basis element $\varphi_{0,0}$ the pyramid algorithm was stopped at level $j = 2$, resulting in the four father basis elements $\varphi_{2,0}, ..., \varphi_{2,3}$. Figure 5 suggests where the name "wavelet" came from. Note that these basis elements give the impression of being "small pieces of waves". But they have special properties, such as orthogonality and smoothness, that cannot be arrived at by simply trying to "slice up" the sinusoids that make up the Fourier basis.

An accessible introduction, from a different viewpoint, to more basic wavelet ideas can be found in Strang (1989). A good source for deeper reading, including historical references, is Benedetto and Frazier (1994). This book gives a good overview of continuous wavelet folklore in Section 1, and a readable presentation of broader and deeper aspects of discrete wavelets in Section 2. The following chapters consider many important variations of wavelet ideas.

2.2.2   Why Wavelets?

A common question is "why should one use the wavelet transform, when the Fourier transform has been such a useful workhorse?" The answer is that wavelet bases allow surprisingly good signal compression of a very wide range of signals. This point is illustrated in Figure 6, where the Symmlet 8 basis, shown in Figure 5, is used to denoise the same two data sets as used in the above figures.

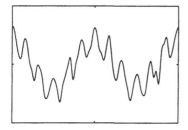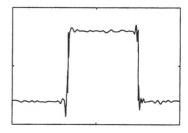

FIGURE 6. Symmlet 8 basis nonparametric regression estimates, using hard thresholding, for the data in, Left panel: the left panel of Figure 1, Right Panel: the left panel of Figure 2.

Note that using this basis, the signal recovery is excellent for *both* of these data sets. In contrast, the Fourier basis gave excellent recovery of the smooth sinusoidal target, but very poor performance for the step target, while the Haar basis gave excellent recovery of the step target, but poor recovery of the sinusoidal target. The key to understanding this is "signal compression". The Fourier basis represents the sinusoid with very few coefficients, but needs many for the step. The Haar represents the step with few coefficients, but needs many for the sinusoid. The power and elegance of smooth wavelet bases, such as the Symmlet 8 comes from being able to represent *both* types of signals in an efficient manner. This is the reason for the excitement that has surrounded the development of wavelet bases.

It is natural to expect that these impressive properties of wavelets come at some price. Perhaps the most serious downside to wavelets is that because they are flexible in both the time and frequency directions, they are rather sparse in each direction. For example, note that the number of "frequencies" represented is only $\log_2 n$ for the wavelet bases, while it is $n/2$ for the Fourier basis. The basis is also sparse in the time direction, as seen by noting that the peaks in the $\psi_{2,k}$ row of Figure 5 appear in only four rather special locations. A signal with peaks that have a similar shape, but different location may not be so well compressed by this basis.

There are methods for addressing these, and other difficulties, although they all come at some cost. Many of these techniques, such as "wavelet frames", are discussed in Bennedetto and Frazier (1994). An approach to the location sparsity problem is the "nondecimated" wavelet transform (also called "stationary" and "shifted") discussed in Nason, Sapatinas and Sawczenko (1997) , which can be viewed as the set of all shifts of the wavelet basis, and is easily computed by eliminating the decimation operators ($\downarrow 2_e$) and ($\downarrow 2_o$) from the wavelet algorithm discussed above.

The already large literature on wavelet nonparametric regression is not surveyed here, but a very few suggested next references are Donoho and Johnstone (1995) and Donoho, Johnstone, Kerkyacharian and Picard (1995). See Marron, Adak, Johnstone, Neumann and Patil (1998) for an "exact risk" analysis of wavelet bases, which provides a different viewpoint on some of these ideas.

Acknowledgments: The organization of the wavelet ideas used here mostly came from an informal presentation by David Donoho at the Oberwolfach meeting "Curves, Images and Massive Computation" in 1993. Sidney Resnick and Peter Müller made a number of helpful comments.

References

Benedetto, J. J. and Frazier, M. W. (1994) *Wavelets: Mathematics and Applications*, CRC Press, Bocan Raton, Florida.

Bloomfield, P. (1976) *Fourier analysis of time series, an introduction*, Wiley, New York.

Daubechies, I. (1992) *Ten Lectures on Wavelets,*. SIAM, Philadelphia.

Donoho, D. L. and Johnstone, I. M. (1995). Adapting to unknown smoothness via Wavelet shrinkage, *J. Amer. Statist. Assoc.*, 90, 1200-1224.

Donoho, D. L., Johnstone, I. M., Kerkyacharian, G. and Picard, D. (1995). Wavelet shrinkage: Asymptopia? *Jour. Roy. Statist. Soc. Ser. B*, 57, 301–369.

Marron, J. S., Adak, S., Johnstone, I. M, Neumann, M. and Patil, P. (1998) Exact risk analysis of wavelet regression, *Journal of Computational and Graphical Statistics*, 7, 278-309.

Nason, G.P., Sapatinas, T. and Sawczenko, A. (1997) Statistical modelling of time series using non-decimated wavelet representations, Technical Report, University of Bristol.

Strang, G. (1989) Wavelets and dilation equations: a brief introduction, *SIAM Review*, 31, 614-627.

3

Bayesian Approach to Wavelet Decomposition and Shrinkage

Felix Abramovich and Theofanis Sapatinas

ABSTRACT We consider Bayesian approach to wavelet decomposition. We show how prior knowledge about a function's regularity can be incorporated into a prior model for its wavelet coefficients by establishing a relationship between the hyperparameters of the proposed model and the parameters of those Besov spaces within which realizations from the prior will fall. Such a relation may be seen as giving insight into the meaning of the Besov space parameters themselves. Furthermore, we consider Bayesian wavelet-based function estimation that gives rise to different types of wavelet shrinkage in non-parametric regression. Finally, we discuss an extension of the proposed Bayesian model by considering random functions generated by an overcomplete wavelet dictionary.

3.1 Introduction

Consider the standard non-parametric regression problem:

$$y_i = g(t_i) + \epsilon_i, \quad i = 1, \ldots, n, \tag{1}$$

and suppose we wish to recover the unknown function g from additive noise ϵ_i given noisy data y_i at discrete points $t_i = i/n$. Only very general assumptions about g are made like that g belongs to a certain class of functions.

One of the basic techniques in non-parametric regression and signal processing is the generalized Fourier series approach. An unknown response function g is expanded in some orthogonal basis $\{\psi_j\}$:

$$g(t) = \sum_j \omega_j \psi_j(t),$$

where $\omega_j = \langle g, \psi_j \rangle$. The key point for the efficiency of such an approach is obviously a proper choice of a basis. A 'good' basis should allow *parsimonious* expansion for a wide variety of possible responses using a relatively small number of basis functions. The original signal then is represented by the set of its few generalized Fourier coefficients ω_j with high accuracy.

Such parsimonious representations may be the key to understanding the basic features of a signal, detecting its regularity, compression, etc. It also plays a crucial role in the non-parametric regression problem (1) - instead of estimating g directly, we estimate its generalized Fourier coefficients ω_j and the resulting estimate is $\hat{g}(t) = \sum_j \hat{\omega}_j \psi_j(t)$. In a way, we transfer the original non-parametric problem to an infinitely parametric one.

For a fixed basis, assuming g to belong to a specific class of possible responses, it implicitly or explicitly yields corresponding assumptions on its generalized Fourier coefficients ω_j. Bayesian approach seems only natural to exhibit these assumptions through putting a prior model on ω_j. Examples of Bayesian orthogonal series estimators based on different bases and priors are well known in the literature. Wahba (1983) proposed a Bayesian model for spline smoothing estimation. It turns out that her prior model for the unknown response function is equivalent to placing a certain prior on its Fourier sine and cosine coefficients. Silverman (1985) obtained similar results for B-spline basis. Steinberg (1990) presented a Bayesian model for the coefficients of a function's expansion in a power series of Hermite polynomials.

Here we discuss Bayesian estimators using orthogonal wavelet series. In Section 2, we show how prior knowledge about a function's regularity can be incorporated into a prior model for its wavelet coefficients. A relationship between the hyperparameters of the proposed model and the parameters of those Besov spaces within which realizations from the prior will fall is established. Such a relation may be seen as giving insight into the meaning of the Besov space parameters themselves. Furthermore, in Section 3, we discuss Bayesian wavelet-based estimation that gives rise to different types of wavelet shrinkage in non-parametric regression. In, particular, for the prior specified, we show that a posterior median is a *bona fide* thresholding rule. Finally, in Section 4, we discuss an extension of the proposed Bayesian model by considering random functions generated by an overcomplete wavelet dictionary.

3.2   Wavelets and Besov spaces

3.2.1   Wavelet series

Orthogonal wavelet series in $L^2(\mathbb{R})$ are generated by dilations and translations of a mother wavelet ψ: $\psi_{jk}(t) = 2^{j/2}\psi(2^j t - k)$, $j, k \in \mathbb{Z}$. In many practical situations, the functions involved are only defined on a compact set, such as the interval $[0, 1]$, and to apply wavelets then requires some modifications. Cohen *et al.* (1993) have obtained the necessary boundary corrections to retain orthonormality. In later sections, however, we confine attention to periodic functions on $\mathbb{R}$ with unit period and work in effect with periodic wavelets (see, for example, Daubechies, 1992, Section 9.3).

The wavelet coefficients w_{jk} of the function are then actually restricted to the resolution and spatial indices $j \geq 0$ and $k = 0, \ldots, 2^j - 1$ respectively, and the function can be expanded in the orthogonal wavelet series as:

$$g(t) = c_0\phi(t) + \sum_{j=0}^{\infty} \sum_{k=0}^{2^j-1} w_{jk}\psi_{jk}(t),$$

where $c_0 = \int g(t)\phi(t)dt$ and $w_{jk} = \int g(t)\psi_{jk}(t)dt$. The function ϕ called the *scaling function* or the *father* wavelet (see any standard text on wavelets).

Wavelets are local in both time (via translations) and frequency/scale (via dilations) domains. This localization allows parsimonious representation for a wide set of different functions in wavelet series – by choosing the scaling function and the mother wavelet with corresponding regularity properties, one can generate an unconditional wavelet basis in a wide set of function spaces, such as Besov spaces (see Section 2.3 below). For detailed comprehensive expositions of the mathematical aspects of wavelets we refer, for example, to Meyer (1992) and Wojtaszczyk (1997).

We shall assume that the scaling function ϕ and the mother wavelet ψ correspond to an r-regular multiresolution analysis, for some integer $r > 0$ (see, for example, Daubechies, 1992). This will imply that ϕ and ψ are members of the Hölder space C^r, and that ψ has vanishing moments up to order r. For examples of mother wavelets with various regularity properties, and with compact support, see Daubechies (1992).

3.2.2   *Prior model*

As we have already mentioned, a large variety of different functions allow parsimonious representation in wavelet series where only a few non-negligible coefficients are present in the expansion. To capture this characteristic feature of wavelet bases, Abramovich, Sapatinas & Silverman (1998a) suggested to place the prior on w_{jk} of the following form:

$$w_{jk} \sim \pi_j N(0, \tau_j^2) + (1 - \pi_j)\delta(0), \quad j \geq 0; \quad k = 0, \ldots, 2^j - 1, \quad (2)$$

where $0 \leq \pi_j \leq 1$, $\delta(0)$ is a point mass at zero, and w_{jk} are independent. To complete the model a vague prior is placed on the scaling coefficient c_0.

According to the prior model (11), every w_{jk} is either zero with probability $1 - \pi_j$ or with probability π_j is normally distributed with zero mean and variance τ_j^2. The probability π_j gives the proportion of non-zero wavelet coefficients at resolution level j while the variance τ_j^2 is a measure of their magnitudes. Note that the prior parameters π_j and τ_j^2 are the same for all coefficients at a given resolution level j.

The hyperparameters of the prior model (2) are assumed to be of the form:

$$\tau_j^2 = c_1 2^{-\alpha j} \quad \text{and} \quad \pi_j = \min(1, c_2 2^{-\beta j}), \quad j \geq 0, \quad (3)$$

where c_1, c_2, α, and β are non-negative constants. Some intuitive under-standing of the model implied by (2), (3) can be found in Abramovich, Sapatinas & Silverman (1998a, Section 4.2).

It is interesting to compare the priors (2) with the three point 'least favourable' priors of the form:

$$\omega_{jk} \sim \frac{\pi_j}{2}\delta(\mu_j) + \frac{\pi_j}{2}\delta(-\mu_j) + (1 - \pi_j)\delta(0) \tag{4}$$

used for derivation minimax wavelet estimators. The expressions for π_j and μ_j are given in Donoho & Johnstone (1994) and Johnstone (1994).

Clyde, Parmigiani & Vidakovic (1998) use a similar formulation to (2) but with different forms for the hyperparameters π_j and τ_j^2. The prior model (2) is also an extreme case of that of Chipman, Kolaczyk & Mc-Culloch (1997). Their prior for each ω_{jk} is the mixture of two normal distributions with zero means but different variances for 'negligible' and 'non-negligible' wavelet coefficients.

3.2.3   Besov spaces on the interval

Before establishing a relation between the hyperparameters of the prior model (2) and the parameters of those Besov spaces within which realiza-tions from the prior will fall, we introduce a brief review of some relevant aspects of the theory of the (inhomogeneous) Besov spaces on the interval that we exploit further. For a more detailed study we refer, for example, to DeVore & Popov (1988), DeVore, Jawerth & Popov (1992), Meyer (1992) and Wojtaszczyk (1997).

Let the r-th difference of a function g be:

$$\Delta_h^{(r)}g(t) = \sum_{k=0}^{r} \binom{r}{k}(-1)^k g(t + kh),$$

and let the r-th modulus of smoothness of g in $L^p[0,1]$ be:

$$\nu_{r,p}(g;t) = \sup_{h \le t} \|\Delta_h^{(r)}g\|_{L^p[0,1-rh]}.$$

Then the Besov seminorm of index (s,p,q) is defined for $r > s$, where $1 \le p, q \le \infty$, by:

$$|g|_{B_{p,q}^s} = \left\{ \int_0^1 \left(\frac{\nu_{r,p}(g;h)}{h^s} \right)^q \frac{dh}{h} \right\}^{1/q}, \quad \text{if } 1 \le q < \infty,$$

and by:

$$|g|_{B_{p,\infty}^s} = \sup_{0 < h < 1} \left\{ \frac{\nu_{r,p}(g;h)}{h^s} \right\}.$$

The Besov space $B_{p,q}^s$ on $[0,1]$ is the class of functions $g : [0,1] \to \mathbb{R}$ for which $g \in L^p[0,1]$ and $|g|_{B_{p,q}^s} < \infty$. The Besov norm is defined then as:

$$\|g\|_{B_{p,q}^s} = \|g\|_{L^p[0,1]} + |g|_{B_{p,q}^s}, \quad 1 \leq q < \infty,$$

$$\|g\|_{B_{p,\infty}^s} = \|g\|_{L^p[0,1]} + |g|_{B_{p,\infty}^s}.$$

The (not necessarily integer) parameter s measures the number of derivatives, where the existence of derivatives is required in an L^p-sense, while the parameter q provides a further finer gradation.

The Besov spaces include, in particular, the well-known Sobolev ($B_{2,2}^m$) and Hölder ($B_{\infty,\infty}^s$) spaces of smooth functions, but in addition less traditional spaces, like the space of functions of bounded variation, sandwiched between $B_{1,1}^1$ and $B_{1,\infty}^1$. The latter functions are of statistical interest because they allow for better models of spatial inhomogeneity (see, for example, Meyer, 1992; Donoho & Johnstone, 1995).

For $j \geq 0$, define w_j to be the vector of wavelet coefficients w_{jk}, $k = 0, 1, \ldots, 2^j - 1$, as defined in Section 2.1. The Besov norm of g is equivalent to the corresponding sequence space norm:

$$\|w\|_{b_{p,q}^s} = |c_0| + \left\{ \sum_{j=0}^{\infty} 2^{js'q} \|w_j\|_p^q \right\}^{1/q}, \quad \text{if } 1 \leq q < \infty, \quad (5)$$

$$\|w\|_{b_{p,\infty}^s} = |c_0| + \sup_{j \geq 0} \left\{ 2^{js'} \|w_j\|_p \right\}, \quad (6)$$

where $s' = s + 1/2 - 1/p$ (see, for example, Meyer, 1992; Donoho et al., 1995).

In Section 2.4, we exploit this equivalence of the norms for relating prior information about the function's regularity to the hyperparameters of our prior model for the wavelet coefficients w_{jk}.

In the particular case $p = q = 1$ the sequence space norm in (5) becomes a weighted sum of the $|w_{jk}|$ and the corresponding Besov space norm is essentially an L^1-norm on the derivatives of g up to order s. This will provide motivation for the loss function we use in Section 3.

3.2.4 A relation between Besov space parameters and hyperparameters of the prior model

In this section we demonstrate how knowledge about regularity properties of an unknown response function can be incorporated into the prior model (2) for its wavelet coefficients by specifying the hyperparameters of the prior. We explore the connections between the parameters α and β in (2) of the prior model (11) and the Besov space parameters s and p.

Note first that it follows from (3) that the prior expected number of non-zero wavelet coefficients on the j-th level is $C_2 2^{j(1-\beta)}$. Then, appealing to the first Borel-Cantelli lemma, in the case $\beta > 1$, the number of non-zero coefficients in the wavelet expansion is finite almost surely and, hence, with probability one, g will belong to the same Besov spaces as the mother wavelet ψ, i.e. those for which $\max(0, 1/p - 1/2) < s < r$, $1 \leq p, q \leq \infty$.

More fruitful and interesting is, therefore, the case $0 \leq \beta \leq 1$. The case $\beta = 0$ corresponds to the prior belief that all coefficients on all levels have the same probability of being non-zero. This characterises self-similar processes such as white noise or Brownian motion, the overall regularity depending on the value of α. The case $\beta = 1$ assumes that the expected number of non-zero wavelet coefficients is the same on each level which is typical, for example, for piecewise polynomial functions (see Abramovich, Sapatinas & Silverman, 1998a for details). In general, for the case $0 \leq \beta \leq 1$, the resulting random functions are fractal (rough) (see, Wang, 1997).

Suppose that g is generated from the prior model (2) with hyperparameters specified by (3). Because of the improper nature of the prior distribution of c_0, we consider the prior distribution of g conditioned on any given value for c_0. The following theorem, proved in Abramovich, Sapatinas & Silverman (1998a), establishes necessary and sufficient conditions for g to fall (with probability one) in any particular Besov space.

Theorem 1 *(Abramovich, Sapatinas & Silverman, 1998a). Let ψ be a mother wavelet that corresponds to an r-regular multiresolution analysis. Consider constants s, p and q such that $\max(0, 1/p - 1/2) < s < r$, $1 \leq p, q \leq \infty$. Let the wavelet coefficients w_{jk} of a function g obey the prior model (2) with $\tau_j^2 = c_1 2^{-\alpha j}$ and $\pi_j = \min(1, c_2 2^{-\beta j})$, where $c_1, c_2, \alpha \geq 0$ and $0 \leq \beta \leq 1$.*

Then $g \in B_{p,q}^s$ almost surely if and only if either:

$$s + 1/2 - \beta/p - \alpha/2 < 0, \tag{7}$$

or:

$$s + 1/2 - \beta/p - \alpha/2 = 0 \quad and \quad 0 \leq \beta < 1, \ 1 \leq p < \infty, \ q = \infty. \tag{8}$$

Remark 2.1. The result of Theorem 1 is true for all values of the Besov space parameter q. This should not be surprising due to the embedding properties of Besov spaces (see, for example, Peetre, 1975). To give some insight on the role of q, Abramovich, Sapatinas & Silverman (1998a) considered a more delicate dependence of the variance τ_j^2 on the level j by adding a third hyperparameter $\gamma \in \mathbb{R} : \tau_j^2 = c_1 2^{-\alpha j} j^\gamma$, and extended the results of Theorem 1 for this case (see their Theorem 2).

Theorem 1 essentially includes several important aspects. It shows how prior knowledge about a function's regularity (measured by a Besov space membership) can be incorporated into the prior model (2) for its wavelet coefficients by choosing the corresponding hyperparameters of their prior distribution. It may also be seen as giving insight into the meaning of the

Besov space parameters themselves and, in a way, attempts to 'exorcise' these 'devilish' spaces for statisticians ('besov' is the literal Russian translation of 'devilish'!). Finally, unlike 'least favourable' realizations implied by the three-point prior (4), the priors (2) may be preferable to generate 'typical' functions of particular Besov spaces (see, Abramovich, Sapatinas & Silverman, 1998a, Section 4.3) for Bayesian simulation procedures that have become very popular in recent years.

3.3   Bayesian wavelet estimators

3.3.1   Wavelet-based thresholding procedure

Before discussing Bayesian wavelet estimators, we review some basic aspects of the wavelet-based thresholding procedure. Recall that according to the original model (1), the unknown response function $g(t), t \in [0, 1]$ corrupted by 'white' noise is observable at n discrete points $t_i = i/n$:

$$y_i = g(t_i) + \epsilon_i, \quad i = 1, \ldots, n,$$

where ϵ_i are independent and identically distributed normal variables with zero mean and variance σ^2.

Given observed discrete data $\mathbf{y} = (y_1, \ldots, y_n)^{\mathrm{T}}$, we may find the vector $\hat{\mathbf{d}}$ of its sample discrete wavelet coefficients by performing the *discrete wavelet transform* (DWT) of $\mathbf{y}$:

$$\hat{\mathbf{d}} = \mathcal{W}\mathbf{y},$$

where $\mathcal{W}$ is the orthogonal DWT-matrix with (jk, i) entry given by:

$$\sqrt{n}\, W_{jk,i} \approx \psi_{jk}(i/n) = 2^{j/2}\psi(2^j i/n - k).$$

As usual, we assume that $n = 2^J$ for some positive integer J. Then the DWT yields $(n - 1)$ sample discrete wavelet coefficients $\hat{d}_{jk}, j = 0, \ldots, J - 1;\ k = 0, \ldots, 2^j - 1$, and one sample scaling coefficient $\hat{c}_0$, which is the sample mean $\bar{y}$ multiplied by $\sqrt{n}$.

Both DWT and inverse DWT are performed by Mallat's (1989) fast algorithm that requires only $O(n)$ operations. Due to the orthogonality of $\mathcal{W}$, the DWT of a white noise is also an array ε_{jk} of independent $N(0, \sigma^2)$, so

$$\hat{d}_{jk} = d_{jk} + \varepsilon_{jk}, \quad j = 0, \ldots, J - 1, \quad k = 0, \ldots, 2^j - 1,$$

where the discrete wavelet coefficients d_{jk} are the DWT of the vector of discrete function values $(g(t_1), \ldots, g(t_n))^{\mathrm{T}}$ and are related to the 'theoretical' wavelet coefficients $w_{jk} = \int g(t)\psi_{jk}(t)\, dt$ by $d_{jk} \approx \sqrt{n}\, w_{jk}$. The $\sqrt{n}$ factor essentially arises from the difference Between continuous and discrete orthogonality conditions. This factor cannot be avoided and, therefore, we use different letters d_{jk} and w_{jk} to clarify the distinction.

As we have discussed before, wavelets allow parsimonious representation for a wide variety of functions so it is reasonable to assume that only a few 'large' $\hat{d}_{jk}$ really contain information about the unknown function g, while the 'small' coefficients are attributed to the noise. The extraction of those 'significant' coefficients can be naturally done by thresholding $\hat{d}_{jk}$'s:

$$\hat{d}^{\star}_{jk} = \hat{d}_{jk}\,\mathrm{I}(|\hat{d}_{jk}| > \lambda) \quad \text{(hard thresholding)} \tag{9}$$

$$\hat{d}^{\star}_{jk} = \mathrm{sign}(\hat{d}_{jk})\max(0, |\hat{d}_{jk}| - \lambda) \quad \text{(soft thresholding)}, \tag{10}$$

where $\lambda \geq 0$ is a threshold value. The hard thresholding is a 'keep' or 'kill' rule, while the soft thresholding is a 'shrink' or 'kill' rule. The resulting coefficients $\hat{d}^{\star}_{jk}$ are then used for selective reconstruction of an estimate by the inverse DWT:

$$\hat{g} = \mathcal{W}^{\mathrm{T}}\hat{d}^{\star}$$

The choice of λ is obviously crucial: small/large threshold values will produce estimates that tend to overfit/underfit the data. Donoho & Johnstone (1994) proposed the *universal* threshold $\lambda_{DJ} = \sigma\sqrt{2\log n}$. Despite the 'triviality' of such a threshold, they showed that the resulting wavelet estimator is asymptotically near-minimax among all estimators within the whole range of Besov spaces. Wang (1996) and Johnstone & Silverman (1997) studied corresponding universal thresholds for the case of 'coloured' noise. Abramovich & Silverman (1998) derived universal thresholds for wavelet estimators based on *indirect* data in inverse problems. However, the universal threshold essentially 'ignores' the data and, hence, it is not 'tuned' to the specific problem at hand.

Several *data-adaptive* thresholding rules have been developed recently. Donoho & Johnstone (1995) proposed the *SureShrink* thresholding rule which is based on minimizing the Stein's unbiased risk estimate (Stein, 1981). Abramovich & Benjamini (1996), Ogden & Parzen (1996a, 1996b) considered thresholding as multiple hypotheses testing procedure. Nason (1996), Jansen, Malfait & Bultheel (1997) adjusted the well known cross-validation approach for choosing λ.

Bayesian approaches to thresholding were recently explored by Chipman, Kolaczyk & McCulloch (1997), Abramovich, Sapatinas & Silverman (1998a), Clyde & George (1998), Clyde, Pargimiani & Vidakovic (1998), Crouse, Nowak & Baraniuk (1998), Johnstone & Silverman (1998) and Vidakovic (1998) among others, and some of them will be discussed in detail below.

3.3.2   Bayesian wavelet shrinkage

In this section we discuss a Bayesian formalism that leads to different types of wavelet shrinkage estimators.

For the discrete wavelet coefficients d_{jk}, the corresponding prior model will be:

$$d_{jk} \sim \pi_j N(0, \tau_j^2) + (1 - \pi_j)\delta(0), \quad j = 0, \ldots, J-1; \quad k = 0, \ldots, 2^j - 1, \quad (11)$$

where the hyperparameters are of the form:

$$\tau_j^2 = C_1 2^{-\alpha j} \quad \text{and} \quad \pi_j = \min(1, C_2 2^{-\beta j}), \quad j = 0, \ldots, J-1, \quad (12)$$

with $C_1 = nc_1, C_2 = c_2$.

Subject to the prior (11), the posterior distribution $d_{jk} | \hat{d}_{jk}$ is also a mixture of a corresponding posterior normal distribution and $\delta(0)$. Letting Φ be the standard normal cumulative distribution function, the posterior cumulative distribution function $F(d_{jk} | \hat{d}_{jk})$ is:

$$F(d_{jk} \mid \hat{d}_{jk}) = \frac{1}{1 + \eta_{jk}} \Phi\left(\frac{d_{jk} - \hat{d}_{jk}\tau_j^2/(\sigma^2 + \tau_j^2)}{\sigma\tau_j/\sqrt{\sigma^2 + \tau_j^2}}\right) + \frac{\eta_{jk}}{1 + \eta_{jk}} I(\hat{d}_{jk} \geq 0),$$

$$(13)$$

where the posterior odds ratio for the component at zero is:

$$\eta_{jk} = \frac{1 - \pi_j}{\pi_j} \frac{\sqrt{\tau_j^2 + \sigma^2}}{\sigma} \exp\left(-\frac{\tau_j^2 \hat{d}_{jk}^2}{2\sigma^2(\tau_j^2 + \sigma^2)}\right). \quad (14)$$

Different losses lead to different Bayesian rules. The traditional Bayes rule usually considered in the literature (see, for example, Chipman, Kolaczyk & McCullagh, 1997; Clyde, Pargimiani & Vidakovic, 1998; Vidakovic, 1998) corresponds to the L^2-loss and yields the posterior mean. Using (13) and (14), we then have:

$$E(d_{jk} \mid \hat{d}_{jk}) = \frac{1}{1 + \eta_{jk}} \frac{\tau_j^2}{\tau_j^2 + \sigma^2} \hat{d}_{jk}. \quad (15)$$

Obviously, such a rule is never a thresholding rule but a (nonlinear) smoothing shrinkage. Instead, Abramovich, Sapatinas & Silverman (1998a) suggested the use of the posterior median that corresponds to the L^1-loss and leads to a *bona fide* thresholding rule. To fix terminology, a *shrinkage* rule shrinks wavelet coefficients towards zero, whilst a *thresholding* rule in addition sets actually to zero all coefficients below a certain threshold. As explained in Section 2.3, L^1-losses on the estimated function and its derivatives, corresponding to $B_{1,1}^s$ norms for the function space loss, will be, for all applicable values of s, equivalent to suitable weighted combinations of L^1-losses on the wavelet coefficients w_{jk}. Thus, whichever weighted combination is used, the corresponding Bayes rule will be obtained by taking the posterior median of each wavelet coefficient. By following Abramovich,

Sapatinas & Silverman (1998a), one gets the following closed form for the posterior medians:

$$\text{Med}(d_{jk} \mid \hat{d}_{jk}) = \text{sign}(\hat{d}_{jk}) \max(0, \zeta_{jk}),$$

where:

$$\zeta_{jk} = \frac{\tau_j^2}{\sigma^2 + \tau_j^2} |\hat{d}_{jk}| - \frac{\tau_j \sigma}{\sqrt{\sigma^2 + \tau_j^2}} \Phi^{-1}\left(\frac{1 + \min(\eta_{jk}, 1)}{2}\right). \qquad (16)$$

The quantity ζ_{jk} is negative for all $\hat{d}_{jk}$ in some implicitly defined interval $[-\lambda_j, \lambda_j]$, and hence $\text{Med}(d_{jk}|\hat{d}_{jk})$ is zero whenever $|\hat{d}_{jk}|$ falls below the threshold λ_j. The posterior median is therefore a level-dependent 'kill' or 'shrink' thresholding rule with thresholds λ_j.

Abramovich, Sapatinas & Silverman (1998a) called this Bayesian thresholding procedure *BayesThresh*. Note that, unlike soft thresholding (10), extent of shrinkage in *BayesThresh* depends on $|\hat{d}_{jk}|$: the larger $|\hat{d}_{jk}|$, the less it is shrinked. For large $\hat{d}_{jk}$ the *BayesThresh* asymptotes to linear shrinkage by a factor of $\tau_j^2/(\sigma^2 + \tau_j^2)$, since the second term in (16) becomes negligible as $|\hat{d}_{jk}| \to \infty$.

Remark 3.1. The universal threshold $\lambda_{DJ} = \sigma\sqrt{2\log n}$ of Donoho & Johnstone (1994) can be also obtained as a particular limiting case of *BayesThresh* rule setting $\alpha = \beta = 0$ and letting $C_1 \to \infty$, $C_2 \to 0$ as n increases in such a way that $\sqrt{C_1}/(C_2 \sigma n) \to 1$. Such a peculiar prior is a direct consequence of its 'least favourable' nature.

Another way to obtain a *bona fide* thresholding rule within a Bayesian framework is via a hypothesis testing approach (see, Vidakovic, 1998). The idea is simple: after observing $\hat{d}_{jk}$, test the hypothesis $H_0 : d_{jk} = 0$ against a two-sided alternative $H_1 : d_{jk} \neq 0$. If the hypothesis H_0 is rejected, d_{jk} is estimated by $\hat{d}_{jk}$, otherwise $d_{jk} = 0$. Such a procedure essentially mimics the hard thresholding rule:

$$\hat{d}_{jk}^\star = \hat{d}_{jk} I(\eta_{jk} < 1), \qquad (17)$$

where $\eta_{jk} = P(H_0 \mid \hat{d}_{jk})/P(H_1 \mid \hat{d}_{jk})$ is the posterior odds ratio. Vidakovic (1998) called this thresholding rule Bayes factor (*BF*) thresholding since the posterior odds ratio is obtained by multiplying the Bayes factor with the prior odds ratio. Thus, a wavelet coefficient $\hat{d}_{jk}$ will be thresholded if the corresponding posterior odds ratio $\eta_{jk} > 1$ and will be kept as it is otherwise, where η_{jk} for our prior model (11), (12) is given by (14).

To compare *BayesThresh* and *BF*, note that *BF* is always a 'keep' or 'kill' hard thresholding, whilst *BayesThresh* is a 'shrink' or 'kill' thresholding, where extend of shrinkage depends on the absolute values of the wavelet coefficients. In addition, *BF* thresholds $\hat{d}_{jk}$ if the corresponding $\eta_{jk} > 1$.

One can verify from (16) that *BayesThresh* will 'kill' those $\hat{d}_{jk}$, whose:

$$\eta_{jk} > 1 - 2\Phi\left(-\tau_j|\hat{d}_{jk}|/\sigma\sqrt{\sigma^2 + \tau_j^2}\right)$$

and, hence, will threshold more coefficients. Figure 1 shows the different Bayesian rules for some choices of the hyperparameters.

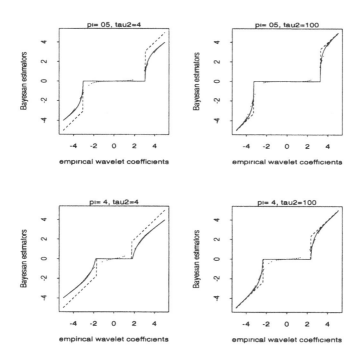

FIGURE 1. The posterior medians (solid lines), the posterior means (dotted lines) and the *BF* (dashed lines) rules as functions of the empirical wavelet coefficients for some choices of the hyperparameters π and τ^2, while σ was fixed at 1.

3.3.3 *Estimation of the hyperparameters*

To apply the Bayesian wavelet-based methods discussed in Section 3.2 in practice, it is necessary first to specify the hyperparameters α, β, C_1 and C_2 in (12). Ideally, the choice of α and β should be made from prior knowledge about regularity properties of the unknown function making use of the results of Theorem 1. Some practical issues for the choice of α and β have been investigated by Abramovich, Sapatinas & Silverman (1998a) and will be briefly discussed further in Section 3.4 below.

In what follows, we assume that α and β have been chosen in advance (or they are known quantities), and estimate C_1 and C_2 by the following procedure in the spirit of empirical Bayes as suggested by Abramovich, Sapatinas & Silverman (1998a).

The set of sample wavelet coefficients $\hat{d}_{jk}$ contains both 'non-negligible' coefficients of the unknown function g and 'negligible' coefficients representing a random noise. Apply the universal threshold of Donoho & Johnstone (1994) $\lambda_{DJ} = \sigma\sqrt{2\log n}$ described above in Section 3.1. Donoho & Johnstone (1994) showed that the probability that even one negligible coefficient will pass the threshold value λ_{DJ} tends to zero, so essentially only non-negligible $\hat{d}_{jk}$ will survive after universal thresholding. Suppose that, on level j, the number of coefficients that pass λ_{DJ} is M_j, and that the values of these coefficients are $x_{j1}, \ldots, x_{jM_j}$. Conditioning on the value M_j, the x_{jm}, $m = 1, \ldots, M_j$, are independent realizations from the tails of the $N(0, \sigma^2 + \tau_j^2)$ distribution beyond the points $\pm\sigma\sqrt{2\log n}$. The log likelihood function is therefore, up to a constant:

$$l(\tau_0^2, \ldots, \tau_{J-1}^2) = -\sum_{j=0}^{J-1} M_j \left\{ \frac{1}{2}\log(\sigma^2 + \tau_j^2) - \log\left(\Phi\left[-\frac{\lambda_{DJ}}{\sqrt{\sigma^2 + \tau_j^2}}\right]\right)\right\}$$
$$-\sum_{j=0}^{J-1}\left\{\frac{1}{2(\sigma^2 + \tau_j^2)}\sum_{m=1}^{M_j} x_{jm}^2\right\}. \tag{18}$$

Substituting $\tau_j^2 = C_1 2^{-\alpha j}$ and $\lambda_{DJ} = \sigma\sqrt{2\log n}$, and given the values of α and σ, we can obtain an estimate of C_1 by a numerical maximization of (18).

The parameter C_2 can be chosen by a cognate procedure. We use the numbers $M_0, \ldots, M_{J-1}$ of coefficients passing the threshold to estimate the π_j. Let $q_j = 2\Phi(-\lambda_{DJ}/\sqrt{\sigma^2 + \tau_j^2})$, the probability conditional on $d_{jk} \neq 0$ that d_{jk} passes the threshold λ_{DJ}. Neglecting the possibility that any $\hat{d}_{jk}$ corresponding to a zero d_{jk} passes the threshold λ_{DJ}, the 'imputed number' of non-zero d_{jk} at level j is M_j/q_j, and the expected value of M_j/q_j is $C_2 2^{(1-\beta)j}$. Given the value of β, a simple method of moments estimate of C_2 based on the total imputed number of non-zero d_{jk} is:

$$\hat{C}_2 = \frac{2^{(1-\beta)} - 1}{2^{(1-\beta)J} - 1}\sum_{j=0}^{J-1}\frac{M_j}{q_j}, \quad \text{if } 0 \leq \beta < 1,$$

$$\hat{C}_2 = \frac{1}{J}\sum_{j=0}^{J-1}\frac{M_j}{q_j}, \quad \text{if } \beta = 1.$$

Note also that if the noise level σ is unknown, it is usual in practice to estimate it robustly by the median absolute deviation of the wavelet coefficients at the finest level, $\hat{d}_{J-1,k} : k = 0, 1, \ldots, 2^{J-1} - 1$, divided by 0.6745 (see, Donoho & Johnstone, 1994). Alternatively, one can adapt a fully Bayesian approach by placing a prior on σ^2 and considering a hierarchical Bayesian model (see, for example, Clyde, Pargimiani & Vidakovic, 1998; Vidakovic, 1998).

We point out that empirical Bayes approaches (conditional maximum likelihood, marginal maximum likelihood) for estimating the hyperparameters π_j and τ_j^2, at each resolution level j separately, in the general form (11) have been recently considered by Clyde & George (1998), Johnstone & Silverman (1998).

3.3.4 Simulations

Abramovich, Sapatinas & Silverman (1998a) performed a comprehensive simulation study to compare *BayesThresh* procedure with standard non-Bayesian thresholding rules. They considered the 'Blocks', 'Bumps', 'Heavisine' and 'Doppler' test functions of Donoho & Johnstone (1994) that caricature spatially variable signals arising in diverse scientific fields and have become standard tests for wavelet estimators.

The results showed that, for various signal-to-noise ratios for all test functions, *BayesThresh* compares favourably with its non-Bayesian counterparts (see, Abramovich, Sapatinas & Silverman, 1998a for details). In particular, for $\alpha = 0.5$ and $\beta = 1$, *BayesThresh* outperformed all non-Bayesian estimators in almost all cases in terms of the mean square error. Abramovich, Sapatinas & Silverman (1998a) suggested to make it a 'standard default' choice for prior hyperparameters when prior knowledge about a function's regularity properties is difficult to elicit.

We have continued the study of Abramovich, Sapatinas & Silverman (1998a). Using the same test functions and for the same signal-to-noise ratios, we have compared different Bayesian wavelet procedures discussed in Section 3.2: posterior means (15), posterior medians (*BayesThresh*) and Bayes Factor (17) using the 'standard' choice $\alpha = 0.5$ and $\beta = 1$. The three Bayesian methods yield quite similar results. Usually, posterior means have a smaller mean square error, *BayesThresh* second with *BF* very close to it. All of them outperformed their non-Bayesian competitors in all cases.

3.4 Stochastic expansions in an overcomplete wavelet dictionary

3.4.1 From bases to dictionaries

In recent years there has been growing interest in the atomic decomposition of functions in overcomplete dictionaries (see, for example, Mallat & Zhang, 1993; Davis, Mallat & Zhang, 1994; Chen, Donoho & Saunders,

1999). Every basis is essentially only a *minimal* necessary dictionary needed to represent a large variety of different functions. Such 'miserly' representation usually causes poor adaptivity (Mallat & Zhang, 1993). The use of *overcomplete* dictionaries increases the adaptivity of the representation, because one can choose now the most suitable one among many available. One can see an interesting analogy with colours. Theoretically, every other colour can be generated by combining three basic colours (green, red and blue) in corresponding proportions. However, a painter would definitely prefer to use the whole available palette (overcomplete dictionary) to get the hues he needs!

In mathematical terms, an atomic decomposition of a function (signal) g is an expression of g as a superposition of a parametric collection of waveforms $(\psi_\lambda)_{\lambda \in \Lambda}$:

$$g(t) = \sum_{\lambda \in \Lambda} \omega_\lambda \psi_\lambda(t).$$

The collection of waveforms $(\psi_\lambda)_{\lambda \in \Lambda}$ is called a *dictionary*, and the waveforms ψ_λ are called *atoms*.

Here, we naturally focus on wavelet dictionaries. The atoms of a wavelet dictionary $\mathcal{D}_\Lambda = \{\psi_\lambda : \lambda \in \Lambda\}$ with the set Λ of indices $\lambda = (a, b)$ are translations and dilations of a single mother wavelet and are of the form:

$$\psi_\lambda(t) = a^{1/2}\psi(a(t - b)), \quad a \geq 1, \quad 0 \leq b \leq 1.$$

In particular, for the orthonormal wavelet dictionary $\Lambda = \{(2^j, k2^{-j}), j \geq 0, k = 0, ..., 2^j - 1\}$. *Overcomplete* wavelet dictionaries are obtained by sampling indices more finely. An important example of overcomplete wavelet dictionaries is the non-decimated (or stationary or translation-invariant) wavelet dictionary (see, for example, Coifman & Donoho, 1995; Nason & Silverman, 1995). Atomic decompositions in overcomplete dictionaries are obviously nonunique and one may think about choosing the 'best' possible representation among many available (see, for example, Mallat & Zhang, 1993; Davis, Mallat & Zhang, 1994; Chen, Donoho & Saunders, 1999).

3.4.2   Prior model

Consider the overcomplete wavelet dictionary where the scales and dilations of wavelet atoms ψ_λ are not dyadic constraints any longer, but arbitrary. To extend the prior model (2) for orthonormal wavelet bases, Abramovich, Sapatinas & Silverman (1998b) modelled the set of the locations of wavelet atoms and their magnitudes as being sampled from a certain marked Poisson process.

More specifically, let the set Λ of indices $\lambda = (a, b)$ be sampled from a Poisson process S on $[1, \infty) \times [0, 1]$ with intensity $\mu(\lambda)$. Conditional on S, the corresponding coefficients ω_λ are assumed to be independent normal random variables:

$$\omega_\lambda \mid S \sim N(0, \tau^2(\lambda)). \tag{19}$$

To complete the model it is assumed that both the variance $\tau^2(\lambda)$ and the intensity $\mu(\lambda)$ depend on the scale a only, and are of the form:

$$\tau_a^2 \propto a^{-\delta} \quad \text{and} \quad \mu_a \propto a^{-\varsigma}, \quad a \geq 1, \tag{20}$$

where $\delta, \varsigma \geq 0$, with $\delta + \varsigma > 0$.

The intuitive basis of the proposed model is an extension of the notion that the orthogonal wavelet series representation of an unknown function is sparse. The parameter ς controls the relative rarity of 'fine-scale' wavelet atoms in the function, while the parameter δ controls the size of the contribution of these atoms when they appear. For example, if ς is small and δ is large, there will be a considerable number of 'fine-scale' atoms but these will each have fairly low contribution, so one might expect the functions to be reasonably smooth and homogeneous. On the other hand, if ς is large and δ is small, there will be occasional large 'fine-scale' effects in the functions.

3.4.3 Regularity properties of random functions

Consider now a random function g generated by the wavelet dictionary $(\psi_\lambda)_{\lambda \in \Lambda}$:

$$g(t) = \sum_{\lambda \in \Lambda} \omega_\lambda \psi_\lambda(t), \tag{21}$$

where the random locations λ of atoms and their random magnitudes ω_λ obey the prior (19), (20). The following Theorem 2 proved in Abramovich, Sapatinas & Silverman (1998b) establishes a relation between the hyperparameters ς and δ of the prior and the parameters s and p of those Besov spaces within which g will fall (with probability one), extending thus Theorem 1 for orthonormal wavelet bases.

Note that, for $\varsigma > 1$, the intensity $\mu_a \propto a^{-\varsigma}$ is integrable over the range of λ for which ψ_λ has support intersecting $[0, 1]$. Therefore, the number of relevant terms in the atomic decomposition (21) is finite almost surely and, hence, with probability one, g will belong to the same Besov spaces as the mother wavelet ψ, namely those for which $\max(0, 1/p - 1/2) < s < r$, $1 \leq p \leq \infty$, $1 \leq q \leq \infty$. The more interesting case is again $0 \leq \varsigma \leq 1$.

Theorem 2 *(Abramovich, Sapatinas & Silverman, 1998b). Let ψ be a compactly supported mother wavelet that corresponds to an r-regular multiresolution analysis. Consider constants s, p and q such that $\max(0, 1/p - 1/2) < s < r$, $1 \leq p, q \leq \infty$. Consider a function g as defined in (21), with the conditional variances $\tau_a^2 \propto a^{-\delta}$ and the intensity of the Poisson process $\mu_a \propto a^{-\varsigma}$. Assume that $\delta \geq 0$, $0 \leq \varsigma \leq 1$, and that $\delta + \varsigma > 0$. Assume also that the wavelets are sufficiently regular that $\delta < 2r + 2\rho - 1$.*

Then $g \in B_{p,q}^s$ almost surely if and only if

$$s + 1/2 - \zeta/p - \delta/2 < 0. \tag{22}$$

Theorem 2 establishes a sufficient and necessary condition for realizations to fall in a particular Besov space. It shows that the function's smoothness measured by the parameter s depends both on the intensity of 'fine-scale' atoms (via ζ) and their magnitudes (via δ). The parameter p can be seen as 'discouraging inhomogeneity', in that the larger the value of p the more emphasis is placed on the parameter δ. For large δ, no matter how many 'fine-scale' atoms there are, they each make a relatively low contribution. On the other hand, if p is small, then there is a trade-off where large weights on 'fine-scale' atoms (small δ) can be tolerated if the corresponding atoms are relatively rare (large ζ).

Theorem 2 makes it possible in principle to incorporate prior knowledge about a function's regularity into a prior model for its atomic wavelet representation. The models considered in this section show how Bayesian ideas can be extended to a broader range of wavelet models, freed from the dyadic positions and scales considered in the classical case. The algorithmic details, probably involving modern Bayesian computational methods, have yet to be worked out in detail and are an interesting subject for future research. The improvement to 'standard' wavelet methods obtained by moving from the discrete (decimated) wavelet transform to the non-decimated wavelet transform (see, for example, Coifman & Donoho, 1995; Nason & Silverman, 1995; Lang et al., 1996; Johnstone & Silverman, 1997) suggest that a Bayesian approach based on a general atomic decomposition may result in yet better performing wavelet shrinkage estimators.

Acknowledgments: This work was started when Dr Theofanis Sapatinas was a Research Associate in the School of Mathematics, University of Bristol, supported by EPSRC grant GR/K70236.

References

Abramovich, F. & Benjamini, Y. (1996). Adaptive thresholding of wavelet coefficients. *Computational Statistics and Data Analysis* **22**, 351-361.

Abramovich, F. & Silverman, B.W. (1998). Wavelet decomposition approaches to statistical inverse problems. *Biometrika* **85**, 115-129.

Abramovich, F., Sapatinas, T. & Silverman, B.W. (1998a). Wavelet thresholding via a Bayesian approach. *J. Roy. Statist. Soc. B* **60**, 725-749.

Abramovich, F., Sapatinas, T. & Silverman, B.W. (1998b). Stochastic expansions in an overcomplete wavelet dictionary. *Probability Theory and Related Fields* (under invited revision).

Chen, S.S.B., Donoho, D.L. & Saunders, M.A. (1999). Atomic decomposition by basis pursuit. *SIAM Journal on Scientific Computing*, **20**, 33-61.

Chipman, H.A., Kolaczyk, E.D. & McCulloch, R.E. (1997). Adaptive Bayesian wavelet shrinkage. *J. Am. Stat. Ass.* **92**, 1413-1421.

Clyde, M. & George, E.I. (1998). Robust empirical Bayes estimation in wavelets. *Discussion Paper* **98-21**, Institute of Statistics and Decision Sciences, Duke University, USA.

Clyde, M., Parmigiani, G. & Vidakovic, B. (1998). Multiple shrinkage and subset selection in wavelets. *Biometrika* **85**, 391-401.

Cohen, A., Daubechies, I., Jawerth, B. & Vial, P. (1993). Wavelets on the interval and fast wavelet transforms. *Applied and Computational Harmonic Analysis* **1**, 54-81.

Coifman, R.R. & Donoho, D.L. (1995). Translation-invariant de-noising. In *Wavelets and Statistics*, Lecture Notes in Statistics **103**, Antoniadis, A. and Oppenheim, G. (Eds.), pp. 125-150, New York: Springer-Verlag.

Coifman, R.R. & Wickerhauser, M.V. (1992). Entropy-based algorithms for best-basis selection. *IEEE Transactions on Information Theory* **38**, 713-718.

Crouse, M., Nowak, R. & Baraniuk, R. (1998). Wavelet-based statistical signal processing using hidden Markov models. *IEEE Transactions on Signal Processing* **46**, 886-902.

Daubechies, I. (1988). Time-frequency localization operators: a geometric phase space approach. *IEEE Transactions on Information Theory* **34**, 605-612.

Daubechies, I. (1992). *Ten Lectures on Wavelets*. Philadelphia: SIAM.

Davis, G., Mallat, S.G. & Zhang, Z. (1994). Adaptive time-frequency approximations with matching pursuit. In *Wavelets: Theory, Algorithms, and Applications*, Chui, C.K., Montefusco, L. and Puccio, L. (Eds.), pp. 271-293, San Diego: Academic Press.

DeVore, R.A., Jawerth, B. & Popov, V. (1992). Compression of wavelet decompositions. *American Journal of Mathematics* **114**, 737-785.

DeVore, R.A. & Popov, V. (1988). Interpolation of Besov Spaces. *Transactions of the American Mathematical Society* **305**, 397-414.

Donoho, D.L. & Johnstone, I.M. (1994). Ideal spatial adaption by wavelet shrinkage. *Biometrika* **81**, 425-455.

Donoho, D.L. & Johnstone, I.M. (1995). Adapting to unknown smoothness via wavelet shrinkage. *J. Am. Stat. Ass.* **90**, 1200-1224.

Jansen, M., Malfait, M. & Bultheel, A. (1997). Generalized cross validation for wavelet thresholding. *Signal Processing* **56**, 33-44.

Johnstone, I.M. (1994). Minimax Bayes, asymptotic minimax and sparse wavelet priors. In *Statistical Decision Theory and Related Topics, V*, Gupta, S.S. and Berger, J.O. (Eds.), pp. 303-326, New York: Springer-Verlag.

Johnstone, I.M. & Silverman, B.W. (1997). Wavelet threshold estimators for data with correlated noise. *J. Roy. Statist. Soc. B* **59**, 319-351.

Johnstone, I.M. & Silverman, B.W. (1998). Empirical Bayes approaches to mixture problems and wavelet regression. *Technical Report*, Department of Mathematics, University of Bristol, UK.

Lang, M., Guo, H., Odegard, J.E., Burrus, C.S. & Wells Jr, R.O. (1996). Noise reduction using an undecimated discrete wavelet transform. *IEEE Signal Processing Letters* **3**, 10-12.

Mallat, S.G. (1989). A theory for multiresolution signal decomposition: the wavelet representation. *IEEE Transactions on Pattern Analysis and Machine Intelligence* **11**, 674–693.

Mallat, S.G. & Zhang, Z. (1993). Matching pursuit in a time-frequency dictionary. *IEEE Transactions on Signal Processing* **41**, 3397-3415.

Meyer, Y. (1992). *Wavelets and Operators.* Cambridge: Cambridge University Press.

Nason, G.P. (1996). Wavelet shrinkage using cross-validation. *J. Roy. Statist. Soc. B* **58** 463-479.

Nason, G.P. & Silverman, B.W. (1994). The discrete wavelet transform in S. *Journal of Computational and Graphical Statistics* **3**, 163-191.

Nason, G.P. & Silverman, B.W. (1995). The stationary wavelet transform and some statistical applications. In *Wavelets and Statistics*, Lecture Notes in Statistics **103**, Antoniadis, A. and Oppenheim, G. (Eds.), pp. 281-300, New York: Springer-Verlag.

Ogden, T. & Parzen, E. (1996a). Data dependent wavelet thresholding in nonparametric regression with change-point applications. *Computational Statistics and Data Analysis* **22**, 53-70.

Ogden, T. & Parzen, E. (1996b). Change-point approach to data analytic wavelet thresholding. *Statistics and Computing* **6**, 93-99.

Peetre, J. (1975). *New Thoughts on Besov Spaces.* Durham: Duke University Press.

Silverman, B.W. (1985). Some aspects of the spline smoothing approach to nonparametric regression curve fitting (with discussion). *J. Roy. Statist. Soc. B* **47**, 1-52.

Stein, C. (1981). Estimation of the mean of a multivariate normal distribution. *Ann. Stat.* **9**, 1135-1151.

Steinberg, D.M. (1990). A Bayesian approach to flexible modeling of multivariate response for functions. *Journal of Multivariate Analysis* **34**, 157-172.

Vidakovic, B. (1998). Non-linear wavelet shrinkage with Bayes rules and Bayes factors. *J. Am. Stat. Ass.* **93**, 173-179.

Wahba, G. (1983). Bayesian 'confidence intervals' for the cross-validated smoothing spline. *J. Roy. Statist. Soc. B* **45**, 133-150.

Wang, Y. (1996). Function estimation via wavelet shrinkage for long-memory data. *Ann. Stat.* **24**, 466-484.

Wang, Y. (1997). Fractal function estimation via wavelet shrinkage. *J. Roy. Statist. Soc. B* **59**, 603-612.

Wojtaszczyk, P. (1997). *A Mathematical Introduction to Wavelets.* Cambridge: Cambridge University Press.

4

Some Observations on the Tractability of Certain Multi-Scale Models.

Eric D. Kolaczyk

ABSTRACT Multi-scale modeling (wavelet-based or otherwise) recently has been found to be an area rich and natural for the use of Bayesian frameworks. However, the tractability of a given approach (e.g., with respect to optimization, computation, simulation, etc.) tends to rely on a sometimes fortuitous interplay between the proposed multi-scale analysis scheme and the probability distributions specified in the model. This interplay is explored here through a series of case studies with recursive dyadic partition (RDP) models, with some attention given to wavelet-based models as well. A graphical modeling perspective proves useful in illustrating the types of relationships between distribution and multi-scale structure that are found to lead to (un)tractable settings. For the RDP models, in which a multi-scale structure is achieved through simple summing of pairs of "children" into "parents," it is observed that a sufficient condition for a particularly tractable model is that the "parent" serve as a *cut* for the distribution of the "children."

4.1   Introduction.

The term "multi-scale modeling" has come to refer broadly to an area of study in which complex objects of interest are represented using structures made up of a simpler component(s), across a variety of scales and positions. An important attribute of a large portion of the multi-scale modeling procedures introduced to date is their use of computationally efficient implementations. Given that many (though by no means all) applications of such procedures involve large data sets, on-line processing, and/or repeated application over sizeable databases, computational issues can play a prominent role in model development. Indeed, it does not seem unreasonable to speculate that the popularity of, for example, wavelet-based methods would have been much slower in spreading, despite their powerful mathematical properties, if not accompanied by "fast" algorithms (e.g., the $O(n)$ orthogonal wavelet transform, etc.). In this paper, a brief study is presented of how the interplay of statistical distributions and multi-scale structure can lead to (un)tractable frameworks, where "tractability" refers loosely to how efficiently tasks such as optimization, simulation, mathematical manipulation, etc. may be accomplished. A series of case studies will be emphasized, using a simple form of multi-scale analysis, wherein the sense of the term "tractable" will be clear.

In considering issues of tractability, a useful attribute to note among multi-scale models is whether the relevant distributional structure(s) is (a) synthesized from, or (b) analyzed by a given dictionary of time-scale / time-frequency elements. For example, the study of multi-scale stochastic processes (e.g., Flandrin (1992), Abry & Sellan (1996)) has tended to fall into the former category, while many of the models for statistical estimation fall into the latter (e.g., Donoho & Johnstone (1994),Donoho, Johnstone, Kerkyacharian & Picard (1994)). Tractability under *synthesis* is something that the researcher often is able to insure through construction, usually by inheritance from the structure of the chosen dictionary. In the case of *analysis,* however, a particular choice of dictionary may yield a (un)tractable framework, depending on the form of both the statistical distribution and the parameterization in multi-scale space induced by the analysis. For the building of *Bayesian* multi-scale models, consideration of both synthesis and analysis can come into play, and in fact interact, e.g., through specification of a prior distribution (synthesis) and a transformation of the data likelihood (analysis).

As a motivating example, consider the estimation of a (discrete) intensity function $\Lambda \equiv (\Lambda_0, \Lambda_1, \dots, \Lambda_{n-1})$ from a time series of Poisson counts $\mathbf{X} \equiv (X_0, X_1, \dots, X_{n-1})$. For a wavelet-based model, statistical dependency will exist among wavelet coefficients of $\mathbf{X}$ whose corresponding wavelets share intervals of support. This dependency is to be contrasted with the independence of coefficients in the case of the canonical "signal plus Gaussian noise" model. As a result of this dependency, the mathematical structure is less tractable in the Poisson case than in the Gaussian case for tasks such as the derivation of shrinkage thresholds or the evaluation of estimator performance (e.g., see Kolaczyk (1998)). An alternative class of multi-scale models for Poisson data was introduced recently by Kolaczyk (1999), one based on *recursive dyadic partitioning* (also see Timmerman & Nowak (1999) for a similar formulation). This framework yields models that are flexible enough to produce estimates with good performance properties, yet possess tractable and interpretable structures – characteristics shared with the standard wavelet-based models for Gaussian data.

Models based on recursive dyadic partitions (RDPs) also have been found to be similarly successful in the canonical Gaussian noise model (e.g., Engel (1994), Donoho (1997)). Combining this observation with the discussion above suggests that how the distributional structure and the elements of the multi-scale analysis interact can play an important role in determining the tractability of the final model structure. An interesting question is whether it is possible to state a set of conditions under which certain forms of tractability follow. Here, as a first step towards addressing this question, an exploratory approach is taken in which a series of case studies are examined. We will concentrate on the RDP as a multi-scale analysis tool, although occasionally wavelet models will enter the discussion as well. In section 4.2, the basic RDP model is introduced in a fairly general sense. A graphical models formalism will then be used to provide a more structured context within which to examine the issue of tractability. Four specific case studies are then presented in section 4.3, in which the Gaussian, Poisson, gamma, and Cauchy distributions are specified for the data. In section 4.4 some general conclusions are drawn regarding the role of *factorization* and *cuts* in achieving tractable multi-scale models based on RDPs. Some additional discussion then follows in section 4.5.

4.2   Recursive Dyadic Partition Models.

We begin with a rather general formulation of a RDP analysis. Throughout the remainder of this article then, our interest will be in exploring and illustrating the effects of the addition of more specific model components.

4.2.1   The Basic Model.

Let $\mathbf{X} \equiv (X_0, X_1, \ldots, X_{n-1})$ be a vector of independent random variables, for $n \equiv 2^J$, where X_i is the i-th variable observed in time, with distribution $\Pr(X_i; \theta_i)$, for unknown scalar parameter θ_i. We will assume that interest is focused on the estimation of $\boldsymbol{\theta}$ from $\mathbf{X}$, through the use of a Bayesian multi-scale model.

For the most part, we will concentrate on the following simple multiscale analysis of $\mathbf{X}$.

$$X_{J,k} \equiv X_k \qquad \text{and} \qquad X_{j,k} = X_{j+1,2k} + X_{j+1,2k+1}, \qquad (1)$$

for scales $j = 0, 1, \ldots, J-1$ and locations $k = 0, 1, \ldots, n_j - 1$, where $n_j \equiv 2^j$. Note that if, for example, $\boldsymbol{\theta} \equiv E[\mathbf{X}]$, then a similar analysis of $\boldsymbol{\theta}$ is induced as well. The process described in (1) will be called a *recursive dyadic partition (RDP) analysis*, since it is produced by recursively splitting the total sum $X_{0,0}$ across a set of dyadic subintervals of the observation period. The basic procedure of (not necessarily dyadic) recursive partitioning of a data-space, of course, has analogues in a variety of settings, under various other names (e.g., the CART algorithm of Breiman *et al.* 1983). Here we use it to effect a particular form of multi-scale analysis, one similar to an analysis with Haar wavelets. [Technically, an RDP analysis possesses a one-to-one correspondence with analysis by a so-called *hereditary* Haar basis (see Engel (1994)), in which the coefficient of a given wavelet may be non-zero only if the coefficient of the corresponding "parent" wavelet is non-zero as well.]

Implicit in pursuing a multi-scale analysis of $\mathbf{X}$ is the hope that it provides a more useful representation of the information in the data than $\mathbf{X}$ itself. Essentially, the RDP analysis may be viewed as examining a collection of histograms of the data, with bin-widths ranging from $1/2^J$ (i.e., the original data, with a unit-length observation period) down to 1 (i.e., the total sum $X_{0,0}$). The random variables $X_{j,k}$ (or histogram bins) containing counts from overlapping intervals will necessarily be correlated. Hence, while the independence of the original observations permitted the data likelihood to be factored over the time index i, which generally allows for more tractable expressions and algorithms, a similarly factored expression for all of the components of the RDP analysis is not forthcoming. However, through selective sub-sampling of the RDP analysis components we find that (under appropriate conditions) the following multi-scale factorization is permitted.

$$\prod_{k=0}^{n-1} \Pr(X_{J,k}) = \Pr(X_{0,0}) \prod_{j=0}^{J-1} \prod_{k=0}^{n_j-1} \Pr(X_{j+1,2k}|X_{j,k}) . \qquad (2)$$

In other words, by recursively conditioning a "child" $X_{j+1,2k}$ on the sum of itself and its "sibling," that is on the "parent," the indexing of the components of the likelihood function changes from one of "time" to one of location/scale. Here, in a slight abuse of notation, we use $\Pr(Y)$ to refer to either a density or a probability mass function, depending on whether Y is a continuous or discrete random variable, and we assume that all of the relevant probabilities are well-defined.

Note that the expression in (2) says nothing about the effect of the RDP analysis on $\boldsymbol{\theta}$. Suppose that the parameterization induced by this analysis is collected in the parameter vector $\boldsymbol{\omega}$, where $dim(\boldsymbol{\omega})$ may or may not be equal to $dim(\boldsymbol{\theta})$. Viewing multi-scale modeling from a Bayesian perspective, it is upon $\boldsymbol{\omega}$ that we would like to impose a prior distribution structure. Hence, a particularly nice occurance would be if we found that

$$\prod_{k=0}^{n-1} \Pr(X_{J,k}; \theta_k) = \Pr(X_{0,0}; \Omega_{0,0}) \prod_{j=0}^{J-1} \prod_{k=0}^{n_j-1} \Pr(X_{j+1,2k}|X_{j,k}; \omega_{j,k}) , \qquad (3)$$

where the re-parameterization $\boldsymbol{\omega} = (\Omega_{0,0}, \{\omega_{j,k}\}_{j,k})$ of $\boldsymbol{\theta}$ is similarly n-length *and* the components of $\boldsymbol{\omega}$ *factor with the likelihood* in a multi-scale fashion.

Given the scenario in (3), if independent prior distributions are placed upon the components of $\boldsymbol{\omega}$, then a multi-scale factorization of the posterior distribution follows as well i.e.,

$$
\begin{aligned}
\Pr(\omega|\mathbf{X}) &= \Pr(\Omega_{0,0}|X_{0,0}) \prod_{j,k} \frac{\Pr(X_{j+1,2k}|X_{j,k},\omega_{j,k})\,\Pr(\omega_{j,k})}{\Pr(X_{j+1,2k}|X_{j,k})} \\
&= \Pr(\Omega_{0,0}|X_{0,0}) \prod_{j,k} \Pr(\omega_{j,k}|X_{j+1,2k}, X_{j+1,2k+1}) ,
\end{aligned}
\tag{4}
$$

under the condition that $\Pr(\omega_{j,k}|X_{j,k}) = \Pr(\omega_{j,k})$. The second line follows from the first in (4) by noting that the terms inside the product of the first can be expressed as $\Pr(\omega_{j,k}|X_{j+1,2k}, X_{j+1,2k+1})\,\Pr(\omega_{j,k})/\Pr(\omega_{j,k}|X_{j,k})$, which are equal to the terms inside the product of the second under the stated condition. This condition dictates that $\omega_{j,k}$ be independent of $X_{j,k}$, for all (j,k), in the sense that $\omega_{j,k}$ in (3) parameterizes the conditional distribution of the $(j+1)$-level children only. In certain contexts (e.g., see the Gaussian and Poisson cases in section 4.3) such a condition is natural to impose, as the $\omega_{j,k}$ only play a role in determining how the $X_{j,k}$ are split into their children, $X_{j+1,2k}$ and $X_{j+1,2k+1}$.

In contexts where (3) and (4) obtain, many tasks pertaining to modeling become quite tractable. For example, maximization of the posterior over over ω is reduced to a set of individual optimization problems for the $\omega_{j,k}$. Also, given the tree-like structure (inherited from the RDP analysis) along which the various components in (2) - (4) separate, simulation methods such as those based on MCMC can be done in an efficient manner.

4.2.2   A Graphical Models Formulation.

The expressions in (2) - (4), ignoring the multi-scale context, are simply the results of a particular specification of conditional distributions among a collection of random variables. As such, it is natural to use the device of graphical models here, as it will help to more clearly illustrate the issues relevant to our discussion of tractable multi-scale models.

We will introduce the use of graphical models through example. A more complete introduction may be found in, for example, Lauritzen (1996) or Frey (1998). For simplicity, consider the case where $n = 4$ i.e., $\mathbf{X} = (X_0, X_1, X_2, X_3)$. Then figure 1 shows a graphical model representation of the RDP analysis described by (1). Random variables at vertices connected by an edge are dependent. Those edges with arrows on one end indicate a "parent-child" relationship, which often can be summarized using a conditional probability statement. For example, with respect to issues of tractability, the graph in figure 1 is most interesting to us when paired with the multi-scale factorization of $\Pr(\mathbf{X})$ in (2). Formally, it is said that the distribution of $\mathbf{X}$ *factorizes* with respect to this graph. This particular factorization follows because of the conditional independence of "children," given their "parents," from the rest of their "ancestors." For example, $X_{2,0}$ is independent of $X_{0,0}$, given $X_{1,0}$.

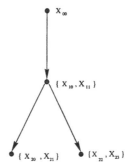

FIGURE 1. Graphical model representation of RDP model.

A graphical model for the probabilistic structures underlying (3) and (4) is shown in figure 2. Here the circles for the $\omega_{j,k}$ have been left unfilled, to indicate the fact that they are unobserved. The graph structure among the $X_{j,k}$'s is simply that of figure 1 laid on its side. The lack of edges between the elements of ω indicates their independence from one another, as does the similar lack of edges between $\omega_{j,k}$ and $X_{j,k}$, for each pair (j, k). In the time domain, the effects of this particular multi-scale structure might be presented as in figure 3, where the elements (X_0, X_1, X_2, X_3) are independent, given θ, but an induced dependency likely exists among the elements of θ (e.g., see Kolaczyk (1999) for an example in the context of Poisson data).

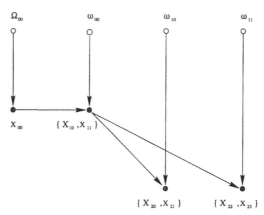

FIGURE 2. Graphical model representation of Bayesian RDP model.

Graphical models also may be used, of course, to represent wavelet-based models. For example, several authors (e.g., Chipman, Kolaczyk & McCulloch (1997), Clyde, Parmigiani & Vidakovic (1998),Abramovich, Sapatinas & Silverman (1998),and Vidakovic (1998)) have introduced Bayesian versions of the standard orthogonal wavelet model for the Gaussian noise problem, in which independent mixture priors are placed on the individual

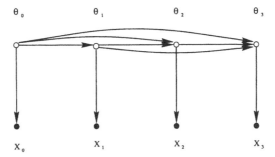

FIGURE 3. Possible graphical model representation of time-domain model generated by the Bayesian RDP model in figure 2.

wavelet coefficients $w_{j,k}$. In addition, conditional on the $w_{j,k}$, the empirical wavelet coefficients, $d_{j,k}$, are modeled as independent. This particular model structure is shown in figure 4, where $c_{0,0}$ is the observed scale coefficient at the zeroth scale. A more general formulation of this idea is given in Crouse, Nowak & Baraniuk (1998), where dependencies are imposed on the wavelet coefficients $w_{j,k}$ (also see the paper of Nowak in this volume). This additional structure would then be reflected in the graphical model in figure 4 through the addition of edges among the $w_{j,k}$ (possibly with arrows, as dictated by the particular form of the dependencies specified).

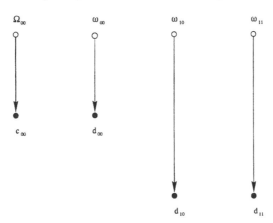

FIGURE 4. Graphical model representation for typical Bayesian wavelet model for canonical Gaussian noise problem.

When wavelets are used in the Poisson model, as mentioned in section 4.1, there are dependencies among many of the empirical wavelet coefficients themselves. Hence, a version of figure 4 would apply, but with edges drawn between various $d_{j,k}$, as dictated by the overlap of support of the corresponding wavelet functions. Qualitatively, such a graph would look quite similar to the graph of the RDP model in figure 2 (possibly without the use of arrows). A crucial difference between the two contexts is that with the wavelets the distribution of $\mathbf{X}$ does not admit of a factorization analogous to (2). The importance of factorization will be addressed further in section 4.4, in conjunction with that of an additional concept (i.e., the concept of a *cut*), the latter of which is motivated by issues arising in the following case studies.

4.3   Case Studies.

In this section, we return our focus to the RDP analysis. As mentioned in section 4.1, this particular vehicle for multi-scale analysis has proven useful with both Gaussian and Poisson noise models. A property shared by both distributions is that they are *reproducing* under summation. With respect to the RDP analysis, having this attribute means that the family of distributions to which the distribution of each "parent" belongs is the same as that of its two "children". Furthermore, in the cases of both models, the conditional distribution of a "child", given its "parent", takes on a particularly nice form. Together, these two facts allow for the components of $\boldsymbol{\omega}$ to factor with the likelihood in the Gaussian and Poisson models, as in equation (3). We demonstrate through example, however, using the gamma and Cauchy models, that the *reproducing* property alone is not a sufficient condition for obtaining this factorization.

4.3.1   Gaussian Model.

Let $X_i \sim \text{Normal}(\theta_i, \sigma^2)$, $i = 0, 1, \ldots, n-1$, with $\sigma > 0$ known. We seek an expression for the multi-scale likelihood factorization in (2). Defining $\theta_{J,k} \equiv \theta_k$ and $\theta_{j,k} = \theta_{j+1,2k} + \theta_{j+1,2k+1}$, for $j = 0, \ldots, J-1$ and $k = 0, 1, \ldots, 2^j - 1$, we have that

$$\begin{pmatrix} X_{j+1,2k} \\ X_{j+1,2k+1} \end{pmatrix} \sim \text{Normal}\begin{pmatrix} \theta_{j+1,2k} \\ \theta_{j+1,2k+1} \end{pmatrix}, 2^{J-j-1}\sigma^2 I_2 \end{pmatrix}, \qquad (5)$$

where I_2 is the 2×2 identity matrix. Using standard properties for the multivariate normal distribution, it then follows that

$$X_{j+1,2k}|X_{j,k} \sim \text{Normal}\left(X_{j,k}/2 + \omega_{j,k}, 2^{J-j-2}\sigma^2\right) \qquad (6)$$

with $\omega_{j,k} \equiv (\theta_{j+1,2k} - \theta_{j+1,2k+1})/2$. As a result, we may write

$$\prod_{i=0}^{n-1} f(X_i; \theta_i, \sigma^2)dX_i \;=\; f(X_{0,0}; \theta_{0,0}, \sigma^2)dX_{0,0} \times \qquad (7)$$

$$\prod_{j=0}^{J-1} \prod_{k=0}^{n_j-1} f(X_{j+1,2k}|X_{j,k}; \omega_{j,k}, \sigma^2)dX_{j+1,2k} \;.$$

In (7) the density for $X_{0,0}$ is Gaussian, with mean $\theta_{0,0}$ and variance σ^2, while the densities within the product are defined according to (6).

Note that the factorization in (7) is of the form shown in (3), in that the multi-scale parameter ω has factored with the likelihood function. The elements of ω are, except for normalization, the coefficients of a Haar transformation of θ. Therefore, it can be seen that in the Gaussian case the parameters in the graphs of figures 2 and 4 are essentially the same, although the (in)dependency of the observable elements (i.e., the $X_{j,k}$ and the $d_{j,k}$, in the RDP and Haar analyses, respectively) differs.

4.3.2 Poisson Model.

Let $X_i \sim \text{Poisson}(\theta_i)$. Due to the reproducing property of the Poisson distribution, it follows that $X_{j,k} \sim \text{Poisson}(\theta_{j,k})$, for any (j,k), where $\theta_{j,k}$ is defined as for the Gaussian case above. Combining this with the well-known result that

$$X_{j+1,2k}|X_{j,k} \sim \text{Binomial}(X_{j,k}; \omega_{j,k}) \tag{8}$$

when $\omega_{j,k} \equiv \theta_{j+1,2k}/\theta_{j,k}$, we arrive at the factorization

$$\prod_{i=0}^{n-1} \text{Pr}(X_i; \theta_i) = \text{Pr}(X_{0,0}; \theta_{0,0}) \prod_{j=0}^{J-1} \prod_{k=0}^{n_j-1} \text{Pr}(X_{j+1,2k}|X_{j,k}; \omega_{j,k}) \ . \tag{9}$$

As in the Gaussian case, the factorization in (9) is of the form in (3). However, note that the entire model (i.e., parameters and observables) is now distinctly different from that obtainable under a Haar transform. Also, whereas in the Gaussian case the RDP analysis produces a "total" parameter $\theta_{0,0}$ that is partitioned into the components of θ via additive adjustments, in the Poisson case the adjustments are now multiplicative.

This structure has been exploited in Kolaczyk (1999) (as has a similar structure in Timmerman & Nowak (1999)) as a foundation for producing a new type of Bayesian multi-scale model for Poisson data. The relevant prior distributions are mixtures of a point-mass at $1/2$ and a symmetric beta distribution. Noting how the parameters $\omega_{j,k}$ enter into the model in (8), it is reasonable to model them as independent from the $X_{j,k}$. In that case, the expression in (4) holds for the posterior distribution of θ, which allows for a computationally efficient, recursive expression (across scales) for the posterior mean. Additionally, an expectation maximization (EM) algorithm is given for computing an empirical Bayes estimate of the scale-dependent mixing parameters in the prior distributions. The marginal distribution of the data admits a factorization that proves fundamental in obtaining simple expressions for the E- and M-steps. Due partly to the fact that this multi-scale model was crafted "within" the framework of the Poisson likelihood (as opposed to being imposed from "without" via, say, a wavelet transform), the mixing parameters have a natural interpretation as the "fraction of homogeneity" in the underlying process; the empirical Bayes estimates thus provide a data-dependent summary of this structure.

4.3.3   Gamma Model.

Let $X_i \sim$ Gamma(θ_i, b), with $b > 0$ known. The gamma distribution is reproducing under summation, when the scale parameter b is common among the summands, and so we have that $X_{j,k} \sim$ Gamma$(\theta_{j,k}, b)$, for any pair (j, k), where $X_{j,k}$ and $\theta_{j,k}$ are defined as above. As a model for non-negative, continuous random variables, the gamma distribution has been found to be quite useful (e.g., see Touzi, Lopes & Bousquet (1988) for an application to synthetic aperature radar imaging), although typically it is not the shape parameters θ that vary with i but rather the scale parameters b. Nevertheless, examination of the particular form of the gamma model specified here is useful in illustrating that the reproducing property alone does not insure that a factorization like that in (3) arises.

Specifically, it can be shown that

$$X_{j+1,2k}|X_{j,k} \sim X_{j,k}B_{j,k} \;, \tag{10}$$

where $B_{j,k} \sim$ Beta$(\theta_{j+1,2k}, \theta_{j+1,2k+1})$. This result follows, for example, by showing that the joint distribution of $(X_{j+1,2k}, X_{j+1,2k+1})$ can be factored into the product of the distributions of $X_{j,k}$ (i.e., gamma) and $X_{j+1,2k}/X_{j,k}$ (i.e., beta) Johnson, Kotz & Balakrishnan (1994, pg. 350). Combining this result with (2), we then have the factorization

$$\prod_{i=0}^{n-1} f(X_i; \theta_i, b)dX_i = f(X_{0,0}; \theta_{0,0}, b)dX_{0,0} \times \tag{11}$$

$$\prod_{j=0}^{J-1} \prod_{k=0}^{n_j-1} f(X_{j+1,2k}|X_{j,k}; \theta_{j+1,2k}, \theta_{j+1,2k+1})dX_{j+1,2k} \;.$$

We see that for the gamma model the parameter $\boldsymbol{\theta}$ does not factor into n separate components with the multi-scale factorization of the likelihood, as in (3), since each of the $n-1$ density functions inside the product in (11) are functions of the two parameters $\theta_{j+1,2k}$ and $\theta_{j+1,2k+1}$. Furthermore, since we have $\theta_{j,k} \equiv \theta_{j+1,2k} + \theta_{j+1,2k+1}$, it does not make sense to place prior distributions on each of the multi-scale parameters, as was the case in the Gaussian and Poisson models, since the values of those at the finest scale (i.e., the original time-domain parameters, θ_i) completely determine those at all coarser scales. Hence, an expression for the posterior distribution like that in (4) will not result, and therefore neither will the particular form of tractability that follows.

A graphical model representing the situation is shown in figure 5. Note that the probabilistic structure on the $X_{j,k}$'s has not changed from that of figure 2. However, the structure corresponding to the multi-scale parameters shows the redundancy just mentioned, where the lines directed from "children" to "parent" reflect the fact that the values of the former define the latter.

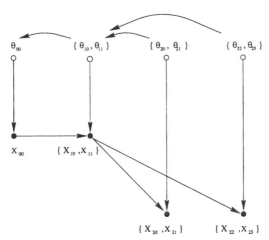

FIGURE 5. Graphical model representation for RDP analysis of gamma model.

4.3.4 Cauchy Model.

For our final case study, consider the following Cauchy model, where $X_i \sim$ Cauchy(θ_i, b), with $b > 0$, i.e.

$$f(X_i; \theta_i, b) = 1/\pi b \left[1 + ((X_i - \theta_i)/b)^2\right]^{-1} . \tag{12}$$

This is a natural example to consider when a "signal plus noise" model is desired in which the noise distribution has much heavier tails than a Gaussian (e.g., with $b \equiv 1$, the X_i are distributed as t random variables with 1 degree of freedom, located at θ_i).

As with the other cases examined above, the Cauchy distribution possesses a reproducing property, and we find that $X_{j,k} \sim$ Cauchy$(\theta_{j,k}, 2^{J-j}b)$. The conditional distribution of $X_{j+1,2k}$, given $X_{j,k}$, can be shown to have the density function

$$f(X_{j+1,2k}|X_{j,k}; \theta_{j+1,2k}, \theta_{j+1,2k+1}, b) = \frac{1}{\pi 2^{J-j-2}b} \times$$

$$\frac{1 + \left[1 + \left(\frac{X_{j,k}-\theta_{j,k}}{2^{J-j}b}\right)^2\right]}{\left[1 + \left(\frac{X_{j+1,2k}-\theta_{j+1,2k}}{2^{J-j-1}b}\right)^2\right]\left[1 + \left(\frac{X_{j,k}-X_{j+1,2k}-\theta_{j+1,2k+1}}{2^{J-j-1}b}\right)^2\right]} \tag{13}$$

and therefore the Cauchy model admits the multi-scale factorization

$$\prod_{i=0}^{n-1} f(X_i; \theta_i, b)dX_i = f(X_{0,0}; \theta_{0,0}, b)dX_{0,0} \times$$

$$\prod_{j=0}^{J-1} \prod_{k=0}^{n_j-1} f(X_{j+1,2k}|X_{j,k}; \theta_{j+1,2k}, \theta_{j+1,2k+1}, b)dX_{j+1,2k}. \quad (14)$$

So the Cauchy model encounters the same trouble under the RDP analysis as the gamma model, and thus may be represented by the graphical model in figure 5 as well. The form of the conditional density of $X_{J+1,2k}$, given $X_{j,k}$, is illustrated in figure 6. In general, this density is symmetric about the point $(\theta_{j+1,2k} + X_{j,k})/2$, with modes at $\theta_{j+1,2k}$ and $X_{j,k} - \theta_{j+1,2k+1}$.

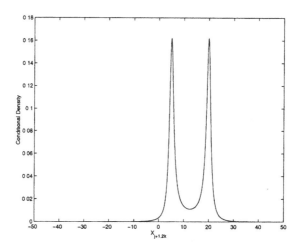

FIGURE 6. Density under the Cauchy model for the conditional distribution of $X_{J+1,2k}$, given $X_{j,k} = 30$, with parameters $\theta_{j+1,2k} = 5.0$, $\theta_{j+1,2k+1} = 10.0$, and $b = 1$.

4.4   Factorization and Cutting.

In section 4.1 it was asked whether conditions could be stated under which tractability of certain multi-scale models follows. For the purposes of this paper, possession of a structure as in (3) has been used as a proxy for the concept of "tractability". Although in this initial inquiry we have contented ourselves with the examination of a set of case studies and examples, these have been sufficiently rich to allow a few (necessarily) general conclusions to be drawn.

4.4.1   The Role of Graphical Models.

Although perhaps not necessary, the graphical models formalism has been found to be a useful device for the visual illustration of individual models, and the comparision of different models. Complex dependency structures often may be represented in a clear, concise fashion, one that serves to unveil relationships that remain more hidden when faced only with a mathematical expression(s).

4.4.2   Factorization.

Focusing on the RDP analysis framework, it should be clear that equation (2) plays a critical role in this paper. Expressions like this have been referred to as *multi-scale factorizations* (Kolaczyk (1999)) of the likelihood of **X**. More formally, equation (2) shows that under fairly general conditions (on the existence of the relevant densities/probabilities), the distribution of **X** *factorizes* with respect to the graph (figure 1) corresponding to the RDP analysis.

This concept of factorizing is a quite general concept in the graphical models literature, the exact definition of which varies slightly with the class of graphs to which a model belongs. Essentially, given a particular choice of graph, a factorization requires that there exist non-negative functions $k_a(\mathbf{X})$, for subsets a of the graph vertices, that depend on **X** only through those elements indexed by vertices in a, such that (with respect to the appropriate product measure) the density of **X** can be written as

$$f(\mathbf{X}) = \prod_a k_a(\mathbf{X}) \ .$$

One class of graphs that is particularly relevant here is the class of directed, acyclic graphs i.e., those that contain an arrow and for which there are no "closed-loop" paths among the vertices (i.e., paths beginning and ending at the same place). Figure 1 is a simple example of a directed, acyclic graph, and for such graphs factorization is said to occur if we may write

$$f(\mathbf{X}) = \prod_{\alpha \in V} k_\alpha \left(X_\alpha, X_{pa(\alpha)} \right) \ . \tag{15}$$

Here the product is over individual vertices α, and the functions $k_\alpha(\cdot)$ typically are the conditional densities of the components X_α, given the "parent(s)" $X_{pa(\alpha)}$.

Hence, to obtain a tractable multi-scale model framework, when the underlying multi-scale analysis admits of a directed, acyclic graph, a necessary condition would seem to be that the distribution of **X** factorize as in (15).

4.4.3 Cuts.

Although factorization is a strong property, the case studies in section 4.3 show that it alone is not a sufficient condition for the type of tractability we have in mind. Specifically, although all four models considered allow for the multi-scale factorization (2), only two yield the further simplification in (3). For the Gaussian and Poisson models, the multi-scale re-parameterizations of the n-length vector $\boldsymbol{\theta}$ (induced by the RDP analysis) take the form of a similarly n-length vector $\boldsymbol{\omega}$ whose elements separate across the multi-scale likelihood, one element per likelihood component; for the gamma and Cauchy models, there is a redundant re-parameterization $\boldsymbol{\omega}$ of length roughly $2n$.

Due to the reproducing property shared by all four of these families of distributions, and the fact that the RDP analysis relies simply on the recursive summation of adjacent pairs of random variables, it is enough to restrict our attention to the expression

$$\Pr\left(X_{j+1,2k}, X_{j+1,2k+1}\right) = \Pr\left(X_{j+1,2k}|X_{j,k}\right) \Pr\left(X_{j,k}\right) \ . \tag{16}$$

On the left-hand side, the distribution is a function of (ignoring nuisance parameters) the pair $(\theta_{j+1,2k}, \theta_{j+1,2k+1})$. On the right-hand side, the distribution of $X_{j,k}$ is a function of $\theta_{j,k}$ in all four cases. However, only in the cases of the Gaussian and Poisson models is the conditional distribution of $X_{j+1,2k}$, given $X_{j,k}$, a function of a single parameter $\omega_{j,k}$.

Formally, this behavior is an example of what Barndorff-Nielsen Barndorff-Nielsen (1978) calls a *cut*: given a random variable pair (Z, W), parameterized by $\chi = (\zeta, \eta)$, where the domain of variation of χ is simply the product of those of ζ and η, W is a cut for Z if we can write

$$\Pr(Z, W; \chi) = \Pr(Z|W; \zeta) \Pr(W; \eta) \ .$$

With respect to (16), in the Gaussian and Poisson models the "parent" $X_{j,k}$ forms a cut for the "children" $(X_{j+1,2k}, X_{j+1,2k+1})$; in the gamma and Cauchy models, it does not.

The concept of a *cut* is quite general, though also rather special. Given a cut, various properties follow, including (by definition) factorization. Hence, regarding the construction of tractable multi-scale models from an RDP analysis, a sufficient condition for tractability is that the "parents" each form a cut for their "children". When the random variables in $\mathbf{X}$ are reproducing, this issue is reduced to the study of a single representative "parent" / "child" pair.

More generally, for a multi-scale analysis that, for example, may be represented using a directed, acyclic graph, it would seem that $X_{pa(\alpha)}$ serving as a cut for X_α, for all vertices α, is a sufficient condition for tractability of the sort exemplified by equation (3).

4.5   Discussion.

In the canonical "signal plus Gaussian noise" model, where the observations are modeled as in section 4.3.1, multi-scale models based on orthogonal wavelet transforms have been the standard to date. The empirical wavelet coefficients may be expressed as $\mathbf{d} = \mathcal{W}\mathbf{X}$, where $\mathcal{W}$ is an $n \times n$ orthogonal matrix. Therefore, the wavelet transform simply represents a certain rotation of the original data, which under the assumptions of independence and Gaussianity leaves everything in the model distribution but the mean θ unchanged. We are left with then, in some sense, the most tractable of situations for which we could hope. This characteristic of invariance under rotation is not, of course, shared by other distributions of key interest, for example the Poisson distribution. Therefore, tractability is not immediate, and the search for a multi-scale analysis that yields tractable expressions for a given model structure then becomes something of interest.

Motivated both by its simplicity and its success in the Gaussian and Poisson contexts, we have in this paper restricted most of our attention to the RDP analyis as a means for building multi-scale models. For the sake of concreteness, we have judged the tractability of a given model by whether or not it achieves a likelihood factorization of the form in equation (3). The exploration of other multi-scale analyses and alternative tractability criteria is, of course, of interest. The graphical models framework that was found to be so useful here is a quite general formalism within which to construct probability models, one which can easily incorporate generalizations and extensions of the work presented herein. Additionally, the concepts of *factorization* and *cuts* are also quite general, and should be similarly useful in such future work. Finally, it should be noted that current work on issues of tractability of complex probabilistic models in fields such as graphical modeling and machine learning (e.g., see Jordan, Ghahramani, Jaakkola & Saul (1999) for a recent overview) shows promise of being similarly useful in the context of multi-scale models.

References

Abramovich, F., Sapatinas, T. & Silverman, B. W. (1998), 'Wavelet thresholding via a Bayesian approach', *Journal of the Royal Statistical Society, Series B* **60**, 725–749.

Abry, P. & Sellan, F. (1996), 'The wavelet based synthesis for fractional Brownian motion proposed by F. Sellan and Y. Meyer: remarks and implementations', *Applied and Computational Harmonic Analysis* **3**, 377–383.

Barndorff-Nielsen, O. (1978), *Information and exponential families in statistical theory*, Wiley and Sons, Inc., New York.

Chipman, H. A., Kolaczyk, E. D. & McCulloch, R. E. (1997), 'Adaptive Bayesian wavelet shrinkage', *Journal of the American Statistical Association* **92**, 1413

– 1421.

Clyde, M., Parmigiani, G. & Vidakovic, B. (1998), 'Multiple shrinkage and subset selection in wavelets', *Biometrika* **85**, 391–402.

Crouse, M. S., Nowak, R. D. & Baraniuk, R. G. (1998), 'Wavelet-based statistical signal processing using hidden Markov models', *IEEE Transactions on Signal Processing* **46**, 886–902.

Donoho, D. L. (1997), 'CART and best-ortho-basis selection: A connection', *The Annals of Statistics* **25**, 1870–1911.

Donoho, D. L. & Johnstone, I. M. (1994), 'Ideal spatial adaptation via wavelet shrinkage', *Biometrika* **81**, 425–455.

Donoho, D. L., Johnstone, I. M., Kerkyacharian, G. & Picard, D. (1994), 'Wavelet shrinkage: Asymptopia ?', *Journal of the Royal Statistical Society, Series B* **57**, 301–370.

Engel, J. (1994), 'A simple wavelet approach to nonparametric regression from recursive partitioning schemes', *Journal of Multivariate Analysis* **49**, 242–254.

Flandrin, P. (1992), 'Wavelet analysis and synthesis of fractional Brownian motions', *IEEE Transactions on Information Theory* **38**, 910–917.

Frey, B. J. (1998), *Graphical Models for Machine Learning and Digital Communication*, The MIT Press, Cambridge, Massachusetts.

Johnson, N. L., Kotz, S. & Balakrishnan, N. (1994), *Continuous Univariate Distributions, Volume 1, Second Edition*, Wiley and Sons, Inc., New York.

Jordan, M. I., Ghahramani, Z., Jaakkola, T. S. & Saul, L. K. (1999), An introduction to variational methods for graphical models, *in* M. I. Jordan, ed., 'Learning in Graphical Models', The MIT Press, Cambridge, Massachusetts.

Kolaczyk, E. D. (1998), 'Wavelet shrinkage estimation of certain Poisson intensity signals using corrected thresholds', *Statistica Sinica* **9**, 119–135.

Kolaczyk, E. D. (1999), 'Bayesian multi-scale models for poisson processes', *Journal of the American Statistical Association*. (in press).

Lauritzen, S. L. (1996), *Graphical Models*, Clarendon Press, Oxford.

Timmerman, K. E. & Nowak, R. D. (1999), 'Multiscale modeling and estimation of Poisson processes with applications to photon-limited imaging', *IEEE Transactions on Information Theory*. (in press).

Touzi, R., Lopes, A. & Bousquet, P. (1988), 'A statistical geometrical edge detector for SAR images', *IEEE Transactions on Geoscience and Remote Sensing* **26**, 764–773.

Vidakovic, B. (1998), 'Nonlinear wavelet shrinkage with Bayes rules and Bayes factors', *Journal of the American Statistical Association* **93**, 173–179.

5

Bayesian Analysis of Change-Point Models

R. Todd Ogden and James D. Lynch

ABSTRACT   A Bayesian analysis based on the empirical wavelet coefficients is developed for the standard change-point problem. This analysis is considered first for the piecewise constant Haar wavelet basis, then extended to using smooth wavelet bases. Although developed initially for use in the standard change-point model, the analysis can be applied to the problem of estimating the location of a discontinuity in an otherwise smooth function by considering only the higher level coefficients in the computations, thereby effectively smoothing the function and analyzing the resulting residuals. The procedure is illustrated by an example using simulated data.

5.1   Change-point models

A common assumption that is tenable in a number of situations is that observations $Y_1, \ldots, Y_n$ are independent and identically distributed (iid) with a particular distributional form with one or more unknown parameters. This paper will focus on modeling violations of the second assumption, that of identical distributions. Generally defined, change-point methodology deals with sets of sequentially ordered observations (as in time) and undertakes to determine whether the fundamental mechanism generating the observations has changed during the time the data have been gathered. Often, the observations are assumed to be mutually independent, the change involving one or more parameters in a specified distributional form.

The classical change-point problem supposes that the data are normally distributed with constant (but unknown) variance and the mean is suspected to undergo an abrupt change in value. This problem has been studied considerably over the years. Some review articles summarizing many years of development on the problems include Csörgő and Horváth (1988) and Krishnaiah and Miao (1988). A thorough theoretical treatment is given in the recent book by Csörgő and Horváth (1997). The problem has been generalized in several different directions: relaxing the assumption of a known parametric form for the data, allowing multiple changes, considering non-abrupt changes, supposing change occurs in several parameters at once, etc.

There are a wide range of applications for these problems. Perhaps the earliest and most enduring scenario is that of quality control, in which a sequence of measurements from a production process is analyzed for a change in, say, the (mean) thickness of a manufactured part. In such situations, *sequential* or *on-line* methods (in which analysis is performed as each new observation is gathered) are often preferred.

One interesting area of study is the extension of change-point methods into the problem of nonparametric regression. In a typical nonparametric regression setting, it is assumed that a smooth curve describes the relationship between a set of x_i's and their corresponding Y_i's. It is of interest in many of these situations to determine whether the regression function is in fact smooth, or whether it contains one or more discontinuities (either in a derivative of the function or in the function itself). The problem of estimating the location of such a discontinuity in a regression function will also be considered here.

This paper will focus initially on nonsequential Bayesian approaches based on the discrete wavelet transform (DWT) to the standard change-point problem with normal data and a single abrupt change in the mean. This standard change-point problem is then generalized quite easily to a nonparametric regression situation, in which the mean of the observations is thought to change smoothly with time, but with one abrupt change. Wavelets might offer little improvement over existing methods in the classical change-point problem, but the more general problem of detecting and estimating jumps in regression functions can benefit a great deal by applying wavelet methods.

This paper is organized as follows: Section 5.2 will specify the basic problem considered and discuss the treatment of the problem in the wavelet domain. Section 5.3 will describe the application of the Haar basis to the problem. Section 5.4 will generalize the approach of Section 5.3 to any orthogonal wavelet basis and also consider the more generalized change-point problem. Section 5.5 will discuss choice of hyperparameters and some other issues of the analysis, and finally, Section 5.6 will present a simple example of the analysis using simulated data.

5.2   Statement of the problem

Let the data consist of ordered observations $Y_1, \ldots, Y_n$ which are independent normal random variables with common variance σ^2 and means $E[Y_i] = \mu_i$. It is convenient to regard the individual means as being generated from a function f defined on $[0, 1]$:

$$\mu_i = f(i/n).$$

In the classical change-point problem, the function f is taken to be piecewise constant with a single jump at the *change-point* $\tau \in (0,1)$; in the more general setting to be considered later, f will be regarded as being smooth everywhere except at τ, where it has a jump discontinuity.

Considering only the standard change-point problem for the moment, let Δ represent the magnitude of the change that occurs at the change-point. Thus, the model becomes:

$$
\begin{aligned}
Y_i &= \mu + \epsilon_i, & i &= 1, \ldots, [n\tau] - 1, \\
Y_i &= \mu + \Delta + \epsilon_i, & i &= [n\tau], \ldots, n,
\end{aligned}
\tag{1}
$$

where $\epsilon_1, \ldots, \epsilon_n$ are a set of independent $N(0, \sigma^2)$ random variables and $[x]$ represents the greatest integer part of x. In the sequel, this model will be referred to as the *standard change-point model*. This standard change-point model corresponds to a function f_τ defined to be

$$
f_\tau(u) = \begin{cases} \mu, & 0 \le u < \tau, \\ \mu + \Delta, & \tau \le u \le 1, \end{cases}
\tag{2}
$$

for some *change-point* $\tau \in (0,1)$. Although the change-point τ is allowed to take on any value in $(0,1)$, with the data from model (1), it is impossible to do more than identify an interval (of length $1/n$) in which τ is thought to lie. It is typical to use the rightmost endpoint of the interval for the estimate of τ; i.e., if it is concluded that the ith observation is the first data point with the changed mean, then $\hat{\tau} = i/n$.

This paper will focus primarily on Bayesian estimation of τ and secondarily on the jump size Δ. The overall noise parameter σ^2 will be regarded as a fixed constant, taking an empirical Bayes approach by estimating σ^2 independently and plugging it into the model. This model could be extended to treating σ^2 as another Bayesian parameter, but this will not be treated in this paper.

As mentioned earlier, the standard change-point problem considered above can be generalized by allowing other possibilities for the mean-generating function f. In particular, one interesting case is that of a smooth function with a jump discontinuity:

$$
f(u) = \begin{cases} g(u), & 0 \le u < \tau, \\ g(u) + \delta(u), & \tau \le u \le 1, \end{cases}
\tag{3}
$$

for a change-point τ and smooth functions g and δ, with $\Delta = \delta(0) \ne 0$. The model specified in (3) will be termed the *generalized change-point model*. This paper focuses primarily on the analysis for the standard change-point model (2); the related analysis for the generalized model is given in Section 5.4.2

For simplicity of development it will be assumed that the sample size n is a power of two, i.e., $n = 2^J$ for some integer $J > 0$. The formulation of the problem will be given first in terms of the Haar wavelet and then extended to a general orthonormal wavelet basis.

The original data are first transformed into the wavelet domain and represented by its set of wavelet coefficients $(w_{j,k}, \ j = 0, \ldots, J - 1, \ k = 0, \ldots, 2^j - 1)$. Denote the vector of these coefficients by w. The distribution of the (j, k)th empirical wavelet coefficient is expressed in terms of a function $q_{j,k}(\tau)$ to be computed below:

$$w_{j,k}|\tau, \Delta \sim N(\Delta q_{j,k}(\tau), \sigma^2) \tag{4}$$

and the $w_{j,k}$'s are mutually (conditionally) independent. Note that if the function is constant, then each of the $n - 1$ coefficients in the set above has mean zero. The final element of the transformation (sometimes denoted $w_{-1,0}$) represents a final smoothing of the data; in the Haar case, this is actually proportional to $\bar{Y}$, the mean of the observations. Thus, the parameter μ is of no concern in this analysis — once the DWT has been applied to the data, the effect of μ is removed.

Applying this fact to the question of computing the mean function of the (j, k)th wavelet coefficient shows that $q_{j,k}(\tau)$ can be computed for any given values of j, k, and τ by applying the wavelet transform to the set of $[n\tau] - 1$ zeroes and $n - [n\tau] + 1$ ones. For the Haar wavelet case, an explicit formula is

$$q_{j,k}(\tau) = \begin{cases} \frac{2^{j/2}}{n}(n2^{-j}k - [n\tau]), & 2^{-j}k \leq \tau < 2^{-j}(k + \frac{1}{2}) \\ -\frac{2^{j/2}}{n}(n2^{-j}(k+1) - [n\tau]), & 2^{-j}(k + \frac{1}{2}) \leq \tau < 2^{-j}(k + 1) \\ 0, & \text{otherwise.} \end{cases}$$

This function is plotted for the Haar wavelet basis in Figure 1. It is shaped like an inverted witch's hat — it is zero outside of the support of the corresponding wavelet and goes to a peak at the midpoint of the support of $\psi_{j,k}$. This function gives some insight into how well a single empirical Haar wavelet coefficient would do at detecting a jump point: If a change occurs outside the support of $\psi_{j,k}$, it has no power to detect it; the power for detecting change is highest when the change occurs near the midpoint of the support of $\psi_{j,k}$.

5.3   Haar wavelet formulation

Since the empirical coefficients are independent,

$$p(w|\tau, \Delta) = \prod_j \prod_k p(w_{j,k}|\tau, \Delta).$$

The joint posterior distribution of τ and Δ is thus

$$p(\tau, \Delta|w) \propto \prod_j \prod_k p(w_{j,k}|\tau, \Delta) \cdot \pi(\tau, \Delta) \tag{5}$$

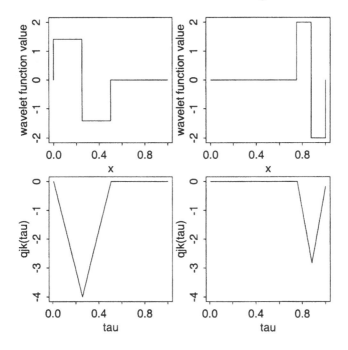

FIGURE 1. Plots of Haar wavelets (top panels) and their corresponding $q_{j,k}(\tau)$ functions (bottom panels). The left-hand plots are for $j = 1$ and $k = 0$; those on the right have $j = 2$ and $k = 3$.

for a joint prior distribution π on τ and Δ.

Applying the distributional result (4) to the expression (5) for the prior distribution of τ and Δ gives

$$
p(\tau, \Delta, |w) \propto \pi(\tau, \Delta) \exp \left(-\frac{1}{2\sigma^2} \sum_j \sum_k (w_{j,k} - \Delta q_{j,k}(\tau))^2 \right) \qquad (6)
$$

Note that if Δ and σ^2 were known and if a uniform prior were applied to τ, then the mode of the posterior distribution (6) would be that value of τ that minimizes the sum of squared differences between the observations and their expectations. Figure 2, taken from Richwine (1996), illustrates this point for a simulated data set in which σ^2 and Δ are both known to be one. The wavelet coefficients are plotted alongside the corresponding values of $q_{j,k}(\tau)$ for two different values of τ: 0.72 and 0.20. The original data had a change at $\tau = 0.72$, and it is easily seen how much better the values line up for the top panel than for the bottom panel.

The marginal posterior distribution for τ may be found applying a uniform prior for τ and an independent $N(\Delta_0, \sigma_\Delta^2)$ prior on Δ, and then inte-

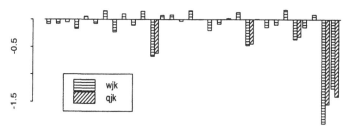

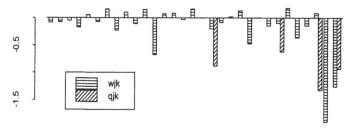

FIGURE 2. Comparing $w_{j,k}$'s with their expectations for two values of τ

grating over Δ:

$$p(\tau|w) \propto G(\tau)^{-1/2} \exp\left\{ -\frac{H_w^2(\tau)}{4G(\tau)} \right\},$$

where

$$G(\tau) = \frac{\sum_j \sum_k q_{j,k}^2(\tau)}{2\sigma^2} + \frac{1}{2\sigma_\Delta^2}$$

and

$$H_w(\tau) = \frac{\sum_j \sum_k w_{j,k} q_{j,k}(\tau)}{\sigma^2} - \frac{\Delta_0}{\sigma_\Delta^2}.$$

The constant σ^2 can be estimated by the usual MAD estimate as in Donoho and Johnstone (1995).

This analysis represents a first-pass approach to the change-point problem, but it doesn't really extend all the way into the wavelet domain. An alternative is to express the uncertainty on τ in a way that has nice conjugacy properties. Consider, for a moment, the problem of estimating the function f which generates the means of the observations. If this is to be done using Bayesian analysis in the wavelet domain, then the resulting analysis is similar to those found in other chapters in this volume. In these analyses, the likelihood is expressed as

$$w_{j,k}|\theta_{j,k} \sim N(\theta_{j,k}, \sigma^2),$$

independent of the other $w_{j,k}$'s. In this more general formulation, the $n-1$ wavelet coefficient parameters take the place of $q_{j,k}(\tau)$ in (4). Prior information is then expressed in terms of the $\theta_{j,k}$'s. One way it could be expressed is with $\theta_{j,k}$'s being mutually independent with

$$\theta_{j,k} \sim N(\tilde{w}_{j,k}, \xi_{j,k}^2).$$

Typically, the $\tilde{w}_{j,k}$'s are all set to zero, resulting in a Bayes estimate that shrinks the empirical wavelet coefficients towards zero. In the case considered here, however, there is more information about the underlying function f (namely, the change-point structure), and so it makes sense to use non-zero $\tilde{w}_{j,k}$'s that will depend on the prior information as to the belief of the location (and magnitude of) the change-point. Carrying this analysis through gives the Bayes estimate of $\theta_{j,k}$ under squared error loss as

$$\hat{\theta}_{j,k} = \frac{\xi_{j,k}^2 w_{j,k} + \sigma^2 \tilde{w}_{j,k}}{\sigma^2 + \xi_{j,k}^2},$$

which is the convex combination of the prior information on $\theta_{j,k}$ and the data. In this more general formulation, the empirical wavelet coefficients are shrunk towards the corresponding $\tilde{w}_{j,k}$'s, rather than always towards zero.

An alternative analysis would be to express the prior information in an analogous form, but to place the uncertainty on the parameter τ directly, rather than on the $\theta_{j,k}$'s. Specifically, replacing the $\theta_{j,k}$'s with the functions $\Delta q_{j,k}(\tau)$ as in (4),

$$\pi(\tau|\Delta) \propto \exp\left\{ -\frac{1}{2} \sum_j \sum_k \frac{(\Delta q_{j,k}(\tau) - \tilde{w}_{j,k})^2}{\xi_{j,k}^2} \right\}. \tag{7}$$

These two approaches (undertainty expressed in terms of τ vs. uncertainty on $\theta_{j,k}$'s) are parallel, and will be developed together in this section. Applying this prior distribution, one may obtain the conditional posterior distribution of τ:

$$p(\tau|\Delta, w) \propto$$

$$\exp\left\{ -\frac{1}{2} \sum_j \sum_k \left[\frac{(w_{j,k} - \Delta q_{j,k}(\tau))^2}{\sigma^2} + \frac{(\Delta q_{j,k}(\tau) - \tilde{w}_{j,k})^2}{\xi_{j,k}^2} \right] \right\}$$

$$= \exp\left\{ -\frac{1}{2} \sum_j \sum_k \left(\frac{1}{\sigma^2} + \frac{1}{\xi_{j,k}^2} \right) \left(\Delta q_{j,k}(\tau) - \frac{\xi_{j,k}^2 w_{j,k} + \sigma^2 \tilde{w}_{j,k}}{\sigma^2 + \xi_{j,k}^2} \right)^2 \right\}$$

$$\times \exp\left\{ -\frac{1}{2} \sum_j \sum_k \left(\frac{1}{\sigma^2 + \xi_{j,k}^2} \right) (w_{j,k} - \tilde{w}_{j,k})^2 \right\}.$$

One problem with this analysis is that the $\tilde{w}_{j,k}$'s (regarded earlier as the means of the corresponding $\theta_{j,k}$'s) depend implicitly on the jump size Δ. Thus, if Δ is not known, the $\tilde{w}_{j,k}$'s can not be computed. To make the dependence of $\tilde{w}_{j,k}$ on Δ explicit, it can be written

$$\tilde{w}_{j,k} = \Delta \tilde{u}_{j,k}.$$

Making this substitution gives the following expression for the (conditional) posterior on τ:

$$p(\tau|\Delta, \boldsymbol{w}) \propto$$

$$\exp\left\{-\frac{1}{2}\sum_j\sum_k\left(\frac{1}{\sigma^2}+\frac{1}{\xi_{j,k}^2}\right)\left(\Delta q_{j,k}(\tau)-\frac{\xi_{j,k}^2 w_{j,k}+\sigma^2\Delta\tilde{u}_{j,k}}{\sigma^2+\xi_{j,k}^2}\right)^2\right\}$$

$$\times\exp\left\{-\frac{1}{2}\sum_j\sum_k\left(\frac{1}{\sigma^2+\xi_{j,k}^2}\right)(w_{j,k}-\Delta\tilde{u}_{j,k})^2\right\}.$$

In the analogous formulation that puts the uncertainty on the $\theta_{j,k}$'s, the conditional Bayes estimator of the scaled (by Δ) $\theta_{j,k}$ would thus be

$$\hat{\theta}_{j,k}|\Delta = \frac{\xi_{j,k}^2 w_{j,k}+\sigma^2\Delta\tilde{u}_{j,k}}{\Delta(\sigma^2+\xi_{j,k}^2)}. \tag{8}$$

The form of (8) requires that the prior for Δ have a finite first negative moment. A family of natural conjugate priors $\pi_{A,B}(\Delta)$ can be defined as follows. First, fix a probability distribution $\pi_0(\Delta)$. For $A \in \mathbb{R}$ and $B > 0$, let

$$K(A, B) = \int_{-\infty}^{\infty} \exp(A\Delta - B\Delta^2/2)\pi_0(\Delta)\, d\Delta.$$

Then, for A and B satisfying $K(A, B) < \infty$, let

$$\pi_{A,B}(\Delta) = \frac{1}{K(A, B)}\exp\left\{A\Delta - \frac{B}{2}\Delta^2\right\}\pi_0(\Delta). \tag{9}$$

If, for instance, $\pi_0(\Delta)$ is a gamma distribution with shape parameter at least two, then $\pi_{A,B}(\Delta)$ will have a finite first negative moment.

The joint posterior is obtained by applying the prior distribution in (9):

$$p(\tau, \Delta|\boldsymbol{w}) \propto$$

$$\exp\left\{-\frac{1}{2}\sum_j\sum_k\left[\frac{(w_{j,k}-\Delta q_{j,k}(\tau))^2}{\sigma^2}+\frac{(\Delta q_{j,k}(\tau)-\Delta\tilde{u}_{j,k})^2}{\xi_{j,k}^2}\right]\right\}$$

$$\times\exp\left\{A\Delta - \frac{B}{2}\Delta^2\right\}\pi_0(\Delta).$$

Grouping terms in Δ and Δ^2 gives that

$$
p(\tau, \Delta | \boldsymbol{w}) \quad \propto \quad \pi_0(\Delta) \exp \left\{ \left[A + \frac{1}{\sigma^2} \sum_j \sum_k w_{j,k} q_{j,k}(\tau) \right] \Delta \right\}
$$

$$
\times \exp \left\{ -\frac{1}{2} \left[B + \sum_j \sum_k \frac{1}{\sigma^2} q_{j,k}^2(\tau) + \frac{1}{\xi_{j,k}^2} (q_{j,k}(\tau) - \tilde{u}_{j,k})^2 \right] \Delta^2 \right\}
$$

$$
= \quad \exp \left\{ A^*(\tau) \Delta - \frac{B^*(\tau)}{2} \Delta^2 \right\} \pi_0(\Delta),
$$

where

$$
A^*(\tau) = A + \frac{1}{\sigma^2} \sum_j \sum_k w_{j,k} q_{j,k}(\tau)
$$

and

$$
B^*(\tau) = B + \sum_j \sum_k \frac{1}{\sigma^2} q_{j,k}^2(\tau) + \frac{1}{\xi_{j,k}^2} (q_{j,k}(\tau) - \tilde{u}_{j,k})^2.
$$

In general, the marginal posterior on τ can then be computed by numerical integration:

$$
p(\tau | \boldsymbol{w}) \propto \frac{1}{K(A^*(\tau), B^*(\tau))}.
$$

5.4   Generalizing the analysis

5.4.1   Smoother wavelet bases

Though the previous analysis was developed initially with the Haar basis in mind, it can be easily adapted to any orthogonal wavelet basis. The only change this would involve is a new definition for the mean function $q_{j,k}(\tau)$ as it appears in (4). For a given wavelet basis, this function can be expressed

$$
q_{j,k}(\tau) = \sqrt{n} \int_\tau^1 \psi_{j,k}(u) \, du + O(n^{-1/2}).
$$

In practice, it is not necessary to compute this integral — the value of the $w_{j,k}(\tau)$'s for all j and k can be computed by applying the discrete wavelet transform algorithm to the vector $(0, \ldots, 0, 1, \ldots, 1)'$ ($[n\tau]$ zeroes, $n - [n\tau]$ ones) for $[n\tau] = 2, \ldots, n$ These $q_{j,k}$ functions for varying wavelet families are plotted in Figure 3 for several values of j and k. These have essentially the same interpretation for smoother wavelets as that for the $q_{j,k}(\tau)$ functions for the Haar wavelets: for values of τ for which $q_{j,k}(\tau)$ is large (in absolute value), the corresponding wavelet coefficient would be effective in describing the change at that location τ; where the function is

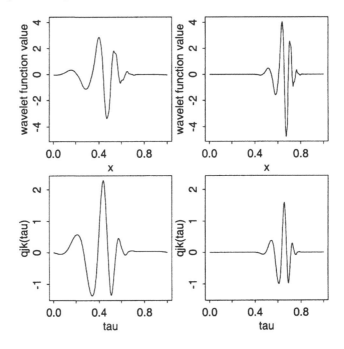

FIGURE 3. Plots of wavelets (top panels) and their corresponding $q_{j,k}(\tau)$ functions (bottom panels) for the Daubechies $N = 5$ Extremal Phase family. The left-hand plots are for $j = 3$ and $k = 3$; those on the right have $j = 4$ and $k = 10$.

near zero, the wavelet is not as useful in representing an abrupt change-point.

Of course, with a different wavelet basis, the prior means (the $\tilde{u}_{j,k}$'s) will also change. These hyperparameters should be chosen while keeping the specific wavelet basis to be used in mind.

5.4.2   Generalizing the change-point model

The analysis discussed so far is appropriate only for the standard change-point model (2), in which the function f is piecewise constant with a single jump. An attractive feature of the framework developed here is that it can be easily extended to more general change-point problems, in particular, in the model (3), in which the underlying function f is smooth except for a single point of abrupt change.

To apply the analysis to this situation, recall that each empirical wavelet coefficient contains information about how the function is changing in a localized region (which is just the support of the corresponding wavelet function). If $w_{j,k}$ is near zero, this indicates that there is not much change in the function over the support of the corresponding wavelet. A smooth function results in a set of $w_{j,k}$'s that are all near zero; introducing an abrupt jump into the function will give rise to several large coefficients, corresponding to wavelets with support overlapping the location of the jump. Since a change-point is a localized phenomenon, it is natural to use only the more localized (higher level) coefficients in the computation of the posterior density, leaving the lower-level coefficients to describe the large-scale features of the data.

To be more precise, define the "raw" estimator of $f(u)$ to be the piecewise constant function

$$f(u) = \begin{cases} Y_i, & \frac{i-1}{n} \le u < \frac{i}{n}, \ i = 1, \ldots, n \\ 0, & u = 1. \end{cases}$$

This function is first projected onto the approximation space V_{j_0} for some suitably chosen integer j_0. This projection is a linear operator (described in terms of kernel methods in Antoniadis, et al. (1994)) which amounts to using only the wavelet coefficients with $j \le j_0$ in the reconstruction. To estimate the location of the abrupt jump, the posterior density is computed using only the wavelet coefficients with dilation index $j > j_0$. Thus, the projection gives a good smooth estimate of f, and the "residual" wavelet coefficients are analyzed to find the probable location of the change-point.

This generalization is readily made; in all the expressions for posterior distributions, the summation indices on j and k were left unspecified. For the analysis associated with the standard change point model, the summation would be over all wavelet coefficients, but to handle the case of locating an abrupt jump in an otherwise smooth function, the double sum $\sum_j \sum_k$ should be replaced by $\sum_{j=j_0+1}^{J-1} \sum_{k=0}^{2^j-1}$.

5.5   Discussion

The choice of hyperparameter values, particularly the $\tilde{u}_{j,k}$'s and the $\xi_{j,k}^2$'s, requires some attention. The conditional prior distribution of τ as defined in (7) is an extremely flexible one; indeed, it requires the choice of $2(n-1)$ hyperparameter values. To make an effective choice for these values, it helps to be able to "think in the wavelet domain."

One strategy for choosing the $\tilde{u}_{j,k}$ values is to first formulate a reasonable prior distribution for τ in the "time" domain, and then transform it into the wavelet domain. For instance, if one has prior probabilities $p(\tau)$ that the change will occur at a particular value of τ, for $\tau = \frac{2}{n}, \frac{3}{n}, \ldots, 1$, then the corresponding $\tilde{u}_{j,k}$'s can be computed by a weighted average of the corresponding transformation. More precisely, the discrete wavelet transform on the data can be represented as

$$w = \mathcal{W}Y,$$

where $Y = (Y_1, \ldots, Y_n)'$ and $\mathcal{W}$ represents the matrix of the DWT operator. the vector $\tilde{u} = (\tilde{u}_{j,k}, \; j = 0, \ldots, J-1, \; k = 0, \ldots, 2^j - 1)$ can be computed by defining z_τ to be the vector of $[n\tau] - 1$ zeroes and $n - [n\tau] + 1$ ones:

$$\tilde{u} = \sum_{n\tau=2}^{n} p(\tau) \mathcal{W} z_\tau.$$

The $p(\tau)$ values can be computed a number of ways. If it is thought that $n\tau$ has an approximately normal distribution with mean $\mu_{n\tau}$ and variance $\sigma_{n\tau}^2$, then the $p(\tau)$ values could be computed according to

$$p(\tau) = \phi \left(\frac{n\tau - \mu_{n\tau}}{\sigma_{n\tau}} \right).$$

The resulting prior on τ (if the $\xi_{j,k}$'s are all chosen to be equal) is a discretized normal distribution.

The next thing to consider is the choice of $\xi_{j,k}$ values. Recall that in the formulation of the problem with $\theta_{j,k}$'s, the updating is done according to

$$\hat{\theta}_{j,k} | \Delta = \frac{\xi_{j,k}^2 w_{j,k} + \sigma^2 \Delta \tilde{u}_{j,k}}{\Delta(\sigma^2 + \xi_{j,k}^2)}.$$

Thus, a large value of $\xi_{j,k}$ will place more weight on the associated empirical coefficient value; choosing a small value for $\xi_{j,k}$ will weight the prior information more heavily. In a typical analysis, there may be some useful prior information on the general location of a change, but it may not be appropriate to weight prior information on higher-level coefficients. Given the multiresolution structure, therefore, it is perhaps natural to set $\xi_{j,k}$ large for large j and small for smaller values of j. A particular choice for these hyperparameters is illustrated in the next section.

Perhaps the greatest criticism for wavelet-based methods is their lack of *translation invariance* (see, e.g., Coifman and Donoho (1995)). It is easy to see, for example, that using the Haar basis, it will be much easier to locate a change at, for instance $\tau = 1/2$, than it will be for a non-dyadic point such as $\tau = 1/3$. The analysis described in this paper could be extended so that it behaves more consistently across a wider range of τ values by applying the stationary wavelet transform as discussed by Nason and Silverman (1995) and by taking into account in the formulation the natural dependencies that result from applying a redundant transformation.

Another practical issue in the generalized analysis is the choice of the integer j_0. Although the choice may often be constrained somewhat by software limitations, the problem of choosing this parameter needs to be addressed. Some data-based choice based on the total energy of empirical coefficients at each level might be appropriate.

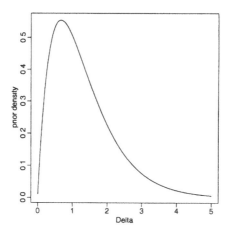

FIGURE 4. The prior density for the jump size Δ

5.6 Example

This section will illustrate the general technique developed in this paper with a simulated data set. For this example, the prior distribution for Δ is chosen according to (9) with $A = -1$, and $B = 0$, where $\pi_0(\Delta)$ is the gamma density with shape parameter $\alpha = 2$ and scale parameter $\beta = 2$. This prior distribution is plotted in Figure 4. The choice of $\beta = 2$ is a useful one, since it allows for direct integration, eliminating the need for numerical quadrature. Even with β set to two, the three parameters remaining in the formulation of the prior distribution for Δ allow for a very flexible structure for expressing prior information.

The $\tilde{u}_{j,k}$'s were generated with $n\tau \sim N(40, 16)$ and the $\xi_{j,k}$'s were chosen according to the following table:

j	$\xi_{j,k}$
0	2
1	5
2	10
3	20
4	50
5	100
6	200

The plot of the conditional prior on τ (using $\Delta = 2$) is given in Figure 5. It is interesting to note the small mode to the right of the main mode; this indicates another general location in which the given prior information on the wavelet structure indicates a change is likely.

The data were generated with $\tau = 0.38$, near the side of the main mode

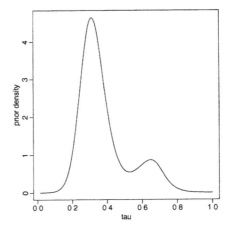

FIGURE 5. Prior density of the change-point τ with $\Delta = 2$

of the prior, and with $\Delta = 1$ and $\sigma^2 = 1$. The data are plotted in Figure 6. The marginal posterior density (having integrated Δ out) is given in Figure 7.

By simple inspection of the raw data in Figure 6, few places where changes may seem to occur. In this simulated example, it is known that the (abrupt) change occurred at 0.38, but the change appears to be somewhat gradual from that point to about 0.5. The posterior distribution for τ seems to find three likely change-points; the first (and smallest) mode occurs at about the right place; the second larger mode occurs a bit after 0.4; and the third a little after 0.5. Looking at the corresponding areas in Figure 6, it can be seen that there are some unusual structures which could be taken as evidence for an abrupt change-point.

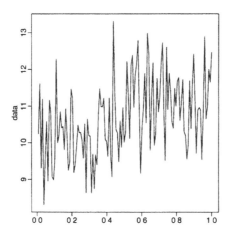

FIGURE 6. Plot of the simulated data

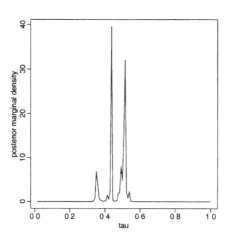

FIGURE 7. Posterior marginal density of the change-point τ

References

A. Antoniadis, G. Gregoire, and I. W. McKeague, "Wavelet methods for curve estimation," *Journal of the American Statistical Association*, vol. 89, pp. 1340–1352, 1994.

R. R. Coifman and D. L. Donoho, "Translation-invariant de-noising," in *Wavelets and Statistics* (A. Antoniadis and G. Oppenheim, eds.), pp. 125–150, New York: Springer-Verlag, 1995.

M. Csorgő and L. Horváth, *Limit Theorems in Change-Point Analysis.* New York: Wiley, 1997.

M. Csorgő and L. Horváth, "Nonparametric methods for changepoint problems," in *Handbook of Statistics, Vol. 7* (P. R. Krishnaiah and C. R. Rao, eds.), pp. 403–425, Amsterdam: Elsevier, 1988.

D. L. Donoho and I. M. Johnstone, "Adapting to unknown smoothness via wavelet shrinkage," *Journal of the American Statistical Association*, vol. 90, pp. 1200–1224, 1995.

P. R. Krishnaiah and B. Q. Miao, "Review about estimation of change-points," in *Handbook of Statistics, Vol. 7* (P. R. Krishnaiah and C. R. Rao, eds.), pp. 375–402, Amsterdam: Elsevier, 1988.

G. P. Nason and B. W. Silverman, "The stationary wavelet transform and some statistical applications," in *Wavelets and Statistics* (A. Antoniadis and G. Oppenheim, eds.), pp. 281–299, New York: Springer-Verlag, 1995.

J. E. Richwine, "Bayesian estimation of change-points using Haar wavelets," 1996. Master's Thesis at the University of South Carolina.

6

Prior Elicitation in the Wavelet Domain

Hugh A. Chipman and Lara J. Wolfson

ABSTRACT Bayesian methods provide an effective tool for shrinkage in wavelet models. An important issue in any Bayesian analysis is the elicitation of a prior distribution. Elicitation in the wavelet domain is considered by first describing the structure of a wavelet model, and examining several prior distributions that are used in a variety of recent articles. Although elicitation has not been directly considered in many of these papers, most do attach some practical interpretation to the hyperparameters which enables empirical Bayes estimation. By considering the interpretations, we indicate how elicitation might proceed in Bayesian wavelet problems.

6.1 Introduction

Bayesian statistical analysis using any statistical method, including wavelets, typically involves specifying priors on components of the problem. Bayesian analysis using wavelets primarily focuses on using priors to determine shrinkage rules – in other words, placing prior distributions in some fashion to indicate which wavelet coefficients, for a given basis, are going to tend towards zero. There are a variety of ways that these priors might be formulated, incorporating increasing levels of complexity and information about the uncertainty inherent in wavelet coefficients at various resolution levels tending towards zero. After using a graphical model to outline some commonly used structures, we then attempt to summarize the different ways that researchers have developed to formulate and apply these priors, focusing particularly on how the parameters for each prior are chosen. In this sense, this is a chapter about the "elicitation" of priors on wavelet coefficients.

When using a Bayesian approach and formulating priors, whether for wavelets or any other problem, one concern to many practitioners of statistics is how the priors are chosen. How should the parametric form, and the actual parameters themselves (usually referred to as the hyperparameters) be determined? The answers, unfortunately, are very difficult to obtain. Considerable attention is given in a paper by Kass and Wasserman (1995) to criteria for choosing a prior; several recent papers (Kadane and Wolfson,

1998; O'Hagan, 1998; and Craig et. al., 1998) attempt to lay out criteria for how the hyperparameters for the prior distribution should be elicited. Each problem tends to be somewhat unique, in that the choice of the prior distribution and its parameters tends to be a function not only of common statistical practice, but of the specific objectives of the statistical analysis being performed.

In wavelets, where subjectivity usually comes in through priors placed on the wavelet coefficients, the process of "eliciting" the values of the hyperparameters is dependent on the form chosen for the prior. One criteria that should therefore be considered in choosing the form of the prior, then, is whether or not the hyperparameters will have meaningful interpretations that will ease the process of elicitation, or if a connection be drawn between the hyperparameters and some quantity that is easier to specify. The existing literature contains examples of both, as shown in Section 3 where we outline the approaches various authors have taken to prior specifications. An alternative, one that seems to be quite popular among those who do research in Bayesian wavelets, is to employ empirical Bayes methods that attempt to determine the hyperparameters of the prior distribution from the data being analyzed. This has some obvious drawbacks, but it can be argued that it eliminates some of the subjectivity from the use of the Bayesian approach.

The wavelet coefficients, however, are not the only subjective area in wavelet analysis. Overlooked in many papers is the actual choice of an appropriate wavelet basis. Most of the papers we examine in this chapter assume that the appropriate basis (Daubechies, Meyer, Haar, etc.) is agreed on in advance, and that the uncertainty lies only in the coefficients estimated for a given basis. Since different bases may more parsimoniously represent signals with different characteristics, priors on the choice of basis seem a natural extension of much of the work on Bayesian model selection. We discuss this further in Section 4.

6.2   The Structure of a wavelet model and priors

In this section, the parameters of a wavelet model are given, and basic choices of prior distributions discussed. Brief references are given to specific papers; for full details, refer to Section 6.3. A general framework is introduced for placing prior distributions on these parameters, via a graphical model. This structure is then used to describe and compare particular choices of priors used in a number of recent papers. Where possible, the notation of Chapter 1 is used.

The vector of observed signal values $Y = (Y_1, Y_2, \ldots, Y_N)$ is assumed to have been generated by the model

$$Y = X\theta + \epsilon \tag{1}$$

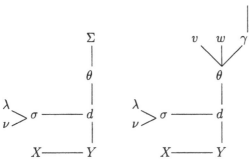

FIGURE 1. Parameters of a wavelet model and corresponding priors. Figure 1(a) (left) involves one level on unobserved wavelet coefficients θ, while 1(b) (right) adds a second hierarchy.

where X is the $(N \times N)$ matrix of wavelet basis functions evaluated at equally spaced $x[i] = i/N$, θ is the $(N \times 1)$ vector of unobserved wavelet coefficients, and ϵ is a N-vector of iid $N(0, \sigma^2)$ errors. Typically $N = 2^n$ with n integer. Because the matrix X is orthogonal, the N-vector of observed wavelet coefficients

$$d = (X'X)^{-1}X'Y = X'Y \tag{2}$$

is also assumed to be an iid vector with normal errors having variance σ^2. That is,

$$d \overset{iid}{\sim} N(\theta, \sigma^2). \tag{3}$$

Orthogonality of X also implies that the observed coefficients d and the observed signal Y are 1-1 functions of one another, and the observed coefficients may be treated as the response, conditional on the choice of a basis X. Orthogonality also gives the simplification of (2). The power of wavelet models is their ability to decompose a signal in terms of both location and scale ("frequency"). Typically the wavelet coefficients d are indexed as d_{jk}, with j indicating the scale and k indicating location.

The parameters (X, θ, σ^2) define the wavelet model. It is on these parameters that priors are to be placed. In what follows, both variables (such as Y), model parameters $((X, \theta, \sigma^2)$, and prior parameters) will be generically referred to as "variables".

These variables and relationships (1) and (3) are represented graphically in Figures 1(a) and 1(b). Figure 1(a) represents the simplest case, and is considered initially. In this graphical model, solid lines are used to indicate dependence between variables (represented by the vertices of the graph). For example, the variable Y depends on (d, X) since $Y = Xd$. Conditional independence is represented as follows: if a variable is conditioned upon, the corresponding vertex of the graph and all lines originating at that vertex are removed. Any variables not connected by a line (directly or through other variables) are conditionally independent. For example, conditional on the observed response, X and d are independent. That is, d and X depend on each other only through the observed response Y.

Note that in some models the (unobserved) wavelet coefficients θ are assumed to have a variance proportional to σ^2, necessitating a dependence between θ and σ. Since some models do not assume this dependence, this relationship is not indicated in Figure 1(a) or 1(b).

Prior distributions may be placed on all variables that are at the periphery of Figure 1(a), e.g. X, σ, θ. Priors on σ are simple enough that they are represented in the main part of the figure. Specifically σ is assumed to follow an inverse gamma distribution with parameters ν, λ, or equivalently $\nu\lambda/\sigma^2 \sim \chi_\nu^2$. Special cases of interest include fixing σ to some value s by taking $\nu = \infty, \lambda = s^2$, and an uninformative prior, taking $\nu = 0$.

The wavelet basis, given by the matrix X is usually assumed to be fixed, rather than placing a prior on it. This is further discussed in Section 4.

Several different priors have been considered for the (unobserved) wavelet coefficients θ. Two classes of priors are represented in Figures 1(a) and 1(b), and described below. Since shrinkage of the wavelet coefficients is achieved via a prior on θ, it is to be expected that a number of different prior structures have been considered. It is in this respect that most of the papers surveyed here differ; here we outline most of the general approaches, and refer the reader to the next section for details.

One straightforward possibility is to assign to θ a multivariate normal prior with mean vector 0 and covariance matrix Σ, i.e. $\theta \sim \text{MVN}(0, \Sigma)$, as in Figure 1(a) (Vannucci and Corradi (this volume), Vidakovic and Muller 1995). Another possibility is to have the prior covariance of θ given as $\sigma^2\Sigma$, indicated by the dashed line between θ and σ in Figure 1(a). An important special case is to assume that Σ is the identity matrix, yielding prior independence of the coefficients. Vidakovic (1998) considers a similar case, but with independent t distributions rather than normals. These approaches achieve shrinkage through the choice of Σ. Dependence between coefficients is also introduced via Σ. Coefficients with strong prior correlations would be those which are "close" to each other - i.e. coefficients at similar locations, or similar resolutions.

Shrinkage of the wavelet coefficients can be achieved by other priors. A popular choice for the prior the coefficients θ is the use of a scale mixture of two distributions. One mixture component would then correspond to "negligible" coefficients, the other to "significant" coefficients. This may be achieved in a number of ways, although most articles to date have used either a scale mixture of two normal distributions or a mixture of one normal and a point mass at 0. The latter case is a limiting case of the first. This parameterization is depicted in Figure 1(b), and is given as

$$\theta_{jk} = \gamma_{jk} N(0, v_{jk}^2) + (1 - \gamma_{jk}) N(0, w_{jk}^2), \qquad \theta_{jk}|\gamma_{jk} \text{ independent} \qquad (4)$$

| Paper | Prior | | |
	θ	σ	shrinkage
Ruggeri & Vidakovic (1995)	various	fixed	threshold
Vannucci & Corradi (1997)	MVN, correlated	IG	shrink
Vidakovic (1998)	T, indep	exp	shrink

TABLE 1. Comparison of techniques using a single prior on wavelet coefficients. Abbreviations are MVN = multivariate normal, T=Student's t, IG = inverse gamma, exp = exponential.

where $v_{jk}^2 \gg w_{jk}^2$ and γ_{jk} is binary $(0/1)$. The greater magnitude of v_{jk}^2 represents the large coefficients, while the mixture component with smaller variance w_{jk}^2 represents the negligible cases. In such cases, v_{jk}^2 and w_{jk}^2 are chosen, and a prior placed on γ_{jk}. The limiting case, in which $w_{jk} = 0$ reduces one mixture component to a point mass at 0.

An important distinction between the use of two normals versus a normal and a point mass is the type of shrinkage obtained. With $w_{jk} > 0$, no coefficient estimate based on the posterior will be exactly equal to zero. If $w_{jk} = 0$, then it is possible that some wavelet coefficient estimates will be exactly zero (i.e. the coefficients are "thresholded"). Abramovich, Sapatinas, and Silverman (1998) achieve this latter result by using the posterior median rather than the posterior mean. Although $w_{jk} > 0$ yields a smoother shrinkage function, it does not compress the data, since all estimated wavelet coefficients are nonzero. If compression is paramount, the point mass prior may be preferred.

Whether a mixture of two normals or a normal and a point mass is used, a prior must still be placed on the binary variables γ_{jk}. The simplest prior is

$$\Pr(\gamma_{jk} = 1) = 1 - \Pr(\gamma_{jk} = 0) = \pi_{jk}, \qquad \gamma_{jk} \text{ independent } \forall j, k \qquad (5)$$

The only article not using such a prior is Crouse, Nowak, and Baraniuk (1998), which instead introduces dependence among the different γ_{jk}, either at similar locations or similar resolutions.

One article related to Figure 1(b), but with an important difference is Holmes and Denison (1998). In this case an infinite number of scale mixtures are used rather than two discrete choices.

6.3   A comparative look at current practice

In this section we give details on the different prior structures as indicated in Tables 1 and 2. Techniques using only one additional hierarchical level on the prior structure of θ (i.e. as in Figure 1(a)) are considered first, followed by the majority of papers, which place at least two levels of prior on θ (as in

	Prior			
Paper	θ	σ	γ	shrinkage
Chipman et.al. (1997)	N-N	fixed	indep.	shrink
Crouse et.al. (1998)	N-N	fixed	depend.	shrink
Abramovich et.al. (1998)	P-N	fixed	indep.	thresh
Clyde & George (1998)	P-N	fixed	indep.	shrink
Clyde et.al. (1998)	P-N	IG	indep.	shrink
Vidakovic (1998)	P-T	exp	indep.	thresh
Holmes & Denison (1998)	N*	IG	—	shrink/thresh

TABLE 2. Comparison of techniques using a mixture prior on wavelet coefficients. Abbreviations are N=Normal, P=point mass, T=student's T, IG=Inverse Gamma. * Holmes and Denison employ a continuous mixture of normal, rather than two mixture elements. Prior independence is still assumed.

Figure 1(b)). The use of differing notations in the original prior necessitates some translation; we use the notation of Section 2.

The papers by Ruggeri and Vidakovic (1995) and Ruggeri (1999) are particularly interesting because they describe structures for priors on wavelet coefficients that differ from the variants on the normal linear model that most authors use. Here, the authors offer a fairly comprehensive catalog of possible distributions for the wavelet coefficients dependent only on a location parameter (the scale parameter is assumed to be known), and suggest priors corresponding to each of the choices. They consider a number of combinations of location parameter priors for θ (listed first) with distributional models for d (listed second): double exponential/normal, normal/normal, double exponential/double exponential, t/t (both with the same and different degrees of freedom), normal/double exponential, t/double exponential, and t/normal. The authors do not focus on how the hyperparameters of the prior should be chosen, other than suggesting empirical Bayes methods.

Vannucci and Corradi (1997, this volume) consider a model with an inverse gamma prior on σ and a multivariate normal prior on θ. They elaborate on an idea introduced in Vidakovic and Muller (1995) that the prior covariance Σ be non-diagonal, i.e. coefficients are correlated. The correlation structure is considered both for coefficients within the same resolution level j and coefficients in different levels. Coefficients are arranged in such a fashion that the covariance matrix can be expressed as a diagonal band outside which correlations are zero. Within the band, the largest correlations are between coefficients that are close in location and scale. The covariance matrix is specified in terms of two hyperparameters, simplifying the elicitation process. Values of these hyperparameters are chosen either by empirical Bayes methods, or by making these parameters a component of the model and simulating their posterior distribution via MCMC. An alternate method for introducing dependence among wavelet coefficients is considered by Crouse et. al. (1998). The two approaches are compared in the description of Crouse et. al. below.

The remaining techniques considered use two or more hierarchical levels of prior on the wavelet coefficients θ. The majority considered use either a mixture of two normals or a mixture of a normal and a point mass. First, we outline methods that use a mixture of two normal distributions for a prior on θ. The main distinction between Chipman, Kolaczyk and McCulloch (1997) and Crouse et. al. (1998) is that the former assume independence in a prior for γ as in (5), while Crouse et. al. introduce dependencies.

In Chipman et. al. (1998), a hierarchical prior (4) is placed on the wavelet coefficients θ and the noise level σ is assumed known. The wavelet coefficients prior is a scale mixture of two normals, each with mean 0, and variances w_{jk}^2 and $v_{jk}^2 \gg w_{jk}^2$. Independence prior (5) on mixture indicators γ is used. Conditional on γ, wavelet coefficients are assumed independent in the prior.

The hyperparameters to be chosen are thus $\sigma, v_{jk}, w_{jk}, \pi_{jk}$. Hyperparameters indexed by jk are assumed to vary with resolution j, but be fixed across location k. This yields level-dependent shrinkage, similar to SureShrink (Donoho and Johnstone 1995). Each of hyperparameters are given an interpretation that allows default values to be estimated from the data. While fixing σ may be theoretically unappealing, this assumption enables rapid closed form calculation of posterior means for wavelet coefficients. The interpretation of π_j is the percentage of coefficients at a resolution level which are expected to be non-negligible. Prior standard deviation w_j represents the magnitude of a negligible coefficient, while v_j gives the largest magnitude of a significant coefficient.

These hyperparameters are not directly elicited. Instead, associations are made between the interpretations given above and

1. The magnitude of an insignificant change in the response Y

2. The magnitude of a substantial change in a given wavelet basis function, and

3. The percent of coefficients at a given resolution level that are above a noise threshold of Donoho, Johnstone, Kerkyacharian and Picard (1995)

Crouse et. al. (1998) consider the same specification of prior for θ as Chipman et. al. (1997) above, except for the innovation of considering dependent priors on the mixture parameters γ_{jk}. Crouse et. al. consider two types of Markov dependence among coefficients. In both cases, the γ_{jk} are arranged in a grid corresponding to the resolution and location of the corresponding wavelet coefficient. The dependency is Markov in the sense that whether or not a specific γ_{jk} is nonzero will depend on the state of its immediate neighbors. These neighbors could either be indicators corresponding to coefficients at the same resolution and location $k \pm 1$, or at an analogous location and at resolution $j \pm 1$. Links within a resolution level are referred to as a "hidden Markov chain", while links across resolutions are a "hidden Markov tree". In essence, if a certain coefficient is significant, then its neighbors (however they are defined) are also likely to be significant. There is an important distinction between this induced correlation and that of Vannucci and Corradi (1997, this volume), where correlation is directly induced on the coefficients via Σ. When correlations are assumed at the γ level, significant coefficients will make neighbors more likely to be significant, but no restrictions are placed on the direction of the significance. A significant coefficient is more likely to have significant neighbors, but they are equally likely to be large with the same or opposite sign. Introducing positive correlations on the coefficients θ themselves via Σ says quite another thing. If a coefficient is large, then a neighboring coefficient is also likely to be large and of the same sign. If the signal to be modeled oscillates rapidly, neighboring wavelet coefficients may be large but of opposite sign, making dependence introduced via γ more appropriate.

The majority of the other mixture-coefficient prior papers rely on a mixture of a point mass and a normal. Abramovich et. al. (1998), Clyde and George (1998), and Clyde, Parmigiani and Vidakovic (1998) all consider such priors. The first two are quite similar, while the last places a prior on σ rather than assuming a fixed value.

In Abramovich et. al. (1998) and Abramovich and Sapatinas (this volume), prior (4) for θ is used, with $w_{ij} = 0$, yielding a mixture of a point mass at zero and a $N(0, v_j^2)$. Independence prior (5) is placed on γ. It is assumed that the prior distribution (i.e., the choice of the hyperparameters v_j^2, π_j) is the same for any given resolution level j, and that the noise level, σ, is known or a good estimate is available. This is a structure similar to that found in Clyde et. al. (1998), and can be viewed as an extreme case of the formulation given in Chipman et. al. (1997). These hyperparameters are assumed to have structure such that $v_j^2 = 2^{-aj}C_1$ and $\pi_j = \min(1, 2^{-Bj}C_2)$, where C_1, C_2, a and B are assumed to be non-negative constants. Some interpretation of these constants is given to explain how they might be derived, particularly in the context of decision-theoretic thresholding rules. For example, in the discrete wavelet transform, this prior structure will correspond to the universal threshold of Donoho & Johnston (1994) when $a = B = 0$, and letting $C_1 \to \infty$, $C_2 \to 0$, such that as n increases, $\sqrt{C_1}/C_2 \sigma n \to 1$. A nice intuitive interpretation of the parameters is also given, in that the prior expected number of non-zero wavelet coefficients on the jth level is $2^{j(1-B)}C_2$. Thus, if $B > 1$, the expected number of non-zero coefficients is finite, and if $B = 0$, this implies a prior belief that all coefficients on all levels have the same probability of being non-zero. If $B = 1$, this assumes that the expected number of non-zero wavelet coefficients is the same on each level. An interesting contribution made by these authors is to connect the parameters a and B, and the Besov space parameters

(s, p), so that if a particular Besov space were chosen to represent prior beliefs, a, B could then be numerically derived. In particular, the authors recommend that a, B be chosen based on prior knowledge about regularity properties of the unknown function be studied. For the parameters C_1 and C_2, the authors propose an empirical-Bayes type procedure that relates C_1 and C_2 to the *VisuShrink* threshold of Donoho and Johnstone (1994). In practice, the authors recommend a "standard" choice of $a = 0.5, B = 1$. Part of the novelty of the approach is the idea that by simulating observations from different Besov spaces, one can elicit the correct space from which to choose the prior. One suggestion made for future research is to consider incorporation of a dependence structure between the wavelet coefficients. A weakness of this approach is that while B has a nice interpretation, the parameters C_1, C_2 don't have good intrinsic interpretability, and so any elicitation of these parameters would be very difficult.

Clyde and George (1998) employ the usual mixture of a point mass and normal distribution as a prior on the wavelet coefficients (i.e. $w_{jk} = 0$ in (4)). They propose an empirical Bayes procedure based on conditional likelihood approximations given the MAD estimate of σ proposed by Donoho et. al. (1995). This doesn't differ a great deal from the approach of Abramovich et. al. (1998).

In Clyde et. al (1998), priors are specified for both the wavelet coefficients θ and the noise standard deviation σ. For θ, (4) is used with $w_{jk} = 0$ and $v_{jk}^2 = c_{jk}\sigma^2$, yielding a mixture of a point mass at zero and a $N(0, c_{jk}\sigma^2)$. Prior independence of different coefficients is assumed. The parameter $\sqrt{c_{jk}}$ indicates the magnitude (relative to σ) of a large (or significant) coefficient at scale j and location k. This parameter is chosen to be constant across scale and location in practice (i.e. $c_{jk} = c$). An inverse gamma prior with parameters ν and λ is placed on σ^2.

An interpretation of c is obtained via linkages to information criteria given in George and Foster (1997). Different choice of c correspond to BIC, AIC, or the risk inflation criterion of Foster and George (1994) and Donoho and Johnstone (1994). The elements of γ are assumed independent. The mixing probability π_{jk} in (5) is chosen to vary across scale j but not location k. The interpretation of π_j is similar to that of Chipman et. al. (1997), although no automatic choices or elicitation recommendations are made.

In Vidakovic (1998), priors similar to both Figure 1(a) and Figure 1(b) are explored. For Figure 1(a), an exponential prior is placed on σ^2 (rather than the inverse gamma used in many other papers considered here), and a scaled t-distribution on θ_{jk}. Prior parameters to be selected are the prior mean of σ^2, the prior variance of θ, and the degrees of freedom for the t prior on θ. No recommendations regarding elicitation are made, although empirical default values are proposed. Shrinkage is achieved with this model via the posterior mean (i.e. Bayes rule). No coefficients are shrunk exactly to zero (i.e. thresholded). To achieve thresholding, the prior on θ_{jk} is modified to be a mixture of a scaled t and a point mass at 0. No choice is given for the mixture probabilities π_{jk}. This differs slightly from (4) as employed by Abramovich et. al. (1998), Clyde and George (1998), and Clyde et. al. (1998), which mix a point mass with a normal. Thresholding is obtained via the Bayes factor.

Holmes and Denison (1998) also assume $\theta \overset{iid}{\sim} N(0,v)$ as in Figure 1(a), and an inverse gamma prior on σ. The difference is that rather than choosing v (they call this λ^{-1}), a hyperprior is placed upon it. This is similar to the other approaches that use a scale mixture of normals, but with the distinction that the prior on the variance is a continuous prior, rather than a prior on two different values. That is, rather than using mixture of two normals, Holmes and Denison use an infinite mixture of normal with different variances. In this way, the paper more closely resembles the approaches using a mixture of two normals or a normal and point mass.

They note that this gives an intuitive interpretation to each component v_i^{-1}, which is to shrink the classical least squares estimate of θ by $(1 + \sigma^2 v_i^{-1})^{-1}$. Because the conjugate inverse-gamma or Wishart prior for v^{-1} would do little to aid elicitation, the authors turn to the fact that beliefs are likely to be most available on the complexity and smoothness on the underlying signal that is being reconstructed. Following the result of Hastie & Tibshirani (1990) on identifying the degrees of freedom of a linear smoother, the authors derive the prior

$$v_i^{-1} \mid \sigma^2 \propto \exp(-c\sum_{i=1}^{N}(1 + \sigma^2 v_i^{-1})^{-1}), \qquad (6)$$

where c is a constant for penalizing model complexity. The hyperparameter c is shown to have a direct relationship with the log model probability, so that $c = \{0, 1, 0.5\log n, \log n\}$ correspond to well-known model choice criteria (Bayes Factor, AIC, BIC, RIC respectively).

This degrees of freedom connection to model complexity allows the elicitation of the prior distribution to capture the effect of both the number of coefficients in the model and the extent of the prior shrinkage on them.

6.4   Future work and conclusions

One component of the wavelet models upon which priors have not been placed is the basis (X in Figures 1(a) and 1(b)). All articles reviewed here have conditioned on the choice of wavelet basis. Choice of an appropriate basis is something of an art form, making formal elicitation difficult. Wavelet bases are often grouped together, and indexed by an integer parameter within each group. For example, Daubechies (1988) has a compactly supported family of wavelets indexed by an integer N, with larger values of N yielding smoother basis functions. Priors might be placed on an indexing parameter within a family of basis functions, or on a range of different families. The first problem seems more approachable in terms of elicitation, provided that a degree of smoothness could be quantified and elicited. Yau and Kohn (1999) consider priors on the basis for purposes of model averaging, but place equal priors weight on each of four bases.

The possibilities for priors and their elicitation will grow as the field of wavelets grows. Interesting possibilities include wavelet packets, a means of constructing a new basis out of several individual bases, and methods that shift either the original data or the wavelet basis.

References

Abramovich, F. and Sapatinas, T. (1999) "Bayesian Approach to Wavelet Decompositions and Shrinkage", this volume.

Abramovich, F., Sapatinas, T., and Silverman, B.W. (1998) "Wavelet Thresholding via a Bayesian Approach", *Journal of the Royal Statistical Society, Series B*, 60, 725–750.

Chipman, H., Kolaczyk, E. and McCulloch, R. (1997) "Adaptive Bayesian Wavelet Shrinkage", *Journal of the American Statistical Association*, 92, 1413–1421.

Clyde, M., and George, E.I. (1998) "Robust Empirical Bayes Estimation in Wavelets", ISDS Discussion Paper 98-21, Duke University, Institute of Statistics and Decision Sciences

Clyde, M., Parmigiani, G., Vidakovic, B. (1998). "Multiple Shrinkage and Subset Selection in Wavelets", Biometrika 85, 391-402.

Crouse, M., Nowak, R. and Baraniuk, R. (1998). "Wavelet-Based Statistical Signal Processing Using Hidden Markov Models". *IEEE Trans. Signal Processing*, 46, 886–902.

Craig, P.S., Goldstein, M., Seheult, A.H., and Smith, J.A. (1998) "Constructing Partial Prior Specifications for Models of Complex Physical Systems", (with discussion) *The Statistician*, 47, 37–54.

Daubechies, I. (1988) "Orthonormal Bases of Compactly Supported Wavelets" *Communications in Pure and Applied Mathematics*, 41, 909-996

Donoho, D. L. and Johnstone, I.M. (1994) "Ideal Spatial Adaptation by Wavelet Shrinkage", *Biometrika*, 81, 425–455.

Donoho, D. L. and Johnstone, I.M. (1995) "Adapting to Unknown Smoothness via Wavelet Shrinkage", *Journal of the American Statistical Association*, 90, 1200–1224.

Donoho, D. L., Johnstone, I. M., Kerkyacharian, G. and Picard, D. (1995) "Wavelet Shrinkage: Asymptopia?" (with discussion), *Journal of the Royal Statistical Society, Series B*, 57, 310-370.

Foster, D. and George, E. (1994) "The Risk Inflation Criterion for Multiple Regression", *Annals of Statistics* 22, 1947–75.

George, E. and Foster, D. (1997) "Empirical Bayes Variable Selection", Technical Report, University of Texas at Austin.

Holmes, C.C., and Denison, D.G.T. (1998) "Bayesian Wavelet Analysis with a Model Complexity Prior", in *Bayesian Statistics 6*, eds. J. Bernardo, J. Berger, A. Dawid, and A. Smith, Oxford University Press.

Kadane, J.B., and Wolfson, L.J. (1998) "Experiences in Elicitation" (with discussion), *The Statistican*, 47, 3–20.

Kass, R.E. and Wasserman, L. (1997) "The Selection of Prior Distribution by Formal Rules", *Journal of the American Statistical Association*, 96, 1343–1370.

O'Hagan, A. (1998) "Eliciting Expert Beliefs in Substantive Practical Applications", (with discussion) *The Statistician*, 47, 21–36.

Ruggeri, F. (1999) "Robust Bayesian and Bayesian Decision Theoretic Wavelet Shrinkage", this volume.

Ruggeri, F. and Vidakovic, B., (1995) "A Bayesian Decision Theoretic Approach to Wavelet Thresholding", ISDS Discussion Paper 95-35, Duke University, Institute of Statistics and Decision Sciences

Vannucci, M. and Corradi, F. (1997) "Some Findings on the Covariance Structure of Wavelet Coefficients: Theory and Methods in a Bayesian Perspective" *Journal of the Royal Statistical Society, Series B*, provisionally accepted.

Vannucci, M. and Corradi, F. (1999) "Modeling Dependence in the Wavelet Domain", this volume.

Vidakovic, B. (1998) "Nonlinear Shrinkage With Bayes Rules and Bayes Factors", *Journal of the American Statistical Association*, 93, 173–179.

Vidakovic, B. and Muller, P. (1995) "Wavelet Shrinkage with Affine Bayes Rules with Applications", Technical Report DP-95-36, ISDS, Duke University.

Yau, P. and Kohn, R. (1999) "Wavelet Nonparametric Regression Using Basis Averaging", this volume.

7

Wavelet Nonparametric Regression Using Basis Averaging

Paul Yau and Robert Kohn

ABSTRACT Wavelet methods for nonparametric regression are fast and spatially adaptive. In particular, Bayesian methods are effective in wavelet estimation. Most wavelet methods use a particular basis to estimate the unknown regression function. In this chapter we use a Bayesian approach that averages over several different bases, and also over the Fourier basis, by weighting the estimate from each basis by the posterior probability of the basis. We show that estimators using basis averaging outperform estimators using a single basis and also estimators that first select the basis having the highest posterior probability and then estimate the unknown regression function using that basis.

7.1 Introduction

Wavelet methods in nonparametric regression are popular because they are fast and spatially adaptive. The spatial adaptivity of wavelet bases is due to the local structure of the wavelet basis terms. The estimators are fast to compute because of the speed of the wavelet transform which takes advantage of the orthogonality of the basis terms. In a series of fundamental papers, Donoho and Johnstone propose several nonparametric wavelet estimators and study their large sample properties; see, for example, Donoho and Johnstone (1994, 1995) and Donoho, Johnstone, Kerkyacherian and Picard (1995). Recent papers by Chipman, Kolaczyk and McCulloch (1997), Vidakovic (1998) and Clyde, Parmigiani and Vidakovic (1998) provide Bayesian estimators for nonparametric wavelet regression and show that the Bayesian approaches compare favourably with the earlier work of Donoho and Johnstone.

The work above assumes that a particular wavelet basis is chosen for the estimation. However, it is clear that no single basis is uniformly best for all regression functions, e.g. a symmlet 8 basis performs well for a great range of functions, but will not do as well for step functions. Kohn, Marron and Yau (1997) propose an estimator which first determines the basis having the highest posterior probability for a given data set and then estimates the regression curve using this basis. They also propose a basis averaging estimator which is a weighted sum of the estimators for each basis, using the posterior probabilities of the individual bases as weights. Kohn et al. (1997) show that the basis averaging estimator compares favourably with the basis selection estimator and the estimators based on fixed bases. This chapter exposits and extends the results of Kohn et al. (1997).

The chapter is organized as follows. Section 2 outlines wavelet shrinkage methods for nonparametric regression. Section 3 describes the empirical Bayes approach taken in this chapter and the basis averaging and basis selection estimators. Section 4 presents the simulation results.

7.2   Wavelet nonparametric regression

Suppose the univariate function $m(x)$ is observed with noise, that is.

$$y_i = m(x_i) + \varepsilon_i, \ i = 1, ..., n,$$

with $E(\varepsilon_i) = 0$. We make the following technical assumptions:

A1 The x_i are equally spaced with $x_i = (i - 1/2)/n$.

A2 The sample size n is a power of two with $n = 2^k$.

A3 The errors $\varepsilon_1, ..., \varepsilon_n$ are independent $N(0, \sigma^2)$.

There are a number of ways to recover the signal vector $m = (m(x_1), ..., m(x_n))'$ from the data vector $y = (y_1, ..., y_n)'$. One way is to assume that $m(x)$ has a particular functional form, for example that $m(x)$ is linear, quadratic or exponential, and use the data to estimate the unknown parameters. This approach is called parametric regression. However, assuming that $m(x)$ is known is often unwarranted and more flexible methods called nonparametric regression estimators assume only that $m(x)$ has a certain degree of smoothness, but do not specify its form. Examples of nonparametric regression estimators are kernel based local linear smoothers (Ruppert, Sheather and Wand, 1995), spline smoothers (Wahba, 1990), polynomial based methods such as those discussed by Stone (1994), and methods based on orthogonal regressors such as wavelet and Fourier regression estimators. Although each of the nonparametric estimators has its strengths, wavelet methods are the fastest and most adaptive spatially.

Orthogonal regression methods for estimating the signal m are described as follows. Suppose $\psi_1, ..., \psi_n$ form an orthonormal basis of n dimensional Euclidean space and let Ψ be the $n \times n$ orthonormal matrix with ith column ψ_i. Then,

$$m = \sum_{i=1}^{n} \beta_i \psi_i = \Psi \beta \qquad (1)$$

for some $n \times 1$ vector $\boldsymbol{\beta}$. Suppose $\boldsymbol{\Psi}$ is known and let $\boldsymbol{w} = \boldsymbol{\Psi}'\boldsymbol{y}$ and $\boldsymbol{e} = \boldsymbol{\Psi}'\boldsymbol{\varepsilon}$. Then $\boldsymbol{w} = \boldsymbol{\beta} + \boldsymbol{e}$, with $\boldsymbol{e} \sim N(0, \sigma^2 I)$. We can regard $\boldsymbol{w}$ as the vector of observations because $\boldsymbol{\Psi}$ is known and the nonparametric regression problem becomes one of estimating n means from n observations. Although this is difficult in general, the problem is made tractable because for many signals $\boldsymbol{m}$, most of the power of the signal (as measured by $\boldsymbol{m}'\boldsymbol{m}$) is captured by just a few of the β_i, with the rest of the β_i close to 0. Therefore, in its simplest form, the nonparametric estimation problem amounts to determining the non-negligible β_i and setting the rest to zero.

Shrinkage estimation generalizes this simple approach for handling orthogonal regressors by writing the estimator of $\boldsymbol{m}$ as

$$\widehat{\boldsymbol{m}} = \sum_{i=1}^{n} \widehat{\beta}_i \psi_i \tag{2}$$

with $\widehat{\beta}_i = \eta_i(w_i)$, and $\eta_i(w_i)$ is a function that shrinks w_i towards zero or even sets $\widehat{\beta}_i$ to zero for small w_i.

The large amount of current interest in wavelets is due to their ability to compress signal effectively for both smooth and periodic signals, as well as for signals that are smooth in most locations, but have some points of non-smoothness. Wavelet bases compress effectively signals with smoothness varying by location because they are adaptive in terms of both scale (this concept is the same as frequency in Fourier analysis) and location. Because of this dual ability to adapt, wavelet bases are most conveniently represented using the following double indexing notation: $j = j_0, ..., \log_2(n/2)$, which indexes scale, and $k = 0, ..., 2^j - 1$, which indexes location. The parameter j_0 is chosen as small as possible for the given basis. This index system has a simple correspondence to the indices $i = 2^{j_0} + 1, ..., n$ used above via

$$i \leftrightarrow (j, k) \text{ as } i = 2^j + k + 1.$$

These two index systems are used interchangeably in the rest of the paper, with the choice of indexing made in terms of convenience. While the Fourier basis does not have a structure requiring double indexing, it is useful to organize it in this way for the empirical Bayes estimators considered in this paper.

It is also useful to write the wavelet bases in terms of father wavelets ψ_i, which are the $i = 1, ..., 2^{j_0}$ elements of the basis, and the mother wavelets $\psi_{j,k}$, which correspond to $i = 2^{j_0} + 1, ..., n$. The wavelet version of the representation (1) becomes

$$\boldsymbol{m} = \sum_{i=1}^{2^{j_0}} \beta_i \psi_i + \sum_{j=j_0}^{\log_2(\frac{n}{2})} \sum_{k=0}^{2^j - 1} \beta_{j,k} \psi_{j,k}.$$

The shrinkage estimator (2) becomes

$$\widehat{m} = \sum_{i=1}^{2^{j_0}} \widehat{\beta}_i \psi_i + \sum_{j=j_0}^{\log_2(\frac{n}{2})} \sum_{k=0}^{2^j - 1} \widehat{\beta}_{j,k} \psi_{j,k},$$

where $\widehat{\beta}_i = w_i$ and $\widehat{\beta}_{j,k} = \eta_j(w_{j,k})$. Note that the shrinkage functions $\eta_j, j \geq j_0$, depend on the level j, but not on the location k. The father terms, $w_i, i = 1, \ldots, 2^{j_0}$, are typically left unthresholded because they represent low frequency terms that usually contain important components of the signal. Thus the parameter j_0 controls how many terms are of this type. The hard thresholding estimator studied by Donoho and Johnstone (1994) has a shrinkage function that can be expressed as

$$\eta_{H,j}(w) = \begin{cases} w & \text{for } |w| \geq \lambda\sigma \\ 0 & \text{for } |w| < \lambda\sigma \end{cases}, \text{ for } j \geq j_0 \qquad (3)$$

for a given value of the threshold λ. The function η_H zeros small values of w and leaves large values of w untouched.

The soft thresholding estimator proposed by Donoho and Johnstone (1994) has the shrinkage function

$$\eta_{S,j}(w) = \begin{cases} sgn(w) \cdot (|w| - \lambda\sigma) & \text{for } |w| \geq \lambda\sigma \\ 0 & \text{for } |w| < \lambda\sigma \end{cases}, \text{ for } j \geq j_0 \qquad (4)$$

and is again independent of k. The function η_S corresponds to moving the w terms $\lambda\sigma$ units towards the origin. We write the hard and soft thresholding estimators as $\widehat{m}_H$ and $\widehat{m}_S$.

The hard and soft thresholding estimators considered in all the examples in this paper have $j_0 = 5$, with $n = 1024$. This choice of j_0 gives good overall performance in our simulations.

The error standard deviation σ is usually unknown, but a good estimator of σ is obtained by using a robust scale estimate based on the highest frequency terms $w_{j,k}$, as suggested by Donoho and Johnstone (1994). Following their suggestion, our scale estimate is based on the median absolute deviation from the median (normalized by dividing by the corresponding standard normal term), which we write as $\widehat{\sigma}$.

We consider two choices of the threshold value λ. Donoho and Johnstone (1994) proposed the denoising threshold,

$$\lambda_D = \sqrt{2 \ln n}, \qquad (5)$$

which corresponds to the largest size of Gaussian pure noise terms, and is well suited for use with η_H. For η_S, a somewhat smaller threshold value is more appropriate which results in more terms in the model to counter the shrinkage effect. Donoho and Johnstone (1994) proposed a minimax optimal value, called λ_{MO}.

7.3   Bayesian estimation

7.3.1   Estimation using a single basis

This section discusses two Bayesian estimators of the vector m that use a single wavelet basis or the Fourier basis. The estimators are empirical Bayes because the parameters for the prior are determined from the data.

For both estimators, let $\gamma_i = 1$ if the basis vector ψ_i is included in the model (2.1) and let $\gamma_i = 0$ if it is not. Let $\gamma = (\gamma_1, \ldots, \gamma_n)$. For a given γ we have from (2.1) that m is a linear function of those vectors ψ_i for which $\gamma_i = 1$. Let $\pi_i = p(\gamma_i = 1)$ be the prior probability that ψ_i is included in the model and let $\pi = (\pi_1, \ldots, \pi_n)$. For $\gamma_i = 1$ the prior for β_i is $N(0, c_i^2)$. For wavelets we use the double index notation and choose $c_i = c_{jk}$ and $\pi_i = \pi_{jk}$ to depend on the scale j only.

The first estimator is almost identical to that of Chipman et al. (1997). The second estimator uses the same prior for the β_i as the first, but estimates the π_i by maximizing the marginal likelihood. There are other priors that we can use, but the two priors below are sufficient to demonstrate the importance of basis averaging.

For the first empirical Bayes estimator, let

$$
\begin{aligned}
c_j &= \|\psi_{j,k}\|_1 \times \max_{i=1,\ldots,n} |y_i|/3 \\
\pi_j &= \#\{k : |y_{j,k}| > \sqrt{2\log n}\}/2^j
\end{aligned}
\tag{6}
$$

for $j \geq 0$ where $\|\psi_{j,k}\|_1$ is the L_1 norm of the basis vectors at scale j; we note that all $\psi_{j,k}$ have the same norm for a given level j. This prior is essentially that of Chipman et al. (1997) who motivate it as follows. Writing $\beta = \Psi' m$, it follows after some algebra that

$$
|\beta_{jk}| \leq \max_{i=1,\ldots,m} |m_i| \|\psi_{j,k}\|_1
$$

A rough approximation to $\max_{i=1,\ldots,n} |m_i|$ is $\max_{i=1,\ldots,n} |y_i|$. In the spirit of the mean plus or minus three standard deviations capturing most of the mass of the Gaussian distribution, $3c_j$ is an approximate bound on $|\beta_{j,k}|$ making $N(0, c_j^2)$ a reasonable prior for $\beta_{j,k}$. The probability π_j is the empirical probability of a term at scale j being larger than λ_D, the denoising threshold defined at (5).

We use the posterior mean of β_i as its estimate and note that the w_i are independent given σ^2. This means that $E(\beta_i|y, \sigma^2, \pi_i, c_i) = E(\beta_i|w_i, \sigma^2, \pi_i, c_i)$. We can show

$$
E(\beta_i|w_i, \sigma^2, \pi_i, c_i) = \eta_i(w_i)
\tag{7}
$$

where

$$
\eta_i(w) = \frac{\pi_i \sigma c_i^2 w}{\pi_i \sigma (c_i^2 + \sigma^2) + (1 - \pi_i)(c_i^2 + \sigma^2)^{3/2} \exp\left(-\frac{w^2}{2\sigma^2}\left(\frac{c_i^2}{c_i^2 + \sigma^2}\right)\right)}
\tag{8}
$$

The notation η_i signifies that the empirical Bayes estimator is a shrinkage estimator similar to the hard and soft thresholding estimators.

For the Fourier basis, we find it convenient to estimate the c_i and π_i by grouping as for the wavelet bases, even though the Fourier basis does not have a structure that requires double indexing. We argue that such a grouping is sensible for the Fourier basis as a rough subdivision of the frequencies, but we are working on a more systematic approach.

The empirical Bayes estimator (8) assumes that the error variance σ^2 is known. In practice it is usually unknown and needs to be estimated from the data. We follow Chipman et al. (1997) and plug-in an estimate $\hat{\sigma}^2$ for σ^2 as we did for the hard and soft thresholding estimators. This is consistent with the empirical Bayes approach used to estimate the c_i and π_i.

The second empirical Bayes estimator determines the c_j as above, but uses the marginal likelihood to estimate the π_j for $j \geq 0$ For a given level j, the marginal likelihood is

$$\prod_{k=0}^{2^j-1} p(w_{j,k}|\pi_j, c_j, \sigma^2) =$$

$$\prod_{k=0}^{2^j-1} \left\{ \frac{\pi_j}{(2\pi(\sigma^2 + c_j^2)^{\frac{1}{2}})} \exp\left(-\frac{w_{j,k}^2}{2(\sigma^2 + c_j^2)}\right) + \frac{(1-\pi_j)}{(2\pi\sigma^2)^{\frac{1}{2}}} \exp\left(-\frac{w_{j,k}^2}{2\sigma^2}\right) \right\}$$

(9)

If the estimate is less than 0 we set π_j to zero and if the estimate is greater than 1 we set $\pi_j = 1$. Once the π_i and c_i are determined, the β_i are estimated by their posterior means as in (7) and (8).

7.3.2   Basis selection and basis averaging

The Bayesian approach also provides a simple method for choosing between different wavelet bases or averaging the curve estimate over different bases. Here we consider a choice among the Symmlet 8, Daubechies 4, Haar and Fourier bases.

Let B denote a generic such basis, and assign to it a prior probability $p(B) = \frac{1}{4}$ so that each basis is equally probable *a priori*. The resulting posterior probability of a basis B is

$$p(B|w, \sigma^2, c, \pi) \propto \frac{1}{4} p(w|\sigma^2, c, \pi, B) = \frac{1}{4} \prod_{i=1}^{n} p(w_i|\sigma^2, c_i, \pi_i, B) \quad (10)$$

because the w_i are independent conditional on σ^2. Each of the densities $p(w_i|\sigma^2, c_i, \pi_i, B)$ is evaluated directly using

$$p(w_i|\sigma^2, c_i, \pi_i, B) = p(w_i|\sigma^2, c_i, B, \gamma_i = 0)(1-\pi_i) + p(w_i|\sigma^2, c_i, B, \gamma_i = 1)\pi_i$$

We assume that if σ^2 is unknown, then its estimate is plugged in.

Using the posterior probabilities for each basis we can form two estimators of m. The first is the estimator using the most probable basis as determined by the posterior probabilities. The second is a weighted average of the posterior mean estimates over the four bases, using the posterior probabilities as weights. We use the notation $\widehat{m}_B$ and $\widehat{m}_A$ for the empirical Bayes estimator using the best (most probable) basis and the basis averaged estimator, respectively, where

$$\widehat{m}_A = \sum_{j=1}^{n} \widehat{m}^j p(B_j | \boldsymbol{w})$$

and $\widehat{m}^j$ is the posterior mean estimate of m for basis B_j.

7.4   Simulation Comparison

7.4.1   Introduction

This section reports the results of a simulation study comparing the performance of single basis estimators, both Bayesian and non-Bayesian, with estimators using basis selection and basis averaging. We use the Symmlet 8 basis for those experiments in which it is necessary to choose a single 'best' basis because it is a good all round performer as shown by the simulation results in section 4.4 and is in accordance with wavelet folklore about these bases.

Figure 1 plots the twelve target curves used in the simulation. The first ten are used by Marron et al. (1998); the 11th is the zero constant function representing no mean structure and the 12th is a mixture of two Gaussian densities representing a smooth function, which requires a variable bandwidth estimator. The design points are $n = 1024$ equally spaced points on $[0, 1]$, and independent Gaussian noise is added to yield the simulated observations $y_i = m(x_i) + \varepsilon_i$. Two noise levels are used; $\sigma = 0.02$ (low noise), and $\sigma = 0.1$ (high noise). Visual impression of these two noise levels is given by the lower right panels of Figure 1. The low noise is close to that often used in the wavelet examples of Donoho and Johnstone.

The estimates of c_i and π_i for the empirical Bayes estimators described in section 3 are different for each basis. To check that our approach to basis selection and basis estimation is sensible we plot the marginal likelihoods of the observations broken down by level for the four bases for 4 different functions. The estimates of c_j and π_j are obtained using the first empirical Bayes method. For each function, the log marginal likelihood is averaged over observations obtained from 10 realizations of the data. The wave, blip, blocks and constant functions are used as test functions. We expect that for the wave function the Fourier basis has the highest marginal likelihood, for

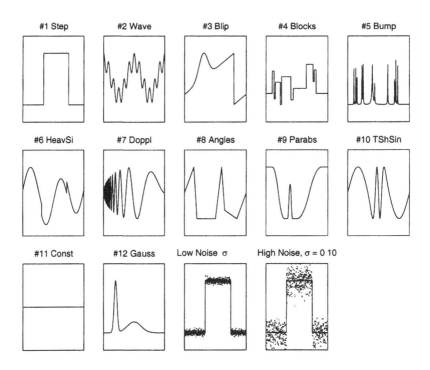

FIGURE 1. *12 target regression curves, plus step with $n = 1024$ pseudo observations show low, $\sigma = 0.02$, and high, $\sigma = 0.1$, Gaussian nose.*

the blip function the Symmlet 8 or the Daubechies 4 basis has the highest marginal likelihood, for the step function the Haar basis has the highest marginal likelihood, and for the constant function all the bases have similar likelihoods. Figure 2 plots the marginal likelihoods broken down by level and confirms our expectations.

In the simulations, the performance of an estimator $\widehat{m}$ is summarized by the Average Squared Error

$$ASE = \frac{1}{n} \sum_{i=1}^{n} \{\widehat{m}(x_i) - m(x_i)\}^2 .$$

For each setting, the various estimators and their ASE values were computed for each of 1000 replicate pseudo data sets. These ASE's are summarized by their average ASE ($AASE$) over the 1000 replicates.

We present the results in three parts, designed to address three issues of interest. Section 4.2 studies whether Bayesian methods are worthwhile, section 4.3 studies the gains from basis selection and basis averaging, and section 4.4 studies how the individual bases perform on the 12 functions.

For each part, and for each of the 24 settings (12 target curves and 2 noise levels), the estimators for that part are compared through the proportion

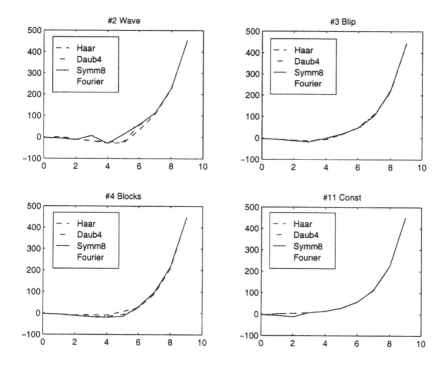

FIGURE 2. *Each panel plots the log marginal likelihood averaged over 10 realizations and broken down by level for each of the 4 bases for a given function.*

$(AASE/AASE_{best}) - 1$, where $AASE_{best}$ is the estimator having smallest AASE for that part and setting. Figures 3, 4 and 5 summarize the results. The figures show the number of times (over the 24 settings) an estimator falls into each of the following 5 categories relative to the other estimators in the group.

1. Best: either the best overall $AASE$, or else within Monte Carlo variability (defined as 2 standard errors) of the best.

2. Excellent: $AASE$ within 10% of the best for that setting.

3. Very Good: $AASE$ between 10% and 20% of the best for that setting.

4. Acceptable: $AASE$ farther than 20% from the best for that setting, but less than 100% from the best.

5. Poor: $AASE$ more than 100% from the best.

7.4.2 Fixed basis Bayesian methods

Figure 3 compares the hard and soft thresholding estimators discussed in section 2 with the two empirical Bayes estimators. All estimators use the

Symmlet 8 basis. The figure shows that the first empirical Bayes method is the best of the 4 estimators and that generally the Bayesian estimators perform favourably relative to the other two estimators. The favourable performance of the Bayesian estimators is consistent with the results of Chipman et al. (1997) and Clyde et al. (1998). Hard thresholding is somewhat better overall than soft thresholding in the sense of $AASE$ which is consistent with the results in Marron et al. (1998).

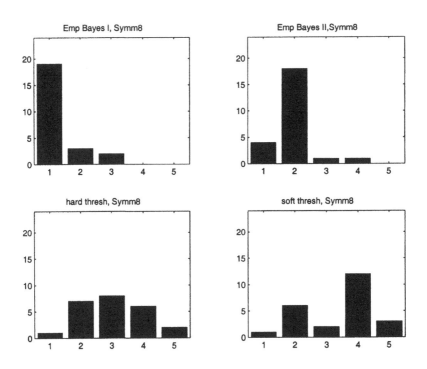

FIGURE 3. *Comparison of classical and Bayesian methods using the Symmlet 8 basis. 1 is best and 5 is worst.*

7.4.3 Basis selection and basis averaging

Figure 4 compares the empirical Bayes estimators using the Symmlet 8 basis with the corresponding estimators using basis selection and basis averaging. The first empirical Bayes method with basis averaging performs best overall. For each of the empirical Bayes approaches the basis averaging method does better than basis selection. The empirical Bayes estimator using the Symmlet 8 basis does well for those functions for which it is the correct basis, but it can perform poorly for those functions for which it is not.

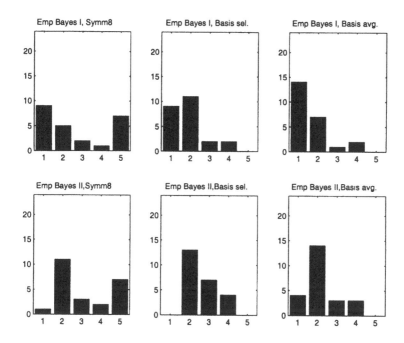

FIGURE 4. *Comparison of the empirical Bayes estimators using the Symmlet 8 basis and the empirical Bayes estimators using basis selection and basis averaging.*

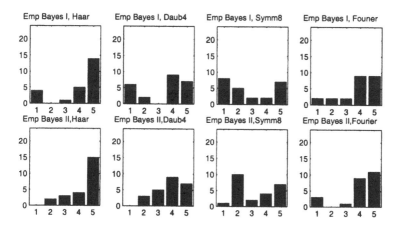

FIGURE 5. *Comparison of the two empirical Bayes approaches for individual bases.*

7.4.4   Empirical Bayes estimation for individual bases

Figure 5 studies the performance of empirical Bayes estimators for the four different bases Rows in Figure 5 correspond to different Bayesian approaches and columns correspond to the two different Bayesian estimators. Comparing across rows shows that the first empirical Bayes estimator generally does better than the second empirical Bayes estimator. Comparing across columns confirms the conventional ideas about wavelet bases: the Symmlet 8 gives generally solid all around performance, but in quite a few of these examples, the Symmlet 8 basis can be much worse than other bases.

Function	Basis			
	Haar	Daub4	Symm8	Fourier
step, lo	1000	0	0	0
step, hi	1000	0	0	0
wave, lo	0	0	0	1000
wave, hi	0	0	0	1000
blip, lo	0	992	8	0
blip, hi	57	692	251	0
bloc, lo	1000	0	0	0
bloc, hi	940	48	12	0
bump, lo	0	999	1	0
bump, hi	82	869	49	0
hvsi, lo	0	3	997	0
hvsi, hi	0	0	462	538
dopp, lo	0	0	1000	0
dopp, hi	0	0	1000	0
angl, lo	0	720	265	15
angl, hi	0	615	235	150
para, lo	0	2	998	0
para, hi	0	18	982	0
tshs, lo	0	0	123	877
tshs, hi	0	1	636	363
cons, lo	182	352	0	466
cons, hi	309	352	0	339
gaus, lo	0	64	936	0
gaus, hi	25	298	677	0

TABLE 1. Basis selection results for first empirical Bayes estimator. Each line shows the number of times out of 1000 that each basis is selected for a given function and noise level. Lo means low noise and hi means high noise.

It is also of some interest to understand the frequency (out of 1000) with which each basis is selected for each function and each noise level. Tables 1 and 2 provide this information.

Function	Basis			
	Haar	Daub4	Symm8	Fourier
step, lo	881	116	3	0
step, hi	987	13	0	0
wave, lo	0	0	52	948
wave, hi	0	0	32	968
blip, lo	6	806	188	0
blip, hi	58	697	245	0
bloc, lo	1000	0	0	0
bloc, hi	935	54	11	0
bump, lo	0	999	1	0
bump, hi	66	897	37	0
hvsi, lo	0	5	995	0
hvsi, hi	0	0	303	697
dopp, lo	0	0	1000	0
dopp, hi	0	0	1000	0
angl, lo	0	629	341	30
angl, hi	0	632	201	167
para, lo	0	1	999	0
para, hi	0	16	984	0
tshs, lo	0	0	81	919
tshs, hi	0	1	600	399
cons, lo	91	310	0	599
cons, hi	204	330	0	466
gaus, lo	0	57	943	0
gaus, hi	18	306	676	0

TABLE 2. Basis selection results for second empirical Bayes estimator. Each line shows the number of times out of 1000 that each basis is selected for a given function and noise level. Lo means low noise and hi means high noise.

7.5   Conclusion

This paper presents an empirical Bayes framework for estimating an unknown regression curve nonparametrically using basis averaging. Unknown parameters are estimated from the data and plugged in. The empirical results show that the basis averaging estimator outperforms single basis estimators.

Acknowledgments: Robert Kohn was partially supported by an Australian Research Council Grant.

References

Clyde, M. Parmigiani, G. and Vidakovic, B. (1997). Multiple shrinkage and subset selection in wavelets. *Biometrika*, 391-402.

Chipman, H. A., Kolaczyk, E. D. and McCulloch, R. E. (1997) Adaptive Bayesian Wavelet Shrinkage. *Journal of the American Statistical Association*, 92, 1413-1421

Daubechies, I. (1992). *Ten Lectures on Wavelets*. SIAM, Philadelphia.

Donoho, D. L. and Johnstone, I. M. (1994) Ideal spatial adaptation via wavelet shrinkage. *Biometrika*, 81, 425-455.

Donoho, D. L. and Johnstone, I. M. (1995). Adapting to unknown smoothness via wavelet shrinkage, *Journal of the American Statistical Association*, 90, 1200-1224.

Donoho, D. L., Johnstone, I. M., Kerkyacherian, G. and Picard, D. (1995) Wavelet shrinkage: asymptopia? (with discussion). *Journal of the Royal Statistical Society*, Ser. B, 57, 301-370.

George, E. I. and Foster, D. (1997). Calibration and empirical Bayes variable selection. Preprint.

Kohn, R., Marron, J. S. and Yau, P. (1997) Wavelet estimation using basis selection and basis averaging. Preprint.

Marron, J. S., Adak, S., Johnstone, I. M., Neumann, M. and Patil, P. (1998) Exact risk analysis of wavelet regression. *Journal of Computational and Graphical Statistics*, 278-309.

Ruppert, D., Sheather, S.J. and Wand, M.P. (1995) An effective bandwidth selector for local least squares regression. *Journal fo the American Statistical Association*, 90, 1257-1270.

Stone, C.J. (1994) "The use of polynomial splines and their tensor products in multivariate function estimation," (with discussion), *Annals of Statistics*, 22, 118-184.

Wahba, G. (1990), *Spline models for observational data,* Philadelphia: SIAM.

Vidakovic, B. (1998) Nonlinear wavelet shrinkage with Bayes rules and Bayes factors. *Journal of the American Statistical Association*, 93, 173-179.

8

An Overview of Wavelet Regularization

Yazhen Wang

ABSTRACT This chapter reviews the construction of wavelet thresholding (shrinkage) estimators by regularization. Both penalty and maximum a posterori approaches are presented and their relation is discussed.

8.1 Introduction

During past several years, there has been great interest in wavelet thresholding/shrinkage methods for function estimation. Consider nonparametric regression

$$y_i = f(x_i) + \varepsilon_i, \qquad i = 1, \cdots, n = 2^J, \tag{1}$$

where $f(x)$ is an unknown function to be estimated, $x_i = i/n \in [0,1]$, and ε_i are i.i.d. Gaussian random variables. Given a wavelet basis $\{\psi_{j,k}(x)\}_{j,k}$, one has the following wavelet expansion,

$$f(x) = \sum_{j,k} \theta_{j,k} \, \psi_{j,k}(x), \tag{2}$$

where $\theta_{j,k}$ are wavelet coefficients of f. The wavelet thresholding estimator developed by Donoho and Johnstone is constructed in three steps. First, from the data (y_i) one computes their empirical wavelet coefficients $y_{j,k}$, which satisfy

$$y_{j,k} = \theta_{j,k} + \varepsilon_{j,k}, \qquad j = 0, \cdots, J-1, \; k = 1, \cdots, 2^j - 1, \tag{3}$$

where $\varepsilon_{j,k}$ are the wavelet coefficients of the noise ε_i. Second, the estimated wavelet coefficients are given by

$$\hat{\theta}_{j,k} = \delta_\lambda(y_{j,k}),$$

where λ is a threshold,

$$\delta_\lambda(x) = x 1_{\{|x| \geq \lambda\}} \tag{4}$$

for hard threshold rule, and

$$\delta_\lambda(x) = sign(x)\,(x - \lambda)_+ \tag{5}$$

for soft threshold rule. Third, the wavelet shrinkage estimator is the reconstruction of $\hat{\theta}_{j,k}$

$$\hat{f}(x) = \sum_{j,k} \hat{\theta}_{j,k}\,\psi_{j,k}(x). \tag{6}$$

The wavelet estimator has been shown to enjoy spatial adaptivity and to achieve minimaxity over a wide range of function classes (Donoho and Johnstone 1998; Donoho, Johnstone, Kerkyacharian and Picard 1995; Wang 1997).

8.2   Estimation by Penalization

The wavelet thresholding estimator can be obtained via penalization. Specifically, the minimizer of

$$\|\mathbf{y} - f\|_2^2 + 2\,\lambda J(f) \tag{7}$$

corresponds to the soft thresholding estimator when the penalty $J(f)$ is the l_1-norm of the wavelet coefficients of f, i.e.

$$J(f) = \sum_{j,k} |\theta_{j,k}|, \tag{8}$$

while for the hard thresholding estimator, the corresponding penalty is the number of non-zero wavelet coefficients, i.e.

$$J(f) = \text{ the number of } \theta_{j,k} \neq 0, \tag{9}$$

where $\theta_{j,k}$ are the wavelet coefficients of f. (See Chen and Donoho 1995, Donoho and Johnstone 1994.)

There is long history in function estimation by regularization. Smoothing splines estimators are constructed by using Sobolev type norm penalty like

$$J(f) = \int [f'(x)]^2\,dx.$$

See Wahba (1990). In general, one may use a loss function and Besov type seminorm penalty to construct an estimator of f, i.e., minimize

$$L(\mathbf{y}, f) + \lambda\,\|f\|_{B_{p,q}^s}, \tag{10}$$

where L is a loss function such as l_2 distance, and $\|f\|_{B^s_{p,q}}$ is a *Besov* seminorm of index (s, p, q) given by

$$|f|_{B^s_{p,q}} = \left(\int_0^1 \left(\frac{w_{r,p}(f; h)}{h^s} \right)^q \frac{dh}{h} \right)^{1/q} \qquad \text{if } q < \infty,$$

$$|f|_{B^s_{p,\infty}} = \sup_{0<h<1} \frac{w_{r,p}(f; h)}{h^s} \qquad \text{if } q = \infty,$$

$$w_{r,p}(f; h) = \|\Delta_h^{(r)} f\|_{L^p[0, 1-rh]},$$

the r-th modulus of smoothness in $L^p[0, 1]$, and

$$\Delta_h^{(r)} f = \sum_{k=0}^r \binom{r}{k} (-1)^k f(t + kh),$$

the r-th difference.

Next section will show that this general approach can accommodate various combination of error distributions and penalties. However, generally regularization estimators involve some thresholding only when $J(f)$ is non-convex or convex but non-differentiable like l_1 norm.

8.3   Estimation by Maximum a Posteriori

Since Donoho and Johnstone's wavelet estimator was developed, variants of shrinkage and threshold rules for constructing wavelet estimators have been proposed. One approach to obtaining shrinkage/threshold rules is though Bayesian methods. It turns out that wavelet thresholding/shrinkage estimators can be obtained from Bayesian estimator by incorporating prior and error distribution, and wavelet regularization estimation can be viewed as maximum a posteriori (MAP) estimation.

For simplicity, consider the prior on f and the error distribution such that $\theta_{j,k}$, $\varepsilon_{j,k}$ [the wavelet coefficients of the function f and the noise ε, see (1)-(3)] all are independent. Thus, the empirical wavelet coefficients $y_{j,k}$ of the data are independent, and one can work coordinate by coordinate and drop the indices j and k.

Suppose that θ has a prior $\pi(\theta)$, and given θ, the empirical wavelet coefficient y has a location model $p(y - \theta)$. Then given the observation y, the posterior distribution of θ is proportional to

$$\pi(\theta|y) \propto p(y - \theta) \times \pi(\theta).$$

The MAP rule $\hat\theta = \hat\theta(y)$ is to maximize the posterior $\pi(\theta|y)$, which is equivalent to minimize

$$- \log p(y - \theta) + J(\theta), \tag{11}$$

where $J(\theta) = -\log \pi(\theta)$.

Consider the normal density $p(y - \theta)$ for which the wavelet coefficients are independent. With $[y|\theta] \sim \mathcal{N}(\theta, \sigma^2)$, the objective functional in (11) becomes

$$\frac{1}{2\sigma^2}|y - \theta|^2 + J(\theta). \tag{12}$$

If $J(\theta)$ is strictly convex and differentiable, the minimizer of (12) is given by

$$\hat{\theta} = h^{-1}(y), \tag{13}$$

where

$$h(u) = u + \sigma^2 J'(u). \tag{14}$$

See Vidakovic (1999).

For double exponential prior $\pi(\theta) = \frac{1}{2}e^{-|\theta|}$, $J(\theta) = |\theta| + \log 2$, $J'(\theta) = sign(\theta)$, and $\hat{\theta}(y) = sign(y) \max(0, |y| - \sigma^2)$, which coincides with the soft thresholding rule defined in (5). In fact, for double exponential prior, the minimization problem (12) corresponds to the coordinatewise version of (7) with penalty given by (8).

Leporini and Pesquet (1998a,b) explore cases for which the prior has an exponential power distribution $\mathcal{EPD}(\alpha, \beta)$

$$\pi(\theta) = \frac{\beta}{2 \alpha \Gamma(1/\beta)} exp\{-(|\theta|/\alpha)^\beta\},$$

and the noise also follows an $\mathcal{EPD}(a, b)$ distribution. The MAP rule minimizes

$$\left(\frac{|y - \theta|}{a}\right)^b + \left(\frac{|\theta|}{\alpha}\right)^\beta, \tag{15}$$

which corresponds to (10) with L and J given by the l_b and l_β norms, respectively. The resulting MAP rule involves thresholding only for $\beta \leq 1$. In this case, $J(\theta) = (|\theta|/\alpha)^\beta$ is non-convex or l_1 norm.

Leporini and Pesquet (1998b) also considered the case with Laplacian prior [i.e. $\mathcal{EPD}(\alpha, 1)$] and the Cauchy noise

$$p(y - \theta) = \frac{a}{\pi} \frac{1}{1 + a^2 (y - \theta)^2}.$$

Now the MAP rule minimizes

$$log\{1 + a^2 (y - \theta)^2\} + |\theta|/\alpha.$$

Due to the nonconvex objective functional, the resulting MAP rule is degenerate for $a\,\alpha \leq 1$ and a double thresholding rule for $a\,\alpha > 1$:

$$\text{if } a\,\alpha \leq 1, \qquad \hat{\theta} = 0,$$

$$\text{if } a\,\alpha > 1, \qquad \hat{\theta} = \begin{cases} sign(y)\,(y - \lambda_L) & y \in (\lambda_L, \lambda_U) \\ 0 & \text{otherwise} \end{cases}$$

where

$$\lambda_L = \alpha - (\alpha^2 - 1/a^2)^{1/2},$$

and λ_U is the unique solution in (λ_L, ∞) of the equation

$$log\left(1 + a^2\,\lambda_U^2\right) = log\left(1 + a^2\,\lambda_L^2\right) + \frac{\lambda_U - \lambda_L}{2}.$$

For Cauchy noise, the MAP rule thresholds not only very small wavelet coefficients but also the extreme large wavelet coefficients. This is quite different from the Gaussian noise case where only small wavelet coefficients are thresholded. The result indicates that for heavy tail distributions like Cauchy, one needs to kill also the very large wavelet coefficients, since it is more likely that the extreme values come from the noise than from the signal.

An interesting problem is to use the method to study robust function estimation for contaminated data. For example, if the errors are $100\,(1 - q)\%$ Gaussian and $100\,q\%$ Cauchy for small $q > 0$, the MAP estimation minimizes

$$(1 - q)\,\frac{|y - \theta|^2}{2\,\sigma^2} + q\,log\{1 + a^2\,(y - \theta)^2\} + \frac{|\theta|}{\alpha}.$$

The resulting MAP rule involves some double thresholding scheme needed to threshold the small wavelet coefficients and to kill the extreme values attributed to the Cauchy noise.

Acknowledgments: I like to thank Brani Vidakovic for suggestions and for providing me with some references.

References

Chen, S. and Donoho, S. (1995). Atomic decomposition by basis pursuit. Manuscript.

Donoho, D. L. and Johnstone, I. M. (1994). Ideal denoising in an orthonormal basis chosen from a library of bases. Comptes Rendus Acad. Sci. Paris A **319**, 1317-1322.

Donoho, D. L. and Johnstone, I. M. (1998). Minimax estimation via Wavelets shrinkage. Ann. Statist. **26**, 879-921.

Donoho, D. L., Johnstone, I. M., Kerkyacharian, G. and Picard, D. (1995). Wavelet shrinkage: Asymptopia ? (with discussion). J. Roy. Statist. Soc. B **57** 301-369.

Leporini, D. and Pesquet, J.C. (1998a). Multiscale regularization in Besov Spaces. Manuscript.

Leporini, D. and Pesquet, J.C. (1998b). Wavelet thresholding for some classes of non-Gaussian noise. Manuscript.

Vidakovic, B. (1999). *Statistical Modeling by Wavelets.* Wiley, N.Y., 381pp.

Wahba, G. (1990). *Spline Models for Observational Data.* Society for Industrial and Applied Mathematics, Philadelphia.

Wang, Y. (1997). Fractal function estimation via wavelet shrinkage. J. Roy. Statist. Soc. B. **59**, 603-613.

9

Minimax Restoration and Deconvolution

Jérôme Kalifa and Stéphane Mallat

ABSTRACT Inverting the distortion of signals and images in presence of additive noise is often numerically unstable. To solve these ill-posed inverse problems, we study linear and non-linear diagonal estimators in an orthogonal basis. General conditions are given to build nearly minimax optimal estimators with a thresholding in an orthogonal basis. The deconvolution of bounded variation signals is studied in further details, with an application to the deblurring of satellite images.

9.1 Introduction

We consider a measurement device that degrades a signal f with a linear operator U and adds a Gaussian white noise W of variance σ^2 :

$$Y = Uf + W .$$

We suppose that U and σ^2 have been calculated through a calibration procedure. Applying the inverse U^{-1} to Y yields an equivalent denoising problem

$$U^{-1}Y = f + U^{-1}W.$$

When the inverse U^{-1} is not bounded, the resulting noise $U^{-1}W$ is amplified by a factor that tends to infinity. Finding an estimate $\tilde{F}$ of the signal f is called an *ill-posed* inverse problem [Ber89, O'S86]. We focus on minimax rather than Bayes estimation because there exists no stochastic models that represents the spatial inhomogeneity of natural images. Section 9.3 compares and relates minimax and Bayes estimation.

Donoho and Johnstone have obtained general minimax optimality results to estimate signals contaminated by white Gaussian noise with thresholding estimators in orthogonal bases [DJ94]. To obtain similar optimality results when estimating signals contaminated by non white noises, one needs to adapt the basis to the covariance properties of the noise. Section 9.4 shows that thresholding estimators are quasi-minimax optimal if the basis nearly diagonalizes the covariance of the noise and if it concentrates the energy of the signal on a few coefficients. This provides a general framework for the resolution of ill-posed inverse problems. Deconvolution is a generic example of instable inverse problem, which is studied in further details in section 9.5, with an application to the removal of blur in satellite images.

9.2   Inverse problems written in a denoising form

A signal f of size N must be estimated from the measured data

$$Y = Uf + W,\tag{1}$$

where W is a Gaussian white noise of known variance σ^2. The degradation U is inverted with the pseudo-inverse. Let $\mathbf{V} = \mathrm{Im}U$ be the image of U and $\mathbf{V}^\perp$ be its orthogonal complement. The pseudo-inverse $\tilde{U}^{-1}$ of U is the left inverse whose restriction to $\mathbf{V}^\perp$ is zero. The restoration is said to be unstable if

$$\lim_{N\to+\infty} \|\tilde{U}^{-1}\|_S^2 = +\infty .$$

where $\|T\|_S$ is the sup operator norm of a linear operator T defined by

$$\|T\|_S = \sup_{f\in\mathbb{C}^N} \frac{\|Tf\|}{\|f\|}.\tag{2}$$

Estimating f from Y is equivalent to estimate it from

$$X = \tilde{U}^{-1}Y = \tilde{U}^{-1}Uf + \tilde{U}^{-1}W.\tag{3}$$

The operator $\tilde{U}^{-1}U = P_{\mathbf{V}}$ is an orthogonal projection on $\mathbf{V}$ so

$$X = P_{\mathbf{V}}f + Z \quad\text{with}\quad Z = \tilde{U}^{-1}W .\tag{4}$$

The noise Z is not white but remains Gaussian because $\tilde{U}^{-1}$ is linear. It is considerably amplified when the problem is unstable. The covariance operator K of Z is

$$K = \sigma^2 \tilde{U}^{-1,*} \tilde{U}^{-1} ,\tag{5}$$

where A^* is the adjoint of an operator A.

To simplify notations, we formally rewrite (4) as a standard denoising problem

$$X = f + Z ,\tag{6}$$

while considering that the projection of Z in $\mathbf{V}^\perp$ is a noise of infinite energy to express the loss of all information concerning the projection of f in $\mathbf{V}^\perp$. It is equivalent to write formally $Z = U^{-1} W$.

Having reformulated the problem in this form leads us to a signal denoising problem. To attenuate the noise while preserving the signal, we need to find a representation that discriminates the signal from the noise. To optimize such a representation and consequently the derived estimators requires taking advantages of prior information on the signals we want to recover. The "Bayes versus minimax" estimation issue depends on the type of prior models available for our signals.

9.3   Bayes versus minimax estimation

Although we may have some prior information, it is rare that we know the probability distribution of complex signals. This prior information often defines a set Θ where the signals are guaranteed to remain, without specifying their probability distribution in Θ. The more prior information the smaller the set Θ. For example, we may know that the signals have at most K discontinuities, with bounded derivatives outside these discontinuities. This defines a particular prior set Θ. Presently, there exists no stochastic model that takes into account the diversity of natural images. However, many images have some form of piecewise regularity, with a bounded total variation. This also specifies a prior set Θ.

The problem is to estimate $f \in \Theta$ from the noisy data

$$X[n] = f[n] + Z[n] .$$

The risk of an estimation $\tilde{F} = DX$ is

$$r(D, f) = \mathsf{E}\{\|DX - f\|^2\}.$$

The expected risk over Θ cannot be computed because we do not know the probability distribution of signals in Θ. To control the risk for any $f \in \Theta$, we thus try to minimize the maximum risk :

$$r(D, \Theta) = \sup_{f \in \Theta} \mathsf{E}\{\|DX - f\|^2\}.$$

Let $\mathcal{O}_n$ be the set of all linear and non-linear operators from $\mathbb{C}^N$ to $\mathbb{C}^N$. The *minimax risk* is the lower bound computed over all operators D :

$$r_n(\Theta) = \inf_{D \in \mathcal{O}_n} r(D, \Theta).$$

In practice, we must find D that is simple to implement and such that $r(D, \Theta)$ is close to the minimax risk $r_n(\Theta)$.

As a first step, one can simplify this problem by restricting D to be a linear operator. Let $\mathcal{O}_l$ be the set of all linear operators from $\mathbb{C}^N$ to $\mathbb{C}^N$. The *linear minimax risk* over Θ is the lower bound :

$$r_l(\Theta) = \inf_{D \in \mathcal{O}_l} r(D, \Theta)$$

This strategy is efficient only if $r_l(\Theta)$ is of the same order as $r_n(\Theta)$.

A Bayes estimator supposes that we know the prior probability distribution π of signals in Θ. If available, this supplement of information can only improve the signal estimation. The central result of game and decision theory shows that minimax estimations are Bayes estimations for a "least favorable" prior.

Let F be the signal random vector, whose probability distribution is given by the prior π. For a decision operator D, the expected risk calculated with respect to π is

$$r(D, \pi) = E_\pi\{r(D, F)\}.$$

The minimum Bayes risks for linear and non-linear operators are defined by :

$$r_l(\pi) = \inf_{D \in \mathcal{O}_l} r(D, \pi) \quad \text{and} \quad r_n(\pi) = \inf_{D \in \mathcal{O}_n} r(D, \pi) .$$

Let Θ^* be the set of all probability distributions of random vectors whose realizations are in Θ. The minimax theorem relates a minimax risk and the maximum Bayes risk calculated for priors in Θ^*.

Theorem 9.1 (Minimax) *For any set Θ of signals of dimension N*

$$r_l(\Theta) = \sup_{\pi \in \Theta^*} r_l(\pi) \quad and \quad r_n(\Theta) = \sup_{\pi \in \Theta^*} r_n(\pi) . \tag{7}$$

A distribution $\tau \in \Theta^*$ such that $r(\tau) = \inf_{\pi \in \Theta^*} r(\pi)$ is called a *least favorable* prior distribution. The minimax theorem proves that the minimax risk is the minimum Bayes risk for a least favorable prior.

In signal processing, minimax calculations are often hidden behind apparently orthodox Bayes estimations. Let us consider an example on images. It has been observed that the histograms of the wavelet coefficients of "natural" images can be modeled with generalized Gaussian distributions [Mal89, Sim99]. This means that natural images belong to certain set Θ but it does not specify a prior distribution over this set. To compensate for the lack of knowledge on the dependency of wavelet coefficients spatially and across scales, one can be tempted to create "a simple probabilistic model" where all wavelet coefficients are considered to be independent. This model is clearly wrong since images have geometrical structures that create strong dependencies both spatially and across scales. However, calculating a Bayes estimator with this prior model may give valuable results to estimate images, because this "simple" prior is often close to a least favorable prior. The resulting estimator and risk are thus good approximations of the minimax optimum. If not chosen carefully, a "simple" prior may yield an optimistic risk evaluation that is not valid for real signals. Understanding the robustness of uncertain priors is what minimax calculations are often about.

9.4   Minimax estimation in Gaussian noise

We saw in section 9.2 that the generic inverse problem of equation (1) can be written as the estimation of a signal f contaminated by an additive Gaussian noise Z :

$$X = f + Z .$$

The random vector Z is characterized by its covariance operator K, and we suppose that $E\{Z[n]\} = 0$. Our goal is to find non-linear estimators which are both close to minimax optimality and simple to implement.

9.4.1   Diagonal Estimation

Donoho and Johnstone [DJ94] proved that diagonal estimators in an orthonormal basis $B = \{g_m\}_{0 \leq m < N}$ are nearly minimax optimal if the basis provides a sparse signal representation, which means that the basis concentrates the energy of the signal on a few coefficients. When the noise is not white, the coefficients of the noise have a variance that depends upon each g_m :

$$\sigma_m^2 = E\{|Z_B[m]|^2\} = \langle Kg_m, g_m \rangle .$$

The basis choice must therefore depend on the covariance K.

We study the risk of estimators that are diagonal in B :

$$\tilde{F} = DX = \sum_{m=0}^{N-1} d_m(X_B[m]) \, g_m . \tag{8}$$

If $d_m(X_B[m]) = a[m] \, X_B[m]$, one can verify that the minimum risk $E\{\|\tilde{F} - f\|^2\}$ is achieved by the following attenuation :

$$a[m] = \frac{|f_B[m]|^2}{|f_B[m]|^2 + \sigma_m^2}, \tag{9}$$

and

$$E\{\|\tilde{F} - f\|^2\} = r_{\inf}(f) = \sum_{m=0}^{N-1} \frac{\sigma_m^2 \, |f_B[m]|^2}{\sigma_m^2 + |f_B[m]|^2} . \tag{10}$$

Over a signal set Θ, the maximum risk of this attenuation is $r_{\inf}(\Theta) = \sup_{f \in \Theta} r_{\inf}(f)$. The attenuation (9) cannot be implemented because $a[m]$ depends upon $|f_B[m]|$ which is not known in practice. It is called an *oracle attenuation* becauses it uses information normally not available. The risk $r_{\inf}(\Theta)$ is thus only a lower bound for the minimax risk of diagonal estimators. We shall see that a simple thresholding estimator has a maximum risk that is close to $r_{\inf}(\Theta)$.

Linear diagonal estimation   We begin by studying linear diagonal estimators D, where each $a[m]$ is a constant. Let $\mathcal{O}_{l,d}$ be the set of all such linear diagonal operators D. Since $\mathcal{O}_{l,d} \subset \mathcal{O}_l$, the *linear diagonal minimax risk* is larger than the linear minimax risk

$$r_{l,d}(\Theta) = \inf_{D \in \mathcal{O}_{l,d}} r(D, \Theta) \geq r_l(\Theta) .$$

We characterize diagonal estimators that achieve the minimax risk $r_{l,d}(\Theta)$. For this purpose, we introduce the notion of quadratic convex hull QH[Θ] of Θ.

The "square" of a set Θ in the basis $\mathcal{B}$ is defined by

$$(\Theta)_\mathcal{B}^2 = \{\tilde{f} \; : \; \tilde{f} = \sum_{m=0}^{N-1} |f_\mathcal{B}[m]|^2 \, g_m \text{ with } f \in \Theta\} . \tag{11}$$

We say that Θ is *quadratically convex* in $\mathcal{B}$ if $(\Theta)_\mathcal{B}^2$ is a convex set. The *quadratic convex hull* QH[Θ] of Θ in the basis $\mathcal{B}$ is defined by

$$\text{QH}[\Theta] = \left\{ f \; : \; \sum_{m=0}^{N-1} |f_\mathcal{B}[m]|^2 \text{ is in the convex hull of } (\Theta)_\mathcal{B}^2 \right\} . \tag{12}$$

It is the largest set whose square $(\text{QH}[\Theta])_\mathcal{B}^2$ is equal to the convex hull of $(\Theta)_\mathcal{B}^2$.

The risk of an oracle attenuation (10) gives a lower bound of the minimax linear diagonal risk $r_{l,d}(\Theta)$:

$$r_{l,d}(\Theta) \geq r_{\inf}(\Theta) \tag{13}$$

The following theorem shows that this inequality is an equality if and only if Θ is quadratically convex [DLG90].

Theorem 9.2 *Let Θ be a closed and bounded set. There exists $x \in$ QH[Θ] such that $r_{\inf}(x) = r_{\inf}(\text{QH}[\Theta])$. If D is the linear operator defined by*

$$a[m] = \frac{|x_\mathcal{B}[m]|^2}{\sigma_m^2 + |x_\mathcal{B}[m]|^2} , \tag{14}$$

then

$$r(D, \Theta) = r_{l,d}(\text{QH}[\Theta]) = r_{\inf}(\text{QH}[\Theta]) . \tag{15}$$

This theorem implies that $r_{l,d}(\Theta) = r_{l,d}(\text{QH}[\Theta])$. To take advantage of the fact that Θ may be much smaller than its quadratic convex hull, it is necessary to use non-linear diagonal estimators.

Non linear diagonal estimation Among non-linear diagonal estimators, we concentrate on thresholding estimators :

$$\tilde{F} = DX = \sum_{m=0}^{N-1} \rho_{T_m}(X_B[m])\, g_m \ , \tag{16}$$

where $\rho_{T_m}(x)$ is a hard thresholding function

$$\rho_{T_m}(x) = \begin{cases} x & \text{if } |x| > T_m \\ 0 & \text{if } |x| \le T_m \end{cases} , \tag{17}$$

or a soft thresholding function

$$\rho_{T_m}(x) = \begin{cases} x - T_m & \text{if } x \ge T_m \\ x + T_m & \text{if } x \le -T_m \\ 0 & \text{if } |x| \le T_m \end{cases} . \tag{18}$$

The risk of this thresholding estimator is

$$r_t(f) = r(D, f) = \sum_{m=0}^{N-1} \mathsf{E}\{|f_B[m] - \rho_{T_m}(X_B[m])|^2\}\, .$$

The threshold T_m is adapted to the noise variance σ_m^2 in the direction of g_m. Donoho and Johnstone compute an upper bound of the risk $r_t(f)$ when $T_m = \sigma_m \sqrt{2\log_e N}$. If the signals belong to a set Θ, the threshold values are improved by considering the maximum of signal coefficients :

$$s_B[m] = \sup_{f \in \Theta} |f_B[m]|\ .$$

If $s_B[m] \le \sigma_m$ then setting $X_B[m]$ to zero yields a risk $|f_B[m]|^2$ that is always smaller than the risk σ_m^2 of keeping it. This is done by choosing $T_m = \infty$ to guarantee that $\rho_{T_m}(X_B[m]) = 0$. Thresholds are thus defined by

$$T_m = \begin{cases} \sigma_m \sqrt{2\log_e N} & \text{if } \sigma_m < s_B[m] \\ \infty & \text{if } \sigma_m \ge s_B[m] \end{cases} . \tag{19}$$

Donoho and Johnstone [DJ94] proved that for these thresholds, the risk of a thresholding is close to the risk of an oracle attenuation

Theorem 9.3 *For the thresholds (19), the risk of a thresholding estimator satisfies for $N \ge 4$*

$$r_t(\Theta) \le (2\log_e N + 1)\left(\bar{\sigma}^2 + r_{\inf}(\Theta)\right) \tag{20}$$

with $\bar{\sigma}^2 = \frac{1}{N}\sum_{\sigma_m < s_B[m]} \sigma_m^2$.

This theorem proves that the risk of a thresholding estimator is not much above $r_{\inf}(\Theta)$. It now remains to understand under which conditions the minimax risk $r_n(\Theta)$ is also of the order of $r_{\inf}(\Theta)$.

9.4.2   Nearly Diagonal Covariance

To estimate efficiently a signal with a diagonal operator, the basis $\mathcal{B} = \{g_m\}_{0 \le m < N}$ must provide a sparse representation of signals in Θ. To analyse the properties of linear and non-linear estimators, we introduce orthosymmetric sets. We say that Θ is *orthosymmetric* in $\mathcal{B}$ if for any $f \in \Theta$ and for any $a[m]$ with $|a[m]| \le 1$ then

$$\sum_{m=0}^{N-1} a[m]\, f_\mathcal{B}[m]\, g_m \in \Theta .$$

This means that the set Θ is elongated along the directions of the vectors g_m of $\mathcal{B}$. The basis $\mathcal{B}$ concentrates the energy of the signals in Θ on a few coefficients if Θ has a star shape in $\mathcal{B}$ as illustrated by figure 1-(a).

We first consider the estimation of signals f in Θ in presence of a white noise Z.

$$X = f + Z$$

We respectively denote $r_l(\Theta)$, $r_{\inf}(\Theta)$ and $r_n(\Theta)$ the corresponding linear, oracle attenuation and non linear minimax risks. In that case, Donoho [DLG90, Don93] showed that if Θ is orthosymmetric in $\mathcal{B}$, then

$$r_l(\Theta) = r_{\inf}(\mathrm{QH}[\Theta]) . \tag{21}$$

and

$$\frac{1}{1.25}\, r_{\inf}(\Theta) \le r_n(\Theta) \tag{22}$$

We want to obtain similar results when the noise Z is not white. In that case, the basis must transform the noise into "nearly" independent coefficients. This approach was studied by Donoho for some specific deconvolution problems where wavelet bases are adapted [Don95]. Section 9.5 shows that it is not the case for *hyperbolic deconvolution* such as deblurring problems. We thus give more general conditions on the orthogonal basis $\mathcal{B}$ to obtain nearly minimax thresholding estimators.

Since the noise Z is Gaussian, to guarantee that the coefficients $Z_\mathcal{B}[m]$ are nearly independent is equivalent to have them nearly uncorrelated, which means that its covariance K is nearly diagonal in $\mathcal{B}$. This approximate diagonalization is measured by preconditioning K with its diagonal. We denote by K_d the diagonal operator in the basis $\mathcal{B}$, whose diagonal is equal to the diagonal of K. The diagonal coefficients of K and K_d are thus $\sigma_m^2 = \mathsf{E}\{|Z_\mathcal{B}[m]|^2\}$. We suppose that K has no eigenvalue equal to zero, because the noise would then be zero in this direction, in which case the estimation is trivial. Let K^{-1} be the inverse of K, and $K_d^{1/2}$ be the diagonal matrix whose coefficients are σ_m. Theorem 9.4 computes lower bounds of the minimax risks with a conditioning factor defined with the operator sup norm $\|\,.\,\|_S$ introduced in (2).

Theorem 9.4 *The conditioning factor satisfies*

$$\lambda_B = \|K_d^{1/2} K^{-1} K_d^{1/2}\|_S \geq 1 . \tag{23}$$

If Θ is orthosymmetric in B then

$$r_l(\Theta) \geq \frac{1}{\lambda_B} r_{\inf}(\mathrm{QH}[\Theta]) , \tag{24}$$

and

$$r_n(\Theta) \geq \frac{1}{1.25\,\lambda_B} r_{\inf}(\Theta) . \tag{25}$$

Proof. We give a sketch of the proof. We consider first the particular case where K is diagonal, which is trivial. If K is diagonal in B then the coefficients $Z_B[m]$ are independent Gaussian random of variance σ_m^2. Estimating $f \in \Theta$ from $X = f + Z$ is equivalent to estimating f_0 from $X_0 = f_0 + Z_0$ where

$$Z_0 = \sum_{m=0}^{N-1} \frac{Z_B[m]}{\sigma_m} g_m \;,\quad X_0 = \sum_{m=0}^{N-1} \frac{X_B[m]}{\sigma_m} g_m \;,\quad f_0 = \sum_{m=0}^{N-1} \frac{f_B[m]}{\sigma_m} g_m . \tag{26}$$

The signal f_0 belongs to an orthosymmetric set Θ_0 and the renormalized noise Z_0 is a Gaussian white noise of variance 1. We thus use equations (21) and (22). By reinserting the value of the renormalized noise and signal coefficients, we derive that

$$r_n(\Theta) \geq \frac{1}{1.25} r_{\inf}(\Theta) \quad\text{and}\quad r_l(\Theta) = r_{\inf}(\mathrm{QH}[\Theta]) . \tag{27}$$

To prove the general case we use inequalities over symmetrical matrices. If A and B are two symmetric matrices, we write $A \geq B$ if the eigenvalues of $A - B$ are positive, which means that $\langle Af, f \rangle \geq \langle Bf, f \rangle$ for all $f \in \mathbb{C}^N$. Since λ_B is the largest eigenvalue of $K_d^{1/2} K^{-1} K_d^{1/2}$, we can derive that

$$K \geq \lambda_B^{-1} K_d . \tag{28}$$

Observe that $\lambda_B \geq 1$ because $\langle K g_m, g_m \rangle = \langle K_d g_m, g_m \rangle$.

One can show that since $K \geq \lambda_B^{-1} K_d$ the estimation problem with the noise Z of covariance K has a minimax risk that is larger than the minimax risk of the estimation problem with a noise of covariance $\lambda_B^{-1} K_d$. But since this covariance is diagonal we can apply (27). The definition of $r_{\inf}(\Theta)$ is the same for a noise of covariance K and for a noise of covariance K_d because $\sigma_m^2 = \langle K g_m, g_m \rangle = \langle K_d g_m, g_m \rangle$. When multiplying K_d by a constant $\lambda_B^{-1} \leq 1$, the value $r_{\inf}(\Theta)$ which appears in (27) is modified into $r'_{\inf}(\Theta)$ with $r'_{\inf}(\Theta) \geq \lambda_B^{-1} r_{\inf}(\Theta)$. We thus derive (25) and (24).

One can verify that $\lambda_B = 1$ if and only if $K = K_d$ and is thus diagonal in B. The closer λ_B is to 1 the more diagonal K. The main difficulty is to find a basis B which nearly diagonalizes the covariance of the noise and provides sparse signal representations so that Θ is orthosymmetric or can be embedded in two close orthosymmetric sets.

An upper bound of $r_l(\Theta)$ is computed in (15) with a linear diagonal operator, and together with (24) we get

$$\frac{1}{\lambda_B} r_{\mathrm{inf}}(\mathrm{QH}[\Theta]) \leq r_l(\Theta) \leq r_{\mathrm{inf}}(\mathrm{QH}[\Theta]). \qquad (29)$$

Similarly, an upper bound of $r_n(\Theta)$ is calculated with the thresholding risk calculated by theorem 9.3. With the lower bound (25) we obtain

$$\frac{1}{1.25\,\lambda_B} r_{\mathrm{inf}}(\Theta) \leq r_n(\Theta) \leq r_t(\Theta) \leq (2\log_e N + 1)\left(\bar{\sigma}^2 + r_{\mathrm{inf}}(\Theta)\right). \qquad (30)$$

If the basis $\mathcal{B}$ nearly diagonalizes K so that λ_B is of the order of 1 then $r_l(\Theta)$ is of the order of $r_{\mathrm{inf}}(\mathrm{QH}[\Theta])$, whereas $r_n(\Theta)$ and $r_t(\Theta)$ are of the order of $r_{\mathrm{inf}}(\Theta)$.

If Θ is quadratically convex then $\Theta = \mathrm{QH}[\Theta]$ so the linear and non-linear minimax risks are close. If Θ is not quadratically convex then its quadratic hull $\mathrm{QH}[\Theta]$ may be much bigger than Θ. This is the case when Θ has a star shape which is elongated in the directions of the basis vectors g_m, as illustrated in Figure 1. In that case, a thresholding estimation in $\mathcal{B}$ may significantly outperform an optimal linear estimation.

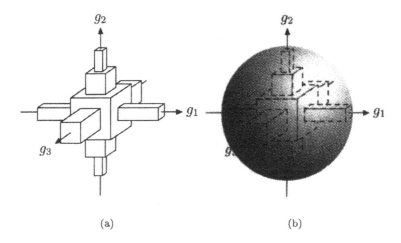

(a)   (b)

FIGURE 1. (a) : Example of orthosymmetric set Θ in 3 dimensions. (b) : The quadratically convex hull $\mathrm{QH}[\Theta]$ is a much larger ellipsoïd.

Thresholding inverse estimation We apply these results to solve inverse problems written in the standard denoising form of equation (6).

$$X = f + Z ,$$

where we write formally $Z = U^{-1} W$.

Let $\mathcal{B} = \{g_m\}_{0 \leq m < N}$ be an orthonormal basis such that a subset of its vectors define a basis of $\mathbf{V} = \mathrm{Im}U$. The coefficients of the noise Z have a variance $\sigma_m^2 = \mathbf{E}\{|Z_B[m]|^2\}$, and we set $\sigma_m = \infty$ if $g_m \in \mathbf{V}^{\perp}$. An oracle attenuation (9) yields a lower bound for the risk

$$r_{\inf}(f) = \sum_{m=0}^{N-1} \frac{\sigma_m^2 \, |f_B[m]|^2}{\sigma_m^2 + |f_B[m]|^2} . \tag{31}$$

The loss of the projection of f in $\mathbf{V}^{\perp}$ appears in the terms

$$\frac{\sigma_m^2 \, |f_B[m]|^2}{\sigma_m^2 + |f_B[m]|^2} = |f_B[m]|^2 \quad \text{if } \sigma_m = \infty .$$

Theorem 9.3 proves that a thresholding estimator in $\mathcal{B}$ yields a risk that is above $r_{\inf}(\Theta)$ by a factor $2 \log_e N$. Theorem 9.4 relates linear and non-linear minimax risk to $r_{\inf}(\Theta)$. Let K_d be the diagonal operator in $\mathcal{B}$, equal to the diagonal of the covariance K defined in (5). The inverse of K is replaced by its pseudo inverse $K^{-1} = \sigma^{-2} U U^*$ and the conditioning number is

$$\lambda_{\mathcal{B}} = \|K_d^{1/2} \, K^{-1} \, K_d^{1/2}\|_S = \sigma^{-2} \, \|K_d^{1/2} \, U\|_S^2 .$$

Thresholding estimators have a risk $r_t(\Theta)$ that is close to $r_n(\Theta)$ if Θ is nearly orthosymmetric in $\mathcal{B}$ and if $\lambda_{\mathcal{B}}$ is of the order of 1. The main difficulty is to find such a basis $\mathcal{B}$.

The thresholds (19) define a projector which is non zero only in the space $\mathbf{V}_0 \subset \mathbf{V}$ generated by the vectors $\{g_m\}_{\sigma_m < s_B[m]}$. This means that the calculation of $X = \tilde{U}^{-1} Y$ in (3) can be replaced by a regularized inverse $X = P_{\mathbf{V}_0} \tilde{U}^{-1} Y$, to avoid numerical instabilities.

9.5   Deconvolution

The restoration of signals degraded by a convolution operator U is a generic inverse problem that is often encountered in signal processing. The convolution is supposed to be circular to avoid border problems. The goal is to estimate f from

$$Y = f \circledast u + W .$$

The circular convolution is diagonal in the discrete Fourier basis $\mathcal{B} = \left\{ g_k[n] = \frac{1}{\sqrt{N}} \exp\left(\frac{i2n\pi k}{N}\right) \right\}_{0 \leq k < N}$. The eigenvalues are equal to the discrete Fourier transform $\hat{u}[k]$, so $\mathbf{V} = \mathrm{Im}U$ is the space generated by the sinusoids g_k such that $\hat{u}[k] \neq 0$. The pseudo inverse of U is $\tilde{U}^{-1}f = f \circledast \tilde{u}^{-1}$ where the discrete Fourier transform of $\tilde{u}^{-1}$ is

$$\widehat{\tilde{u}^{-1}}[k] = \begin{cases} \frac{1}{\hat{u}[k]} & \text{if } \hat{u}[k] \neq 0 \\ 0 & \text{if } \hat{u}[k] = 0 \end{cases} .$$

The deconvolved data is

$$X = \tilde{U}^{-1}Y = Y \circledast \tilde{u}^{-1} = f + Z. \tag{32}$$

The noise $Z = \tilde{U}^{-1}W$ is circular stationary. Its covariance K is a circular convolution with $\sigma^2 \tilde{u}^{-1} \circledast \overline{\tilde{u}}^{-1}$, where $\overline{\tilde{u}}^{-1}[n] = \tilde{u}^{-1}[-n]$. The Karhunen-Loève basis which diagonalizes K is therefore the discrete Fourier basis $\mathcal{B}$. The eigenvalues of K are $\sigma_k^2 = \sigma^2 |\hat{u}[k]|^{-2}$. When $\hat{u}[k] = 0$ we formally set $\sigma_k^2 = \infty$.

When the convolution filter is a low-pass filter with a zero at high frequency, the deconvolution problem is highly unstable. Suppose that the discrete Fourier transform $\hat{u}[k]$ has a zero of order $p \geq 1$ at the highest frequency $k = \pm N/2$

$$|\hat{u}[k]| \sim \left| \frac{2k}{N} - 1 \right|^p. \tag{33}$$

The noise variance σ_k^2 has a hyperbolic growth when the frequency k is in the neighborhood of $\pm N/2$. This is called a *hyperbolic deconvolution problem of degree p*.

9.5.1 Linear Deconvolution

In many deconvolution problems the set Θ is translation invariant, which means that if $g \in \Theta$ then any translation of g modulo N also belongs to Θ. Since the amplified noise Z is circular stationary the whole estimation problem is translation invariant. In this case, the following theorem [Mal99] proves that the linear estimator that achieves the minimax linear risk is diagonal in the discrete Fourier basis. It is therefore a circular convolution. In the discrete Fourier basis, the coefficients of f are proportionnal to its discrete Fourier transform

$$f_B[k] = \langle f, g_k \rangle = \frac{1}{\sqrt{N}} \sum_{n=0}^{N-1} f[n] \exp\left(\frac{-i2n\pi k}{N} \right) = \frac{1}{\sqrt{N}} \hat{f}[k].$$

Hence the oracle risk (10) is rewritten

$$r_{\inf}(f) = \sum_{k=0}^{N-1} \frac{\sigma_k^2 N^{-1} |\hat{f}[k]|^2}{\sigma_k^2 + N^{-1} |\hat{f}[k]|^2}. \tag{34}$$

We denote by QH[Θ] the quadratic convex hull of Θ in the discrete Fourier basis.

Theorem 9.5 *Let Θ be a translation invariant set. The minimax linear risk for estimating f from $X = f + Z$ is reached by circular convolutions and*

$$r_l(\Theta) = r_{\inf}(\text{QH}[\Theta]) . \tag{35}$$

If Θ is closed and bounded, then there exists $x \in \text{QH}[\Theta]$ such that $r_{\inf}(x) = r_{\inf}(\text{QH}[\Theta])$. The minimax linear risk for estimating f from $X = f + Z$ is then achieved by a filter whose transfer function $\hat{d}_1[k]$ is specified by (14). The resulting estimator is

$$\tilde{F} = D_1 X = d_1 \circledast X = d_1 \circledast \tilde{u}^{-1} \circledast Y .$$

So $\tilde{F} = DY = d \circledast Y$, and one can verify that

$$\hat{d}[k] = \frac{N^{-1} |\hat{x}[k]|^2 \, \hat{u}^*[k]}{\sigma^2 + N^{-1} |\hat{x}[k]|^2 \, |\hat{u}[k]|^2} . \tag{36}$$

If $\sigma_k^2 = \sigma^2 |\hat{u}[k]|^{-2} \ll N^{-1} |\hat{x}[k]|^2$ then $\hat{d}[k] \approx \hat{u}^{-1}[k]$, but if $\sigma_k^2 \gg N^{-1} |\hat{x}[k]|^2$ then $\hat{d}[k] \approx 0$. The filter d is thus a regularized inverse of u.

Total variation The total variation of a discrete signal f of size N is defined with

$$\|f\|_V = \sum_{n=0}^{N-1} |f[n] - f[n-1]| . \tag{37}$$

The total variation measures the amplitude of all signal oscillations and is well suited to model the spatial inhomogeneity of piece-wise regular signals. Bounded variation signals may include sharp transitions such as discontinuities. A set Θ_V of bounded variation signals of period N is defined by :

$$\Theta_V = \left\{ f \; : \; \|f\|_V = \sum_{n=0}^{N-1} \left| f[n] - f[n-1] \right| \leq C \right\} .$$

Theorem 9.5 can be applied to the set Θ_V which is indeed translation invariant.

Proposition 9.1 *For a hyperbolic deconvolution of degree p, if $1 \leq C/\sigma \leq N$ then*

$$\frac{r_l(\Theta_V)}{N\sigma^2} \sim \left(\frac{C}{N^{1/2}\sigma} \right)^{(2p-1)/p} . \tag{38}$$

Proof. Since Θ_V is translation invariant, Theorem 9.5 proves that $r_l(\Theta_V) = r_{\inf}(\text{QH}[\Theta_V])$. For all $f \in \Theta_V$, let us define $b[n] = f[n] - f[n-1]$,

$$|\hat{b}[k]| = |\hat{f}[k]| \left| 1 - \exp\left(\frac{-i2\pi k}{N} \right) \right| = 2 |\hat{f}[k]| \left| \sin \frac{\pi k}{N} \right| . \tag{39}$$

Since $\sum_{n=0}^{N-1} |b[n]| \leq C$, necessarily $|\hat{b}[k]| \leq C$ so

$$|\hat{f}[k]|^2 \leq \frac{C^2}{4 |\sin \frac{\pi k}{N}|^2} = |\hat{x}[k]|^2 , \tag{40}$$

Hence Θ_V is included in the set

$$\mathcal{R}_x = \{f \; : |\hat{f}[k]| \le |\hat{x}[k]| \text{ for } 0 \le k < N\}.$$

The convex hull $\text{QH}[\Theta_V]$ is thus also included in $\mathcal{R}_x$ which is quadratically convex, and one can verify that

$$r_{\inf}(\text{QH}[\Theta]_V) \le r_{\inf}(\mathcal{R}_x) \le 2\, r_{\inf}(\text{QH}[\Theta]_V) \;. \tag{41}$$

The value $r_{\inf}(\mathcal{R}_x)$ is calculated by inserting (40) with $\sigma_k^{-2} = \sigma^{-2}\,|\hat{u}[k]|^2$ in (34)

$$r_{\inf}(\mathcal{R}_x) = \sum_{k=0}^{N-1} \frac{N^{-1}\,C^2\,\sigma^2}{4\,\sigma^2\,|\sin\frac{\pi k}{N}|^2 + N^{-1}\,C^2\,|\hat{u}[k]|^2} \;. \tag{42}$$

For $|\hat{u}[k]| \sim |2k\,N^{-1} - 1|^p$, if $1 \le C/\sigma \le N$ then an algebraic calculation gives $r_{\inf}(\mathcal{R}_x) \sim (C\,N^{-1/2}\,\sigma^{-1})^{(2p-1)/p}$. So $r_l(\Theta_V) = r_{\inf}(\text{QH}[\Theta_V])$ satisfies (38).

For a constant signal to noise ratio $C^2/(N\,\sigma^2) \sim 1$, (38) implies that

$$\frac{r_l(\Theta_V)}{N\sigma^2} \sim 1 \;. \tag{43}$$

Despite the fact that σ decreases and N increases the normalized linear minimax risk remains of the order of 1.

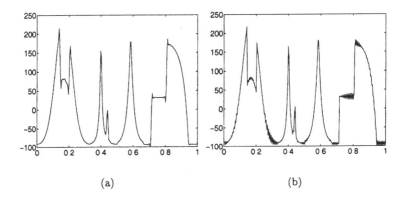

(a)   (b)

FIGURE 2. (a) : Degraded data Y, blurred with the filter (44) and contaminated by a Gaussian white noise (SNR = 25.0). (b) Deconvolution calculated with a circular convolution estimator whose risk is close to the linear minimax risk over bounded variation signals (SNR = 25.8db).

Example 9.1 Figure 2(a) is a signal Y obtained by smoothing a signal f with the low-pass filter

$$\hat{u}[k] = \cos^2\left(\frac{\pi\,k}{N}\right) \;. \tag{44}$$

This filter has a zero of order $p = 2$ at $\pm N/2$. Figure 2(b) shows the estimation $\tilde{F} = Y \circledast d$ calculated with the transfer function $\hat{d}[k]$ obtained by inserting (40) in (36). The maximum risk over Θ_V of this estimator is within a factor 2 of the linear minimax risk $r_l(\Theta_V)$.

9.5.2 Thresholding Deconvolution

An efficient thresholding estimator is implemented in a basis $\mathcal{B}$ which defines a sparse representation of signals in Θ_V and which nearly diagonalizes K. As mentioned in section 9.4.2, this approach was introduced by Donoho [Don95] to study inverse problems such as Radon transforms. We concentrate on more unstable hyperbolic deconvolutions.

The covariance operator K is diagonalized in the discrete Fourier basis and its eigenvalues are

$$\sigma_k^2 = \frac{\sigma^2}{|\hat{u}[k]|^2} \sim \sigma^2 \left| \frac{2k}{N} - 1 \right|^{-2p}. \qquad (45)$$

Yet the discrete Fourier basis is not appropriate for the thresholding algorithm because it does not approximate efficiently bounded variation signals. Periodic wavelet bases provide efficient approximations of bounded variation signals. The signal support is normalized to $[0, 1]$ and its samples have a distance of N^{-1}. The wavelet basis is thus defined on scales $1 \geq 2^j > N^{-1}$. Let $\psi_{0,0}[n] = \frac{1}{\sqrt{N}}$. For $L = -\log_2(N)$ a discrete and periodic orthonormal wavelet basis can be written

$$\mathcal{B} = \{\psi_{j,m}\}_{L < j \leq 0, \, 0 \leq m < 2^{-j}}. \qquad (46)$$

where $\psi_{j,m}[n] = \psi_j[n - m N 2^j]$. However, we shall see that this basis fails to approximatively diagonalize K.

The discrete Fourier transform $\hat{\psi}_{j,m}[k]$ of a wavelet has an energy mostly concentrated in the interval $[2^{-j-1}, 2^{-j}]$, as illustrated by Figure 3. If $2^j < 2N^{-1}$ then over this frequency interval (45) shows that the eigenvalues σ_k^2 remain of the order of σ^2. These wavelets are therefore approximate eigenvectors of K. At the finest scale $2^l = 2N^{-1}$, $|\hat{\psi}_{l,m}[k]|$ has an energy mainly concentrated in the higher frequency band $[N/4, N/2]$, where σ_k^2 varies by a huge factor of the order of N^{2r}. These fine scale wavelets are thus far from approximating eigenvectors of K.

To construct a basis of approximate eigenvectors of K, the finest scale wavelets must be replaced by vectors that have a Fourier transform concentrated in subintervals of $[N/4, N/2]$ where σ_k^2 varies by a factor that does not grow with N.

An orthogonal wavelet transform is implemented by cascading conjugate mirror filters $\bar{h}$ and $\bar{g}$ in a predetermined order [Mal89], as illustrated in figure 3. Wavelet packet bases [CMW92] extend this construction by

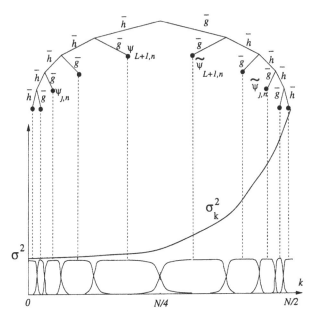

FIGURE 3. A mirror wavelet basis is computed with a wavelet packet filter bank tree, where each branch corresponds to a convolution with a filter $\bar{h}$ or $\bar{g}$ followed by a subsampling. The resulting wavelets $\psi_{j,m}$ and mirror wavelets $\tilde{\psi}_{j,m}$ have a Fourier transform shown below. The variance σ_k^2 of the noise has a hyperbolic growth but varies by a bounded factor on the frequency support of each mirror wavelet.

cascading conjugate mirror filters in an arbitrary order depending on the segmentation of the frequency axis one wants to obtain. We replace the finest scale wavelets by wavelet packets whose discrete Fourier transform support decrease exponentially at high frequencies.

These wavelet packets must also have a small spatial support to efficiently approximate piecewise regular signals, and hence the largest possible frequency support. The optimal trade-off is obtained with wavelet packets that we denote $\tilde{\psi}_{j,m}$, which have a discrete Fourier transform $\widehat{\tilde{\psi}}_{j,m}[k]$ mostly concentrated in $[N/2 - 2^{-j}, N/2 - 2^{-j-1}]$, as illustrated by Figure 3. This basis is constructed with a wavelet packet filtering tree which sub-decomposes the space of the finest scale wavelets. These particular wavelet packets are called *mirror wavelet* because

$$|\widehat{\tilde{\psi}}_{j,m}[k]| = |\hat{\psi}_{j,m}[N/2 - k]| \ .$$

A mirror wavelet basis is a wavelet packet basis composed of wavelets $\psi_{j,m}$ at scales $2^j < 2^l = 2\,N^{-1}$ and mirror wavelets to replace the finest scale wavelets

$$\mathcal{B} = \left\{ \psi_{j,m}, \tilde{\psi}_{j,m} \right\}_{0 \le m < 2^{-j}, l < j \le 0} \ .$$

To prove that the covariance K is "almost diagonalized" in $\mathcal{B}$ for all N, the asymptotic behavior of the discrete wavelets and mirror wavelets must be controlled. The following theorem thus supposes that these wavelets and wavelet packets are constructed with a conjugate mirror filter which yields a continuous time wavelet $\psi(t)$ that has $q > p$ vanishing moments and which is $\mathbf{C}^q$. The near diagonalization is verified to prove that a thresholding estimator in a mirror wavelet basis has a risk whose decay is equivalent to the non-linear minimax risk.

Theorem 9.6 *Let $\mathcal{B}$ a mirror wavelet basis constructed with a conjugate mirror filter that defines a wavelet that is $\mathbf{C}^q$ with q vanishing moments. For a hyperbolic deconvolution of degree $p < q$, if $1 \leq C/\sigma \leq N^{p+\frac{1}{2}}$ then*

$$\frac{r_n(\Theta_V)}{N\sigma^2} \sim \frac{r_t(\Theta_V)}{N\sigma^2} \sim \left(\frac{C}{\sigma}\right)^{4p/(2p+1)} \frac{(\log_e N)^{1/(2p+1)}}{N} . \tag{47}$$

Proof. The main ideas of the proof are outlined. We must first verify that there exists λ such that for all $N > 0$

$$\|K_d^{1/2} K^{-1} K_d^{1/2}\|_S \leq \lambda . \tag{48}$$

The operator $K^{-1} = \sigma^{-2} U U^*$ is a circular convolution whose transfer function is $\sigma^{-2} |\hat{u}[k]|^2 \sim \sigma^2 |2k/N - 1|^{2p}$. The matrix of this operator in the mirror wavelet basis is identical to the matrix of a circular convolution operator of transfer function $\sigma^{-2} |\hat{u}[k + N/2]|^2 \sim \sigma^{-2} |2k/N|^{2p}$ in the discrete wavelet basis. This operator is a discretized and periodized version of a convolution operators in $\mathbf{L}^2(\mathbb{R})$ of transfer function $\hat{u}(\omega) \sim \sigma^{-2} N^{-2p} |\omega|^{2p}$. One can prove [Jaf92, Mey92] that this operator is preconditioned by its diagonal in a wavelet basis of $\mathbf{L}^2(\mathbb{R})$ if the wavelet has $q > p$ vanishing moments and is $\mathbf{C}^q$. We can thus derive that in the finite case, when N grows $\|K_d^{1/2} K^{-1} K_d^{1/2}\|_S$ remains bounded.

The minimax and thresholding risk cannot be calculated directly with the inequalities (30) because the set of bounded variation signals Θ_V is not orthosymmetric in the mirror wavelet basis $\mathcal{B}$.

We first show that we can compute an upper bound and a lower bound of $\|f\|_V$ from the absolute value of the decomposition coefficients of f in the mirror wavelet basis $\mathcal{B}$. The resulting inequalities construct two orthosymmetric sets Θ_1 and Θ_2 such that $\Theta_1 \subset \Theta_V \subset \Theta_2$. A refinement of the inequalities (30) shows that over these sets the minimax and thresholding risks are equivalent, with no loss of a $\log_e N$ factor. The risk over Θ_1 and Θ_2 is calculated by evaluating $r_{\inf}(\Theta_1)$ and $r_{\inf}(\Theta_2)$, from which we derive (47).

This theorem proves that a thresholding estimator in a mirror wavelet basis yields a quasi-minimax deconvolution estimator for bounded variation signals. If we suppose that the signal to noise ratio $C^2/(N\sigma^2) \sim 1$ then

$$\frac{r_n(\Theta_V)}{N\sigma^2} \sim \frac{r_t(\Theta_V)}{N\sigma^2} \sim \left(\frac{\log_e N}{N}\right)^{1/(2p+1)} \tag{49}$$

As opposed to the normalized linear minimax risk (43) which remains of the order of 1, the thresholding risk in a mirror wavelet basis converges to zero as N increases. The larger the number p of zeros of the low-pass filter $\hat{u}[k]$ at $k = \pm N/2$ the slower the risk decay.

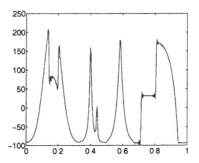

FIGURE 4. Deconvolution of the signal in Figure 2(a) with a thresholding in a mirror wavelet basis (SNR = 29.2 db).

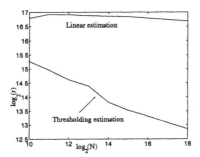

FIGURE 5. When $C^2/(N\sigma^2) = 1$, the risk of a nearly optimal linear estimation remains nearly constant whereas it decays linearly for a thresholding deconvolution in a mirror wavelet basis.

Example 9.2 Figure 2(a) shows an example of signal Y degraded by a convolution with a low pass filter $\hat{u}[k] = \cos^2\left(\frac{\pi k}{N}\right)$. The result of the deconvolution and denoising with a thresholding in the mirror wavelet basis is shown in Figure 4. A translation invariant thresholding, which means that the signal is decomposed in all the translated wavelet bases, is performed to reduce the risk [CD95]. The SNR is 29.2 db whereas it was 25.8 db in the linear restoration of Figure 2(b). Figure 5 shows the risk of the restoration as a function of $\log_2 N$ for the deconvolution of this same signal, with $C^2/(N\sigma^2) = 1$. As expected from (43) the risk of the linear estimator remains nearly constant. For a thresholding estimator in a mirror wavelet basis, (49) shows that $\log_2(r_t(\Theta_V)) \sim -\frac{1}{2p+1}\log_2 N$. In figure 5, p is chosen as $p = 1$, and the measured slope of the logarithm of the thresholding estimation error is approximately $-0,32$ while (49) predicts $-\frac{1}{3} \sim -0.33$. If we chose $p = 2$, (49) predicts $-\frac{1}{5} = -0,2$, and the measured slope is $-0,18$. Numerical results are thus close to theoretical predictions.

9.5.3 Deconvolution of Satellite Images

Nearly optimal deconvolution of bounded variation images can be calculated with a separable extension of the deconvolution estimator in a mirror wavelet basis. Such a restoration algorithm is used by the French Spatial Agency (CNES) for the production of satellite images.

The exposition time of the satellite photoreceptors cannot be reduced too much because the light intensity reaching the satellite is small and must not be dominated by electronic noises. The satellite movement thus produces a blur and the imperfection of the optics combines another blur. The electronic of the photoreceptors adds a Gaussian white noise. When the satellite is in orbit, a calibration procedure measures the impulse response u of the blur and the noise variance σ^2. The image 7(b) is a simulated satellite image provided by the CNES, which is calculated from an airplane image shown in Figure 7(a). The impulse response is a separable low-pass filter

$$U f[n_1, n_2] = f \circledast u[n_1, n_2] \quad \text{with} \quad u[n_1, n_2] = u_1[n_1] u_2[n_2] .$$

The discrete Fourier transform of u_1 and u_2 have respectively a zero of order p_1 and p_2 at $\pm N/2$

$$\hat{u}_1[k_1] \sim \left| \frac{2k_1}{N} - 1 \right|^{p_1} \quad \text{and} \quad \hat{u}_2[k_2] \sim \left| \frac{2k_2}{N} - 1 \right|^{p_2} .$$

Image Total Variation   Most satellite images are well modeled by bounded variation images. For a square discrete image of N^2 pixels, the total variation is defined by

$$\|f\|_V = \frac{1}{N} \sum_{n_1=0}^{N-1} \sum_{n_2=0}^{N-1} \left(\left| f[n_1, n_2] - f[n_1-1, n_2] \right|^2 + \left| f[n_1, n_2] - f[n_1, n_2-1] \right|^2 \right)^{\frac{1}{2}} .$$

We say that an image has a bounded variation if $\|f\|_V$ is bounded by a constant independent of the resolution N. Let Θ_V be the set of images that have a total variation bounded by C

$$\Theta_V = \left\{ f \; : \; \|f\|_V \leq C \right\} .$$

Bounded variation plays an important role in image processing, where its value depends on the length of the image level sets.

The deconvolved noise has a covariance K that is diagonalized in a two-dimensional discrete Fourier basis. The eigenvalues are

$$\sigma_{k_1,k_2}^2 = \frac{\sigma^2}{|\hat{u}_1[k_1]|^2\,|\hat{u}_2[k_2]|^2} \sim \sigma^2 \left|\frac{2k_1}{N} - 1\right|^{-2p_1} \left|\frac{2k_2}{N} - 1\right|^{-2p_2}. \qquad (50)$$

The main difficulty is again to find an orthonormal basis which provides a sparse representation of bounded variation images and which nearly diagonalizes the noise covariance K. Each vector of such a basis should have a Fourier transform whose energy is concentrated in a frequency domain where the eigenvectors σ_{k_1,k_2}^2 vary at most by a constant factor. Rougé [Rou97] has demonstrated numerically that efficient deconvolution estimations can be performed with a thresholding in a wavelet packet basis. This algorithm is inspired by his approach although the chosen basis is different.

At low frequencies $(k_1, k_2) \in [0, N/4]^2$ the eigenvalues remain approximatively constant $\sigma_{k_1,k_2}^2 \sim \sigma^2$. This frequency square can thus be covered with a separable discrete wavelet basis. We derive from the one-dimensionnal wavelets three two-dimensional separable wavelets $\{\psi_{j,m_1,m_2}^l\}_{l=1,2,3}$ which are translated on a square grid

$$\psi_{j,m_1,m_2}^l[n_1, n_2] = \psi_j^l[n_1 - m_1\,N\,2^j, n_2 - m_2\,N\,2^j]$$

and the resulting basis is

$$\{\psi_{j,m_1,m_2}^l[n_1, n_2]\}_{l=1,2,3,\,L+1<j\leq 0,\,0\leq m_1,m_2<2^{-j}}.$$

The remaining high frequency annulus is covered by two-dimensional mirror wavelets that are separable products of two one-dimensional mirror wavelets

$$\{\tilde{\psi}_{j_1,m_1}[n_1]\,\tilde{\psi}_{j_2,m_2}[n_2]\}_{L+1\leq j_1,j_2\leq 0,\,0\leq m_1<2^{-j_1},\,0\leq m_2<2^{-j_2}}.$$

One can verify that the union of these two families define an orthonormal basis of images of N^2 pixels

$$\mathcal{B} = \left\{\{\psi_{j,m}^l[n_1, n_2]\}_{j,m,l}\,,\; \{\tilde{\psi}_{j_1,m_1}[n_1]\,\tilde{\psi}_{j_2,m_2}[n_2]\}_{j_1,j_2,m_1,m_2}\right\}. \qquad (51)$$

This two-dimensional mirror wavelet basis segments the Fourier plane as illustrated in Figure 6. It is an anisotropic wavelet packet basis as defined in [Wic94]. Decomposing a signal in this basis with a filter bank requires $O(N^2)$ operations [Wic94].

To formally prove that a thresholding estimator in $\mathcal{B}$ has a risk $r_t(\Theta_V)$ that is close to the non-linear minimax risk $r_n(\Theta_V)$, one must prove that there exists λ such that $\|K_d^{1/2}\,K^{-1}\,K_d^{1/2}\|_S \leq \lambda$ and that Θ_V can be embedded in two close sets that are orthosymmetric in $\mathcal{B}$. The following theorem computes the risk in a particular configuration of the signal to noise ratio, to simplify the formula. More general results can be found in [KM99].

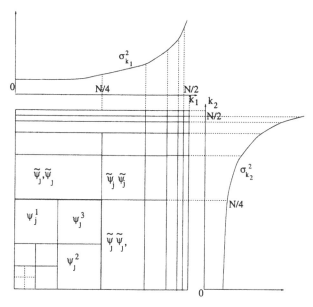

FIGURE 6. The mirror wavelet basis (51) segments the frequency plane (k_1, k_2) into rectangles over which the noise variance $\sigma^2_{k_1,k_2} = \sigma^2_{k_1} \sigma^2_{k_2}$ varies by a bounded factor. The lower frequencies are covered by separable wavelets ψ^l_j, and the higher frequencies are coverd by separable mirror wavelets $\tilde{\psi}_j \tilde{\psi}_{j'}$.

Theorem 9.7 *For a separable hyperbolic deconvolution of degree $p \geq 3/2$ with $p = \max(p_1, p_2)$, if $C^2/(N^2 \sigma^2) \sim 1$ then*

$$\frac{r_l(\Theta_V)}{N^2 \sigma^2} \sim 1 \tag{52}$$

and

$$\frac{r_n(\Theta_V)}{N^2 \sigma^2} \sim \frac{r_t(\Theta_V)}{N^2 \sigma^2} \sim \left(\frac{\log_e N}{N^2} \right)^{\frac{1}{2p+1}} . \tag{53}$$

The theorem proves that the linear minimax estimator does not reduce the original noise energy $N^2 \sigma^2$ by more than a constant. On the contrary, the thresholding estimator in a separable mirror wavelet basis has a quasi-minimax risk that converges to zero as N increases.

Figure 7(c) shows an example of deconvolution calculated in the mirror wavelet basis. The thresholding is performed with a translation invariant algorithm. This can be compared with the linear estimation in Figure 7(d), calculated with a circular convolution estimator whose maximum risk over bounded variation images is close to the minimax linear risk. Like in one dimension, the linear deconvolution sharpens the image but leaves a visible noise in the regular parts of the image. The thresholding algorithm removes completely the noise in these regions while improving the restoration of edges and oscillatory parts. This algorithm was chosen among several competing algorithms by photointerpreters of the French spatial agency (CNES) to perform the deconvolution of satellite images. Deconvolution procedures with regularization using total variation norms [RS94], Markov random field models [BI96], and others, have been tested. The thresholding algorithm obtained both the best metrical and perceptual notations, and it is now integrated in the CNES satellite image acquisition channel.

9.6   Conclusion

We have built a theoretical framework for minimax optimal restoration of signals and images in the case of ill-posed inverse problems. We focused on minimax rather than Bayes optimality because the choice of a set Θ as prior for our signals is better suited for complex signal such as images, as we don't know any prior probability distribution that can model complex signals such as natural images.

One can perform an optimal restoration if one can find an orthogonal basis which can both compress the signal to estimate on a few coefficients and nearly diagonalize the covariance of the non-white Gaussian noise obtained after applying the inverse of the degradation operator. The use of this approach to solve hyperbolic deconvolution of signals and images leads to the creation of mirror wavelet bases in which a simple thresholding procedure on the coefficients of the decomposition yields previously unobtained minimax optimality results. A competition set by the French spatial agency showed that this type of algorithms gives the best numerical results among all competing algorithms.

9.7   REFERENCES

[Ber89]   M. Bertero. *Advances in Electronics and Electron Physics*, chapter "Linear inverse and ill-posed problems". Academic Press, 1989.

[BI96]   S. Brette and J. Idier. Optimized single site update algorithms for image deblurring. In *IEEE ICIP*, Lausanne, Switzerland, 1996.

[CD95]   R. R. Coifman and D. Donoho. Translation invariant de-noising. Technical Report 475, Dept. of Statistics, Stanford University, May 1995.

[CMW92] R. R. Coifman, Y. Meyer, and M. V. Wickerhauser. Wavelet analysis and signal processing. In *Wavelets and their Appli-*

cations, pages 153–178, Boston, 1992. Jones and Barlett. B. Ruskai et al. eds.

[DJ94]   D. Donoho and I. Johnstone. Ideal spatial adaptation via wavelet shrinkage. *Biometrika*, 81 :425–455, December 1994.

[DLG90]   D.L. Donoho, R.C. Liu, and K.B. Gibbon. Minimax risk over hyperrectangles and implications. *Annals of statistcs*, (18) :1416–1437, 1990.

[Don93]   D. Donoho. Unconditional bases are optimal bases for data compression and for statistical estimation. *J. of Appl. and Comput. Harmonic Analysis*, 1 :100–115, 1993.

[Don95]   D. Donoho. Nonlinear solution of linear inverse problems by wavelets-vaguelette decompositions. *J. of Appl. and Comput. Harmonic Analysis*, 2(2) :101–126, 1995.

[Jaf92]   S. Jaffard. Wavelet methods for fast resolution of elliptic problems. *SIAM Journal of Numerical Analysis*, 29, 1992.

[KM99]   J. Kalifa and S. Mallat. Minimax deconvolution in mirror wavelet bases. Technical report, CMAP, Ecole Polytechnique, 1999.

[Mal89]   S. Mallat. A theory for multiresolution signal decomposition : the wavelet representation. *IEEE Trans. Patt. Recog. and Mach. Intell.*, 11(7) :674–693, July 1989.

[Mal99]   S. Mallat. *A Wavelet Tour of Signal Processing* (new edition to appear). Academic Press, 1999.

[Mey92]   Y. Meyer. *Wavelets and Operators*. Advanced mathematics. Cambridge university press, 1992.

[O'S86]   F. O'Sullivan. A statistical perspective on ill-posed inverse problems. *Statist. Sci.*, pages 502–527, 1986.

[Rou97]   B. Rougé. *Théorie de la chaine image optique et restauration*. PhD thesis, Université Paris-Dauphine, 1997. Thèse d'habilitation.

[RS94]   L. Rudin and S.Osher. Total variation based image resoration with free local constraints. In *Proc. IEEE ICIP*, volume I, pages 31–35, Austin-texas, USA, Nov. 1994.

[Sim99]   E. P. Simoncelli. *Bayesian Inference in Wavelet Based Models*, chapter "Bayesian Denoising of Visual Images in the Wavelet Domain". Springer-Verlag, 1999. P. Müller and B. Vidakovic, eds.

[Wic94]   M. V. Wickerhauser. *Adapted Wavelet Analysis from Theory to Software*. AK Peters, 1994.

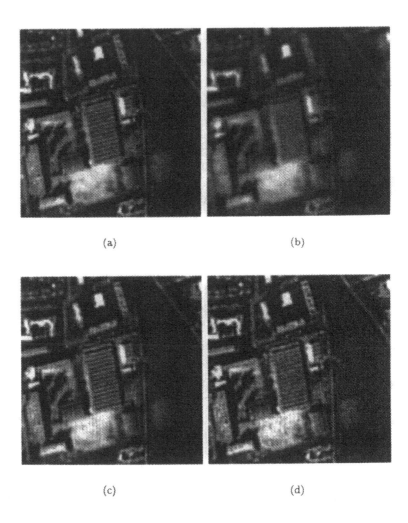

(a)   (b)

(c)   (d)

FIGURE 7. (a) : Original airplane image. (b) : Simulation of a satellite image
provided by the CNES (SNR = 31.1db). (c) : Deconvolution with a translation
invariant thresholding in a mirror wavelet basis (SNR = 34.1db). (d) : Deconvolution calculated with a circular convolution, which yields a nearly minimax risk
for bounded variation images (SNR = 32.7db).

10

Robust Bayesian and Bayesian Decision Theoretic Wavelet Shrinkage

Fabrizio Ruggeri

ABSTRACT Recent results on the application of Bayesian decision theory to wavelet thresholding are reviewed and expanded. Choices of commonly used models for wavelets coefficients are discussed, both for location and scale parameter families. Convenient priors are suggested to achieve non-trivial thresholding rules. Besides, the choice of priors are discussed from a robust Bayesian viewpoint, looking for conditions ensuring shrinkage of the estimates of the wavelet coefficients.

10.1 Introduction

Donoho and Johnstone proposed *wavelet shrinkage,* a class of simple and efficient procedures for nonparametric estimation of functions and densities based on thresholding of wavelet coefficients. An overview is given by Donoho and Johnstone (1994, 1995).

Wavelet shrinkage can be described as a three-step procedure:

1. Data (a noisy signal, measurements, blurred image pixels, etc.) are transformed by a discrete wavelet transformation from the "time domain" to the "wavelet domain".

2. The transformed data are shrunk.

3. The processed data are transformed back to the "time domain".

The choice of a shrinkage rule in step **2** is important. Many different thresholding methods have recently been proposed. An excellent overview of shrinkage rules based on thresholding is given by Nason (1995).

Here we are mainly interested in the hard thresholding rules δ^{hard}. Note that δ^{hard} is fully specified by the threshold λ. Ruggeri and Vidakovic (1999) found an "optimal" threshold λ by minimising the (integrated) Bayes risk over the class of hard thresholding rules δ^{hard}. Instead of finding unrestricted Bayes shrinkage rules which never threshold under the squared error loss, they found the rules that are Bayes in the class of δ^{hard}-type rules. In this respect, their approach is "restricted Bayes." When the risk-minimising λ^* exists, the rule $\delta^{hard}(x, \lambda^*)$ will be called *Bayesian decision* *theoretic (BDT) rule*, and the induced shrinkage estimator, the *BDT estimator*. For unrestricted Bayesian approaches see, e.g., Vidakovic (1998), Chipman *et al.* (1997) and Clyde *et al.* (1998).

The interest relies not only in finding the optimal rules with good denoising properties, but also in identifying the model-prior pairs which produce nontrivial risk-minimising values for λ. Both problems are important to provide both a sound theoretical background (based on decision theory) and practical guidelines on which model-based shrinkage should be performed. The choice of models and priors for wavelet thresholding is based, as well, on the use of prior information on the unknown signal of interest and the level of noise.

The choice of a prior is crucial in order to achieve wavelet shrinkage. If the prior knowledge can be expressed only by a class Γ of distributions (e.g. all symmetric and unimodal ones), then robust Bayesian techniques can be used to discuss if shrinkage is achieved regardless of the choice of the prior in Γ.

The paper is organized as follows. In Section 2, we give the decision theoretic background necessary for optimality considerations. In Sections 3 and 4 we discuss the search of nontrivial minimisers of the Bayes risk under location and scale 0 families, respectively. Hard thresholding rules minimising the posterior expected loss are examined in Section 5, whereas shrinkage under classes of priors is discussed in Section 6.

10.2   Decision Theoretic Setup

Given the real space $\mathcal{R}$ and its subset Θ, let X be a random variable on a dominated statistical space denoted by $(\mathcal{R}, \mathcal{B}_X, \{P_\theta, \theta \in (\Theta, \mathcal{B}_\Theta)\})$, with density $f(x|\theta)$. Let Π denote a probability measure on the parameter space $(\Theta, \mathcal{B}_\Theta)$, with density $\pi(\theta)$.

The goal is to make an inference about the parameter θ, given an observation X. A solution is a *decision rule* $\delta(x)$, that identifies particular inference for each value of x that can be observed. Let $\mathcal{A}$ be the class of all possible realisations of $\delta(x)$, i.e. *actions*. The *loss function* $L(\theta, a)$ maps $\Theta \times \mathcal{A}$ into the set of real numbers and defines a cost to the statistician when he/she takes the action a and the value of the parameter is θ. A *risk function* $R(\theta, \delta)$ characterises the performance of the rule δ for each value of the parameter $\theta \in \Theta$. The risk is defined in terms of the underlying loss function $L(\theta, a)$ as $R(\theta, \delta) = E^{X|\theta} L(\theta, \delta(X))$ where $E^{X|\theta}$ is the expectation with respect to (w.r.t.) P_θ. Since the risk function is defined as an average loss w.r.t. a sample space, it is called the *frequentist risk*. Let $\mathcal{D}$ be the collection of all decision rules under consideration. There are several principles for assigning preference among the rules in $\mathcal{D}$ (e.g. Bayes principle, minimax principle, and Γ-minimax principle). We will present only the first one, referring to Berger (1985) for a thorough presentation of them.

Under the Bayes principle, the prior distribution π is specified on the parameter space Θ. Any rule δ is characterised by its *Bayes risk* $r(\pi, \delta) = \int R(\theta, \delta)\pi(d\theta) = E^\pi R(\theta, \delta)$. The rule δ_π that minimises the Bayes risk is called *Bayes rule*, i.e. $\delta_\pi = arg \inf_{\delta \in \mathcal{D}} r(\pi, \delta)$. The *Bayes risk of the prior distribution π (Bayes envelope function)* is $r(\pi) = r(\pi, \delta_\pi)$. Here we consider some classes $\mathcal{D}$ of decision rules dependent on a parameter λ, usually a threshold. Therefore, we will be interested in the search for the optimal parameter λ^*, corresponding to the rule $\delta_\pi \in \mathcal{D}$ that minimises the Bayes risk. We will mainly consider the class $\mathcal{D}$ of decision rules $\mathcal{D} = \{\delta_\lambda(x) = x \cdot \mathbf{1}(|x| > \lambda), \ \lambda \in [0, \infty)\}$ which corresponds, in the wavelet world, to the class of hard thresholding rules. Note that $\mathbf{1}(A)$ is the indicator function of the set A.

We will show that, under adequate conditions on model and prior, the Bayes risk might be the same for all λ's; it might even happen that $\lambda^* = 0$ or ∞, which correspond, respectively, to keep any x unchanged or changed into 0. Therefore, our interest is focused on the search of models and/or priors leading to nontrivial λ^*, i.e. $0 < \lambda^* < \infty$. We consider both location and scale parameter models, focusing mainly on the former.

10.3   Location Parameter

Suppose the observed data $\underline{y}$ is the sum of an unknown signal $\underline{s}$ and random noise $\underline{\epsilon}$. Coordinate-wise,

$$y_i = s_i + \epsilon_i, \ i = 1, \ldots, n. \tag{1}$$

In the wavelet domain (after applying a linear and orthogonal wavelet transformation W), expression (1) becomes $x_i = \theta_i + \eta_i, \ i = 1, \ldots, n$, where x_i, θ_i, and η_i are the i-th coordinates of $W\underline{y}, W\underline{s}$ and $W\underline{\epsilon}$, respectively. For notational simplicity, the double indexing typical in discrete wavelet representations is replaced by a single index.

Assuming a location model $[x_i|\theta_i] \sim f(x_i - \theta_i)$ and a prior $[\theta_i] \sim \pi(\theta_i)$, we find the Bayes risk of the decision rule δ^{hard}.

10.3.1   *General results*

Suppose that $f(x|\theta) = f(x - \theta)$ for any $(x, \theta) \in \mathcal{B}_X \otimes \mathcal{B}_\Theta$, and that both f and π are symmetric functions. Consider a symmetric loss function $L(\theta, a) = L(\theta - a)$ and the hard thresholding decision rule $\delta_\lambda(x)$ from the class $\mathcal{D}$. It follows that $L(\theta, \delta_\lambda) = L(x - \theta)$ (for $|x| > \lambda$) or $L(\theta)$ (for $|x| \leq \lambda$). Furthermore, the risk function and the Bayes risk, w.r.t. the prior π, of the decision rule δ_λ are given, respectively, by $R(\theta, \delta_\lambda) =$

$C + \int_{-\lambda}^{\lambda} [L(\theta) - L(\theta - x)]f(x - \theta)dx$ and $r(\pi, \delta_\lambda) = C + \int_{\mathcal{R}} [\int_{\theta-\lambda}^{\theta+\lambda} [L(\theta) - L(t)]f(t)dt]\pi(\theta)d\theta$, where $C = \int_{\mathcal{R}} L(t)f(t)dt$ is the risk function (and the Bayes risk) when considering the decision rule $\delta(x) = x$ for any real x. It is worth looking for λ^* which improves upon the Bayes risks $r(\pi, \delta_0) = C$ and $r(\pi, \delta_\infty) = \lim_{\lambda \to \infty} r(\pi, \delta_\lambda) = \int_{\mathcal{R}} L(t)\pi(t)dt$. Note that these Bayes risks coincide when $f \equiv \pi$.

Ruggeri and Vidakovic (1999) considered the derivative of $r(\pi, \delta_\lambda)$ to find the optimal λ^*, i.e. the hard thresholding rule δ_{λ^*} which minimises the Bayes risk. They proved the following lemma under the (very general, indeed) conditions for the differentiability under the integral sign.

Lemma 1

$$\frac{\partial r(\pi, \delta_\lambda)}{\partial \lambda} = r'(\lambda) = 2 \int_{\mathcal{R}} [L(\theta) - L(\theta - \lambda)]f(\theta - \lambda)\pi(\theta)d\theta. \qquad (2)$$

We now suppose that f and L are continuous functions so that r' is continuous as well and $r'(\lambda) = 0$ at non null stationary points. By dividing the right hand side of (2) by $\int_{\mathcal{R}} f(\theta - \lambda)\pi(\theta)d\theta$, we can prove the following result, more understandable than the one in Ruggeri and Vidakovic (1999).

Theorem 1 *The optimal decision rule λ^*, if it exists, is a non null solution to*

$$E^{\theta|\lambda}L(\theta) = E^{\theta|\lambda}L(\theta - \lambda), \qquad (3)$$

where $E^{\theta|\lambda}$ denotes the posterior expectation under the prior π, the model f and the "observation" λ.

As a consequence of the previous Theorem, (3) becomes $\lambda = 2E^{\theta|\lambda}\theta$, when considering the squared loss function $L(t) = t^2$.

Corollary 1 *If (3) has no solution, then $\lambda^* = 0$ or ∞, depending on the sign of r'.*

Example 1 *Consider $X \sim \mathcal{N}(\theta, \sigma^2)$ and $\theta \sim \mathcal{N}(0, \tau^2)$. Condition (3) becomes $\lambda = \frac{2\tau^2\lambda}{\tau^2 + \sigma^2}$, which has no solution when $\tau \neq \sigma$, whereas it is solved by all nonnegative λ when $\tau = \sigma$. As proved by Ruggeri and Vidakovic (1999), and mentioned later in this paper, the former case corresponds to $\lambda^* = 0$ or ∞, whereas the latter corresponds to constant Bayes risk, regardless of the choice of the threshold.*

10.3.2 Case $f = \pi$

The next theorem, due to Ruggeri and Vidakovic (1999), suggests of not choosing models which are functionally equal to the prior distribution.

Theorem 2 *Given* $f(x|\theta) = f(x - \theta)$, *let* $f(t) = \pi(t)$ *for all* $t \in \mathcal{R}$. *It follows that the Bayes risk is constant, i.e.* $r(\pi, \delta_\lambda) = C$, *where* $C = \int_{\mathcal{R}} L(t) f(t) dt$.

Under the above conditions, any number in $[0, \infty)$ can be chosen as λ^* so that it is impossible to discriminate among the hard thresholding rules. This behaviour is exhibited by the following examples involving normal model and prior and, similarly, double exponential and t pairs. We will say that the random variable X has a double-exponential $\mathcal{DE}(\theta, \beta)$ distribution if the density of X has the form: $f(x) = \beta \exp\{-\beta|x - \theta|\}/2$, $\beta > 0$.

Example 2 (Normal) *Assume that* $X \sim \mathcal{N}(\theta, \sigma^2)$, $\theta \sim \mathcal{N}(0, \sigma^2)$, *with* σ^2 *known. It can be shown that* $r(\pi, \delta_\lambda) = \sigma^2$ *for* $L(t) = t^2$. *Similarly,* $r(\pi, \delta_\lambda)$ *equals* $\sqrt{2/\pi}\sigma$ *for* $L(t) = |t|$ *and* $2\Phi(-\mu)$ *for* $L(t) = \mathbf{1}(|t| > \mu)$.

Example 3 (Double Exponential) *Assume that* $X \sim \mathcal{DE}(\theta, \sigma)$ *and* $\theta \sim \mathcal{DE}(0, \sigma)$ *and let* σ *be a given a nonnegative number. It can be shown that* $r(\pi, \delta_\lambda) = 4\sigma^3$ *for* $L(t) = t^2$. *Similarly,* $r(\pi, \delta_\lambda)$ *equals* $2\sigma^2$ *for* $L(t) = |t|$ *and* $\sigma e^{-\mu/\sigma}$ *for* $L(t) = \mathbf{1}(|t| > \mu)$.

Example 4 (t, with s d.f.) *Assume that* $X \sim t_s(\theta, \sigma^2)$ *and* $\theta \sim t_s(0, \sigma^2)$ *and let* σ *be a given nonnegative number. It can be shown that, for* $s > 2$ *and* $L(t) = t^2$, $r(\pi, \delta_\lambda) = \dfrac{s\sigma^2}{s - 2}$.

10.3.3 Case $f \neq \pi$, same functional form

Consider the case when the model $f(x - \theta)$ and the prior $\pi(\theta)$ have the same functional form, apart from a scale parameter. Ruggeri and Vidakovic (1999) have shown that λ^* cannot be a positive, finite number when the ratio of model and prior is decreasing on $\mathcal{R}^+$, as it happens for some standard distributions.

Theorem 3 *Consider a density* $f(x|\theta) = f(x - \theta)$, *the prior* $\pi(\theta)$ *and a symmetric loss function* L *such that* $L(t)$ *is nondecreasing for nonnegative t. If* $f(t)/\pi(t)$ *is decreasing for nonnegative t, then* $\lambda^* = 0$; *if it is increasing, then* $\lambda^* = \infty$.

The next example considers normal model and prior and shows that $\lambda^* = 0$ or ∞, depending upon whether the variance of the model is smaller or larger than the variance of the prior. Similar result holds for double exponential and t distributions.

Example 5 (Normal - Normal) *Let* $f(x - \theta) \propto \exp\{-(x - \theta)^2/(2\sigma_f^2)\}$ *and* $\pi(\theta) \propto \exp\{-\theta^2/(2\sigma_\pi^2)\}$. *For any* $0 < x_1 < x_2$, *it follows that*

$$f(x_1)\pi(x_2) - f(x_2)\pi(x_1)$$

$$\propto \exp\{-x_1^2/(2\sigma_f^2) - x_2^2/(2\sigma_\pi^2)\} - \exp\{-x_2^2/(2\sigma_f^2) - x_1^2/(2\sigma_\pi^2)\},$$

which is positive if and only if $\sigma_f^2 < \sigma_\pi^2$. Therefore, applying Theorems 2 and 3, it follows that λ^* equals 0 for $\sigma_f^2 < \sigma_\pi^2$, ∞ for $\sigma_f^2 > \sigma_\pi^2$ or it can take any value for $\sigma_f^2 = \sigma_\pi^2$.

10.3.4   Case $f \neq \pi$, different functional form

As discussed in Ruggeri and Vidakovic (1999), nontrivial thresholding might be obtained only if the model f and the prior π have different functional forms. Their interest in normal, double exponential and t models/priors is motivated by the wavelet context.

If the noise $\underline{\epsilon}$ in $\underline{y} = \underline{s} + \underline{\epsilon}$ is i.i.d. normal, then orthogonality and linearity of wavelet transformation ensure that the noise $\underline{\eta}$ in $\underline{x} = \underline{\theta} + \underline{\eta}$ is i.i.d. normal as well. When σ^2 in $x_i \sim N(\theta_i, \sigma^2)$ is known or estimable, one may proceed by specifying a prior on θ_i. However, if σ^2 is not known, it should be integrated out by eliciting an appropriate prior $\pi(\sigma^2)$, before proceeding with the marginal likelihood $f(x_i|\theta_i) = \int \phi_\sigma(x_i - \theta_i)\pi(\sigma^2)d\sigma^2$. The function $\phi_\sigma(x_i - \theta_i)$ is the density function of the normal $N(\theta_i, \sigma^2)$ law and $\pi(\sigma^2)$ is a prior on σ^2.

For instance, if we want to be non informative about σ^2, then it is convenient to choose an exponential distribution for σ^2, so that the resulting marginal likelihood $f(x_i - \theta_i)$ is a double exponential; whereas if we have more information about σ^2, then the choice of an inverse gamma prior implies that $f(x_i - \theta_i)$ is distributed as t.

The choice of different models and priors could give nontrivial λ^*, as in the case of normal model and double exponential prior, thoroughly studied in Ruggeri and Vidakovic (1999); the pairs t/double exponential and t/normal can be treated similarly. Based on both the existence of an optimal, 1 and bounded λ^* and the knowledge on σ^2 (described above), Ruggeri and Vidakovic thoroughly discussed the choice of pairs of models and priors which are reasonable from their viewpoint, i.e. based on the Bayesian decision theoretic rule.

Example 6 (Normal - Double Exponential) *Consider the normal model $X \sim N(\theta, \sigma^2)$, in which σ^2 is assumed known (or estimable), and the double exponential prior $\theta \sim \mathcal{DE}(0, \beta)$. For any $0 < x_1 < x_2$, it follows that*

$$f(x_1)\pi(x_2) - f(x_2)\pi(x_1) \propto \exp\{-x_1^2/(2\sigma^2) - \beta x_2\} - \exp\{-x_2^2/(2\sigma^2) - \beta x_1\},$$

which is positive if and only if $x_1 + x_2 > 2\sigma^2\beta$. Therefore, the ratio $f(t)/\pi(t)$ is not strictly monotone and it is possible that $0 < \lambda^* < \infty$, because of Theorem 3. Ruggeri and Vidakovic (1999) considered $\sigma = 1$ and a squared loss function, finding out that there exists nontrivial λ^* when $\beta > 0.90$

(approximately) (e.g. $\lambda^* = 0.94055$ when $\beta = 1$). Furthermore, they found that $\hat{\lambda} = 2\beta\sigma^2$ is a useful approximation to λ^* for large β.

Consider the squared loss function $L(t) = t^2$; then λ^*, if it exists, can be found solving

$$\exp\{\beta\lambda\}\{-2\sigma\phi(-\lambda/\sigma - \beta\sigma) + (2\beta\sigma^2 + \lambda)\Phi(-\lambda/\sigma - \beta\sigma)\}+$$

$$+ \exp\{-\beta\lambda+\}\{2\sigma\phi(-\lambda/\sigma + \beta\sigma) + (-2\beta\sigma^2 + \lambda)[1 - \Phi(-\lambda/\sigma + \beta\sigma)]\} = 0.$$

Ruggeri and Vidakovic (1999) noticed that λ^* is sensitive to the choice of the prior and warned the user to be careful in choosing the parameter values. Sensitivity to a prior is a well studied problem in Bayesian analysis, but neglected in the Bayesian literature on wavelets. The choice of the parameters β and σ must be careful, as shown by Figure 1.

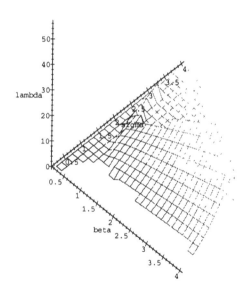

FIGURE 1. λ^* for different β and σ

No thresholding is performed either for small β or for pairs of large β and σ, while severe oversmoothing occurs for large β and σ in the range $[1, 2]$. As a consequence, already pointed out by Ruggeri and Vidakovic, the *elicitation of extreme (small or large) values of β may cause unsatisfactory MSE performance of BDT rules.*

A heuristic rule for eliciting the parameter β is given in Ruggeri and Vidakovic (1999). Whenever the signal-to-noise ratio, i.e. the ratio of standard deviations of the signal and the noise, is moderate, they suggest using β = √0.4 log N /σ̂, where N is the sample size and σ̂ is an estimate of the standard deviation of the noise. The corresponding threshold λ ≈ √1.6 log N σ̂ is smaller than the standard Donoho and Johnstone threshold λ^U, in agreement with findings of Bruce and Gao (1996) in the context of optimal minimax thresholding.

*Ruggeri and Vidakovic (1999) performed an extensive numerical study for this pairs of model and prior. They considered BDT thresholding estimators for the standard test signals (*bumps, blocks, doppler *and* heavisine*). They evaluated the performance of BDT estimators by comparing the averaged mean squared errors (AMSE) on 10 simulated runs with the AMSE of other common estimators: nonadaptive shrinkage methods of Donoho and Johnstone (1994, 1995) (hard thresholding with the universal threshold) and that of Bruce and Gao (1996) (hard thresholding with an optimal minimax threshold), besides the adaptive Bayesian wavelet shrinkage method of Chipman, McCulloch and Kolaczyk (1997).*

Finally, we want to mention the improper priors which are often considered in literature, even though they are strongly questioned by many Bayesians. A typical choice of an improper prior about the location parameter θ, defined all over $\mathcal{R}$, is given by $\pi(\theta) = 1$ (or any other constant). We show that such a prior leads to $\lambda^* = \infty$.

Example 7 *Let X have symmetric density $f(x - \theta)$ and let $L(t)$ be any symmetric, increasing in $|t|$, loss function. Consider $\pi(\theta) = 1$ for any real θ. From*

$$r'(\lambda) = \int_{\mathcal{R}} [L(\theta) - L(\theta - \lambda)] f(\theta - \lambda) d\theta = - \int_{\mathcal{R}} [L(\theta) - L(\theta - \lambda)] f(\theta) d\theta,$$

it follows that

$$r''(\lambda) = - \int_{\mathcal{R}} L'(\theta - \lambda) f(\theta) d\theta = - \int_0^\infty [L'(\theta + \lambda) + L'(\theta - \lambda)] f(\theta) d\theta < 0,$$

since $L'(\theta + \lambda)$ and $L'(\theta - \lambda)$ are positive for $\theta \geq \lambda$, and

$$L'(\theta + \lambda) + L'(\theta - \lambda) = L'(\theta + \lambda) - L'(\lambda - \theta) > 0,$$

for $0 \leq \theta < \lambda$. Since $r'(0) = 0$, it follows that $\lambda^* = \infty$.

10.4   Scale Parameter

Here we consider the distribution of X dependent on a scale parameter. Unlike the location parameter case, here we suppose that the observed data $\underline{y}$ is the sum of a *known* signal $\underline{s}$ and random noise $\underline{\epsilon}$. In the wavelet domain, each coordinate $z_i = \theta_i + \eta_i$, $i = 1, \ldots, n$, can be modelled by $[x_i|\theta_i] \sim \theta_i f(x_i\theta_i)$, where $x_i = z_i - \theta_i$.

Therefore, we consider the symmetric density $f(x|\theta) = \theta g(|x|\theta)$ for any $(x, \theta) \in \mathcal{B}_X \otimes \mathcal{B}_\Theta$, such that $\int_\mathcal{R} g(t)dt = 1$. Let $\pi(\theta)$ be a prior on $\theta \in \mathcal{R}^+$ and $L(\theta, a)$ be a symmetric (in a) loss function such that $L(\theta, a) = L(|a|\theta)$ (e.g. $(|a|\theta - 1)^2$).

The risk function and the Bayes risk, w.r.t. the prior Π, of the hard thresholding rule δ_λ are given, respectively, by $R(\theta, \delta_\lambda) = C + 2 \int_0^{\lambda\theta} [L(0) - L(t)]g(t)dt$ and $r(\pi, \delta_\lambda) = C + 2 \int_{\mathcal{R}^+} [\int_0^{\lambda\theta} [L(0) - L(t)]g(t)dt]\pi(\theta)d\theta$, where $C = \int_\mathcal{R} L(t)g(t)dt$. Note that $r(\pi, \delta_0) = C$ and $r(\pi, \delta_\infty) = \lim_{\lambda\to\infty} r(\pi, \delta_\lambda) = L(0)$, so that the Bayes risk cannot be constant when $C \neq L(0)$.

For simplicity, we suppose that r' is a continuous function of λ, as it happens in many common situations, including those presented in this paper; otherwise standard techniques should be applied if the function r' were not differentiable at λ^*.

Theorem 2 in Ruggeri and Vidakovic (1996) can be reformulated into a more useful theorem.

Theorem 4 *If there exists an optimal* λ^*, $0 < \lambda^* < \infty$, *then* λ^* *is a solution of* $E^{\theta|\lambda}L(\lambda\theta) = L(0)$, *where* $E^{\theta|\lambda}$ *denotes the posterior expectation under the prior* π, *the model* f *and the "observation"* λ.

We consider now some examples under the loss function $(|a|\theta - 1)^2$ such that $L(0) = 1$. It can be shown that

$$\lambda^* = 2m_1/m_2, \tag{4}$$

where m_i, $i = 1, 2$, is the i-th posterior moment.

We will present now both normal and double exponential models, considering very flexible and convenient priors and computing λ^*.

Result 1. Let $X \sim \mathcal{N}(0, \theta)$, where θ is the precision, i.e. the reciprocal of the standard deviation. Let $\pi(\theta) \propto \theta^{\alpha-1}e^{-\beta\theta^2/2}$. After simple computations, it follows that $m_1 = \dfrac{\Gamma(\alpha/2 + 1)}{\Gamma(\alpha/2 + 1/2)\sqrt{(\lambda^2 + \beta)/2}}$ and $m_2 = \dfrac{\Gamma(\alpha/2 + 3/2)}{\Gamma(\alpha/2 + 1/2)(\lambda^2 + \beta)/2}$. It can be proved that $\lambda^* = \delta\sqrt{\dfrac{2\beta}{1 - 2\delta^2}}$, with $\delta = \Gamma(\alpha/2 + 1)/\Gamma(\alpha/2 + 3/2)$, is the solution to (4) when $\alpha > 2.4695$; for smallest values of α, it follows that $\lambda^* = 0$.

Result 2. Let $X \sim \mathcal{DE}(0,\theta)$, i.e. double exponentially distributed with density $f(x|\theta) = \theta e^{-\theta|x|}/2$. Let $\theta \sim \mathcal{G}(\alpha,\beta)$, i.e. Gamma distributed. It follows that $m_1 = \dfrac{\alpha+1}{\lambda+\beta}$ and $m_2 = \dfrac{(\alpha+1)(\alpha+2)}{(\lambda+\beta)^2}$. It follows that $\lambda^* = 2\beta/\alpha$, i.e. twice the reciprocal of the prior mean on θ.

10.5 Posterior Expected Loss

Consider the class of decision rules $\mathcal{D} = \{\delta_\lambda, \lambda \geq 0\}$ such that the following properties hold for any $x \in \mathcal{R}$:

1. $\delta_0(x) = x$;

2. $\lim\limits_{\lambda \to \infty} \delta_\lambda(x) = 0$;

3. $\delta_\lambda(x)$ is a continuous strictly decreasing function for $0 \leq \lambda \leq \lambda_x$, where λ_x (possibly ∞) is such that, for any $x \neq 0$, $\delta_\lambda(x) > 0$ for $0 \leq \lambda < \lambda_x$ and $\delta_\lambda(x) = 0$ for $\lambda \geq \lambda_x$.

The previous properties are satisfied by many thresholding rules considered in literature (but excluding the hard thresholding ones), like the following ones:

Soft-thresholding : $\delta_\lambda^{soft}(x) = (x - \lambda\,\mathrm{sign}(x))\mathbf{1}(|x| > \lambda)$;

Rational : $\delta_\lambda^r(x) = \dfrac{x^{2n+1}}{(\lambda^{2n} + x^{2n})}$;

Hyperbola : $\delta_\lambda^{hi}(x) = \mathrm{sign}(x)\sqrt{x^2 - \lambda^2}\,\mathbf{1}(|x| \geq \lambda)$;

Garrote : $\delta_\lambda^g(x) = (x - \lambda^2/x)\mathbf{1}(|x| > \lambda)$.

We consider a convex loss function $L(\theta, a)$ (in a), so that the posterior expected loss is convex as well and there is a unique Bayes action $a_L(x)$. We consider positive x, whereas negative ones lead to similar results.

Because of the properties of $\mathcal{D}$ and the convexity of $L(\theta, a)$, it follows that the decision rules minimising the posterior expected loss are given by

$$\delta_0(x) = x \quad \text{if} \quad a_L(x) > x,$$
$$\delta_{\hat{\lambda}}(x) = a_L(x) \quad \text{if} \quad 0 \leq a_L(x) \leq x,$$
$$\delta_\infty(x) = 0 \quad \text{if} \quad a_L(x) < 0.$$

Note that $a_L(x)$ cannot be negative under the assumptions on f, π and L made in this paper.

Lemma 2 *Let f and π be symmetric functions (around 0) such that $f(x|\theta) = f(x - \theta)$. Consider a loss function $L(\theta, a)$, increasing in $|\theta - a|$, such that $L(-\theta, -a) = L(\theta, a)$ (e.g. $L(\theta, a) = L(|\theta - a|))$. Given the sample $x \geq 0$, then the Bayes action $a_L(x)$ is nonnegative.*

PROOF. The symmetry conditions imply, for any positive θ, that $\pi(\theta|x) \geq \pi(-\theta|x)$ since $f(x - \theta)\pi(\theta) \geq f(x - (-\theta))\pi(-\theta)$.

Consider a positive a; it can be shown that $EL(\theta, a) \leq EL(\theta, -a)$ if and only if

$$\int_0^\infty [L(\theta, a) - L(\theta, -a)][\pi(\theta|x) - \pi(-\theta|x)]d\theta \leq 0.$$

The previous condition is satisfied since $\pi(\theta|x) \geq \pi(-\theta|x)$, as seen above, and $|\theta - a| \leq |\theta + a|$ implies $L(\theta, a) \leq L(\theta, -a)$. As a consequence, the Bayes action has to be nonnegative. $\triangle$

Under the previous assumptions on x, $\mathcal{D}$, f, π and L, the following theorem holds.

Theorem 5 *The optimal restricted rule in $\mathcal{D}$ corresponds to the action a^* such that $a^* = \min\{x, a_L(x)\}$.*

A different behaviour is exhibited by the hard thresholding rules. Consider, for illustration, a squared loss function and the corresponding Bayes action, the posterior mean $E^{\theta|X}\theta$; it follows that the coefficient is unchanged if and only if $0 < x < 2E^{\theta|X}\theta$, while it is chosen equal to 0 otherwise.

The use of the posterior expected loss instead of the (frequentist) Bayes risk might be more acceptable to many Bayesians, since the latter approach averages over all possible wavelet coefficients instead of considering only the actual observed ones. We believe that the former approach is justified from a practical Bayesian viewpoint, specially when we are looking for a unique "optimal" λ w.r.t. all observed coefficients in the same level of the wavelet decomposition. The number of such coefficients might be thousands, so that there is no relevant difference in searching for an "optimal" λ either for all the possible coefficients or just the observed ones. See Berger (1984) for a more detailed discussion on the use of frequentist measures in Bayesian analysis. Moreover, the approach based on the posterior expected loss is computationally more intensive since it requires the computations of Bayes actions for all the wavelet coefficients x instead of a unique λ^*.

10.6   Bayesian Robustness

Vidakovic (1998) considered the estimation of the wavelet coefficients from a Bayesian viewpoint. He considered the wavelet coefficient x to be normally distributed, i.e. $X|\theta, \sigma^2 \sim \mathcal{N}(\theta, \sigma^2)$, with both θ and σ^2 unknown. Vidakovic integrated out σ^2 considering an exponential distribution $\mathcal{E}(\mu)$

on it. Therefore, he obtained the marginal model $X|\theta \sim \mathcal{DE}(\theta, 1/\sqrt{2\mu})$. He considered a t prior distribution on θ, such that $\theta \sim t_n(0, \tau)$, and considered the posterior mean of θ, i.e. the Bayes rule under squared loss, as a replacement for the observed wavelet coefficient. He plotted the Bayes rules for some choices of n, τ and μ and found that the estimators were shrinking the coefficients. Motivated by testing the hypothesis $\theta = 0$ versus $\theta \neq 0$, Vidakovic introduced the prior $\pi(\theta) = (1 - \epsilon)\delta_0(\theta) + \epsilon q(\theta)$, where δ_0 is a point mass at 0 and $q(\theta)$ is a prior that describes the spread of θ when $\theta \neq 0$.

The choice of $q(\theta)$ could be a difficult task and, actually, a class of priors could better reflect the prior knowledge. The estimators are therefore changing in a set whose range might be large. The uncertainty in the prior raises a robustness issue, which has been thoroughly studied in literature; see Berger (1994) and Berger et al. (1996) for its presentation and a wide list of references. Despite being a crucial issue, robustness has not been addressed in papers about Bayesian wavelet estimation. The illustration of robust Bayesian techniques is well beyond the goals of this paper so that the interested reader is referred to Berger (1990, 1994) where those techniques are thoroughly presented.

In this paper, we simply focus on shrinkage of estimators under some classes of priors. As discussed in this volume, shrinkage is an important issue in wavelet estimation since shrinkage in the wavelet domain is intimately connected with smoothing in the time domain, as shown by Jaffard (1991) and Meyer (1992). Therefore, it is worth exploring the shrinkage properties of Bayes rules under quite general models in order to help practitioners to select an effective model that will induce desirable smoothing in the time domain.

Stemming from previous work by DasGupta and Rubin (1993), Vidakovic and Ruggeri (1999) studied shrinkage properties of Bayes rules corresponding to priors from several important classes, assuming a symmetric and unimodal distribution for $X|\theta$, with a continuous density $f(x|\theta)$ on $\mathcal{R}$. Their main results are given by the following theorems.

Theorem 6 *Let $f(x|\theta) = f(x - \theta)$ be any symmetric and unimodal density defined for any $(x, \theta) \in \mathcal{R} \otimes \mathfrak{R}$ and let the prior distribution $\pi(\theta)$ be in the class Γ_S of all symmetric (about 0) distributions. Then, for any real $x \neq 0$,*

$$\sup_{\pi \in \Gamma_S} \left| \frac{d_\pi(X)}{X} \right| > 1. \tag{5}$$

Note that the supremum in (5) can actually be infinite for any $x \neq 0$. As an example, consider $f(x|\theta)$ to be a $\mathcal{N}(\theta, 1)$ density, so that

$$\sup_{\pi \in \Gamma_S} \left| \frac{d_\pi(x)}{x} \right| = \sup_{\mu \geq 0} \frac{\mu}{x} \frac{e^{\mu x} - e^{-\mu x}}{e^{\mu x} + e^{-\mu x}} = \infty.$$

The supremum becomes bounded when considering prior distributions with a point mass p at 0, like the ones considered in Vidakovic (1998).

Theorem 7 *Let $f(x|\theta) = f(x-\theta)$ be any symmetric and unimodal density defined for any $(x,\theta) \in \mathcal{R} \otimes \mathfrak{R}$ and let the prior distribution $\pi(\theta)$ be in the class Γ^p_S of all symmetric (about 0) distributions, with given point mass p,*
$$0 < p \le 1, \text{ at } 0. \text{ Then, for any real } x \ne 0, \text{ it follows that } \sup_{\pi \in \Gamma^p_S} |\frac{d_\pi(X)}{X}| < \infty.$$

Some examples illustrated in Vidakovic and Ruggeri (1999) demonstrate that even large point mass at 0 may not guarantee shrunk rules. Shrinkage is guaranteed by adding the unimodality constraint, as shown by the next two theorems.

Theorem 8 *Let $f(x|\theta) = f(x-\theta)$ be any symmetric and unimodal density defined for any $(x,\theta) \in \mathcal{R} \otimes \mathfrak{R}$ and let the prior distribution $\pi(\theta)$ be in the class Γ_{SU} of all symmetric (about 0) unimodal distributions. Then, for any*
$$\text{real } x \ne 0, \quad \sup_{\pi \in \Gamma_{SU}} |\frac{d_\pi(X)}{X}| = 1.$$

Theorem 9 *Let $f(x|\theta) = f(x - \theta)$ be any symmetric and unimodal density defined for any $(x,\theta) \in \mathcal{R} \otimes \mathfrak{R}$ and let the prior distribution $\pi(\theta)$ be in the class Γ^p_{SU} of all symmetric (about 0) unimodal distributions, with a given point mass p, $0 < p \le 1$, at 0. For any $x \ne 0$, it follows that*
$$\sup_{\pi \in \Gamma^p_{SU}} |\frac{d_\pi(X)}{X}| < 1.$$

As a consequence of the previous theorems, the approach by Vidakovic (1998) can lead to expansion, instead of shrinkage, when considering normal, double exponential or t models, provided that a symmetric prior (around 0) is chosen for the location parameter.

We now briefly illustrate the results in two cases: the first similar to the one considered in Vidakovic (1998) and the second based on a different classes of priors, the generalised moments class considered by Betrò et al. (1994).

Example 8 *Let $f(x|\theta)$ be $\mathcal{DE}(\theta,1)$, and let the distribution for θ belong to the class Γ_{SU} of all symmetric (about 0) unimodal distributions. The Bayes rules is a shrinker for all realisations of X and μ. Besides, $\sup_{\pi \in \Gamma_{SU}} |\frac{d_\mu(X)}{X}|$ is achieved for a uniform prior on $[-\mu,\mu]$, with the resulting Bayes rule*

$$d_\mu(x) = \begin{cases} 1 - \mu \coth(\mu), & x < -\mu \\ \frac{x-(\mu+1)\exp(-\mu)\sinh(x)}{1-\exp-\mu\cosh(x)}, & -\mu \le x \le \mu \\ \mu \coth(\mu) - 1, & x > \mu \end{cases} .$$

Example 9 *Suppose, like in Betrò et al. (1994), that the probability distribution for the data X is $\mathcal{N}(\theta,1)$ and that the prior Π_0 on θ is $\mathcal{N}(0,2)$. Consider the ε-contamination class $\Gamma_\varepsilon = \{\Pi : \Pi = (1-\varepsilon)\Pi_0 + \varepsilon q(\theta), q \in \mathcal{Q}\}$, where $\mathcal{Q}$ contains all the contaminating priors such that the marginal distribution on X have the same median, 0, as the one from the prior Π_0.*

Given $\varepsilon = 0.2$, we consider a grid of values x in $[0,10]$ and plot (Fig. 2) the supremum of the posterior mean $d_\pi(x)$ w.r.t. the sample x.

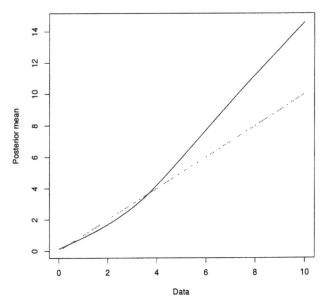

FIGURE 2. Posterior mean in the generalised moments class

We can see that δ_π, the Bayes rule under squared loss function, is a shrinker for "moderate" values of x, but it is not for "large" or very small ones. It can be easily shown, applying the results in Betrò et al. (1994), that $\sup_{\Gamma_\varepsilon} d_\pi(X)$ equals

$$\sup_{p,\theta_1 \leq 0, \theta_2 \geq 0} \frac{(1-\varepsilon)2x\phi_3(x)/3 + \varepsilon[p\theta_1\phi_1(x-\theta_1) + (1-p)\theta_2\phi_1(x-\theta_2)]}{(1-\varepsilon)\phi_3(x) + \varepsilon[p\phi_1(x-\theta_1) + (1-p)\phi_1(x-\theta_2)]},$$

where p satisfies the generalised moments constraint $p = \dfrac{.5 - \Phi_1(-\theta_2)}{\Phi_1(-\theta_1) - \Phi_1(-\theta_2)}$ and ϕ_τ (Φ_τ) denotes the df (cdf) of a $\mathcal{N}(0,\tau)$ distribution.

10.7   Discussion and Conclusions

In this paper, we have outlined two approaches for Bayesian wavelet shrinkage based on the decision theoretic paradigm. Under the first approach, based on the Bayes risk, we have presented a hard thresholding rule which can be applied to all the detail wavelet coefficients. Conditions on models and priors are discussed such that a unique hard thresholding rule exists, under both location and scale parameter models. The second approach deals with the posterior expected loss and different (restricted) Bayes actions are computed for all the detail wavelet coefficients. Robustness issues have been discussed as well, both for their influence on the choice of an optimal threshold and the possibility of choosing priors, compatible with the prior knowledge (e.g. symmetry), leading to expansion instead of shrinkage.

References

Berger, J. (1984). The robust Bayesian viewpoint (with discussion). In *Robustness of Bayesian Analyses.* (J. Kadane ed.). North-Holland: Amsterdam.

Berger, J. (1985). *Statistical Decision Theory and Bayesian Analysis.* (2nd Edition). Springer-Verlag: New York.

Berger, J. (1990). Robust Bayesian analysis: sensitivity to the prior. *J. Statist. Plann. Inference*, 25, 303-328.

Berger, J. (1994). An overview of robust Bayesian analysis (with discussion). *Test*, 3, 5-124.

Berger, J., Betrò, B., Moreno, E., Pericchi, L., Ruggeri, F., Salinetti, G. and Wasserman, L. (eds.) (1996). *Bayesian Robustness.* Institute of Mathematical Statistics: Hayward.

Betrò, B., Męczarski, M. and Ruggeri, F. (1994) Robust Bayesian analysis under generalized moments conditions. *J. Statist. Plann. Inference*, 41, 257-266.

Bruce, A. and Gao, H-Y. (1996). Understanding WaveShrink: Variance and bias estimation. *Biometrika,* 83, 727-745.

Chipman, H., McCulloch, R. and Kolaczyk, E. (1997). Adaptive Bayesian Shrinkage. *J. Amer. Statist. Assoc.,* 92, 440.

Clyde, M., Parmigiani, G. and Vidakovic, B. (1998). Multiple Shrinkage and Subset Selection in Wavelets. *Biometrika,* 85, 391-402.

DasGupta, A. and Rubin, H. (1993). Bayes Estimators as Expanders in One and Two Dimensions. Technical Report # 93-38, Purdue University.

Donoho, D. and Johnstone, I. (1994). Ideal spatial adaptation by wavelet shrinkage. *Biometrika*, 81, 425-455.

Donoho, D. and Johnstone, I. (1995). Adapting to unknown smoothness via wavelet shrinkage. *J. Amer. Statist. Assoc.*, 90, 1200-1224.

Jaffard, S. (1991). Pointwise smoothness, two-microlocalization and wavelet coefficients. *Publ. Mat.*, 35, 155-168.

Meyer, Y. (1992). *Wavelets and Operators*. Cambridge University Press: Cambridge.

Nason, G. (1995). Choice of the threshold parameter in wavelet function estimation. *Wavelets in Statistics,* (A. Antoniadis and G. Oppenheim, eds). Springer-Verlag: New York.

Ruggeri, F. and Vidakovic, B. (1996). A Bayesian decision theoretic approach to wavelet thresholding: scale parameter models. *Proceedings Joint Statistical Meetings*, Chicago, IL, August 4 - 8, 1996.

Ruggeri, F. and Vidakovic, B. (1999). A Bayesian decision theoretic approach to the choice of thresholding parameter. *Statistica Sinica*, 9, 183-197.

Vidakovic, B. (1998). Nonlinear wavelet shrinkage with Bayes rules and Bayes Factors. *J. Amer. Statist. Assoc.*, 93, 173-179.

Vidakovic, B. and Ruggeri, F. (1999). Expansion estimation by Bayes rules. To appear in *J. Stat. Plann. Infer.*

11

Best Basis Representations with Prior Statistical Models

David Leporini, Jean-Christophe Pesquet, and Hamid Krim

ABSTRACT Wavelet packets and local trigonometric bases provide an efficient framework and fast algorithms to obtain a "best representation" of a deterministic signal. Applying these deterministic search techniques to stochastic signals may, however, lead to statistically unreliable results. In this chapter, we revisit this problem and introduce prior models on the underlying signal in noise. We propose several techniques to derive the prior parameters and develop a Bayesian-based approach to the best basis problem. As illustrated by applications to signal denoising, this leads to reduced estimation errors while preserving the classical tree search algorithm.

11.1 Introduction

Basing multiscale representations of signals on statistical approaches has recently been of great research interest (see *e.g.* [CW92, DJ94a, KP95, KMDW95]). Particular emphasis has been on optimizing representations of observed processes typically consisting of an unknown (deterministic) signal embedded in random noise. While this model is adequate for a number of applications, oftentimes the signal falls subject to a variety of random effects (*e.g.* those of the underwater environment reflected in noise corrupted sonar signals), and it becomes essential to view it as a random process itself. These adverse effects coupled with the fact that prior information about a signal (or its transformation) is often known/given, make it compelling to adopt an alternative approach which naturally account for this valuable prior information.

The optimization of representation, as first described in [CW92], uses the structure of a dictionary of orthonormal bases to perform an efficient basis search which in turn is used to estimate an underlying signal of interest. In this setting, the orthonormality property affords one to discriminate between the signal components from those of the noise, subsequently leading to an efficient reconstruction of the signal [DJ94a, Sai94, KP95, KMDW95]. These have been demonstrated to lead to good results in relatively moderate noise scenarios and have been successfully applied in a variety of applications. They are, however, generally based upon a selected threshold [DJ94a] which presents two drawbacks:

a- It is directly dependent upon the noise variance without regard to the signal statistics.

b- It grows unbounded with the data record length.

In some applications, these shortcomings may greatly reduce the performance of threshold-based reconstruction algorithms. Our goal in this chapter is to investigate the impact of accounting for signal prior information in regularizing this "ill-conditioning" of the estimation problem, and mitigating the existing limitations. The so-called Bayesian framework [Ber85] will thus be the basis of our newly proposed approach.

One should note that wedding the Bayesian framework with the multiscale reconstruction techniques was independently proposed by [Vid98, CKM97, CPV98, ASS98, CNB98, PKLH96]. In contrast with most of these works, we focus herein on using a probabilistic prior to optimize a representation of a signal in a basis, i.e. select a Best Basis (BB) for carrying out the reconstruction/estimation.

In Section 2, we give some background relevant to the later developments of the chapter. In Section 3, we address the BB search of a signal in noise taking into account available statistical prior information and derive corresponding solutions. In Section 4, we propose some possible prior models for the components of a signal in its BB and address the problem of the BB search when the parameters of the model are assumed to be known. In Section 5, we describe a Maximum Likelihood approach we have used to successfully estimate these parameters. In Section 6 we substantiate the proposed methods by way of simulations on synthesized as well as collected real data.

11.2   Background

As first described in [CW92], and further discussed in [KMDW95], the efficient search for a BB on a binary tree representing the entire dictionary of possible bases, hinges upon minimizing an additive criterion. Variability in selecting a basis [CW92] and failure of existing multiscale denoising techniques in low Signal to Noise Ratio (SNR) are limitations which result from a deterministic hypothesis testing of a random-valued entropy and from overlooking the statistical structure of the signal of interest. In [KP95] the former problem was mitigated by utilizing the statistical properties of the entropy variables, and our goal here, is to rather account for any available prior probabilistic information about the underlying signal. For efficiency and speed of search for the BB, the dictionary $\mathcal{D}$ of possible bases is structured according to a binary tree of depth $J \in \mathbb{N}^*$. Each node

(j, m) (with $j \in \{0, \ldots, J\}$ and $m \in \{0, \ldots, 2^j - 1\}$) of the tree then corresponds to a given orthonormal basis $\mathcal{B}_{j,m}$ of a vector subspace of the space $\mathbb{R}^K$ of real discrete-time signals supported on $\{1, \ldots, K\}$,[1] $K \in \mathbb{N}^*$. An orthonormal basis of $\mathbb{R}^K$ is then $\mathcal{B}_{\mathcal{P}} = \cup_{(j,m)/I_{j,m} \in \mathcal{P}} \mathcal{B}_{j,m}$ where $\mathcal{P}$ is a partition of $[0, 1[$ in intervals $I_{j,m} = [2^{-j}m, 2^{-j}(m + 1)[$. To be more concrete, we consider the particular case of discrete-time wavelet packets which are defined recursively, for all $l \in \mathbb{Z}$, by

$$W_{0,0}(l) = \delta[l], \tag{1}$$

and (away from the domain boundaries)

$$W_{j+1,2m}(l) = \sum_{k \in \mathbb{Z}} h_0(k) W_{j,m}(l - 2^j k), \tag{2}$$

$$W_{j+1,2m+1}(l) = \sum_{k \in \mathbb{Z}} h_1(k) W_{j,m}(l - 2^j k), \tag{3}$$

where the sequences $\big(h_0(k)\big)_{k \in \mathbb{Z}}$ and $\big(h_1(k)\big)_{k \in \mathbb{Z}}$ correspond to the (finite-length) impulse responses of a paraunitary perfect reconstruction filter bank. The involved subspaces subsequently follows from the initial definition $\mathcal{B}_{0,0} = \{W_{0,0}(\cdot - k), k \in \{1, \ldots, K\}\}$ as described in [CW92]. By taking advantage of the property

$$\text{Span}\{\mathcal{B}_{j,m}\} = \text{Span}\{\mathcal{B}_{j+1,2m}\} \overset{\perp}{\oplus} \text{Span}\{\mathcal{B}_{j+1,2m+1}\},$$

a fast bottom-up tree search algorithm was developed in [CW92] to optimize the partition $\mathcal{P}$. For the sake of simplicity, we shall index-number each possible partition with an integer $n \in \{1, \ldots, N\}$ where $N \in \mathbb{N}^*$ is the size of $\mathcal{D}$, and resort back to the $\mathcal{B}_{j,m}$ notation whenever a node dependent property is needed.

11.3   Problem Statement

Consider the standard additive noise model

$$y_k = x_k + \epsilon_k, \quad k = 1, \ldots, K,$$

where ϵ_k are independent and identically distributed normal variables with zero mean and finite variance σ^2. Estimation/recovery of the underlying *unknown* signal x_k is of interest. A natural way to incorporate available prior knowledge about x_k is provided by the Bayesian statistical framework, and further discussed next with various possible techniques.

---

[1] Discrete decompositions on the interval [CDV93, HV94] are used here.

11.3.1 Bayesian approach

Let $\mathbf{y}_n$, $\mathbf{x}_n$ and ϵ_n respectively denote the K-dimensional vector of components of the observed, original and noise signals in some basis $\mathcal{B}_n = \cup_{(j,m)/I_{j,m} \in \mathcal{P}_n} \mathcal{B}_{j,m}$. Using the linearity property of wavelet packets and local cosine transforms, we have

$$\mathbf{y}_n = \mathbf{x}_n + \epsilon_n.$$

The signals $\mathbf{x}$ and ϵ are assumed to be two mutually independent stochastic processes. We will further assume that there exists a $n_0 \in \{1, \dots, N\}$ such that the Probability Density Function (PDF) of $\mathbf{x}_{n_0}$ is $f_{n_0}(\cdot)$ and that of ϵ_{n_0} is $g_{n_0}(\cdot)$. The functions $f_{n_0}(\cdot)$ and $g_{n_0}(\cdot)$ are subsequently assumed to belong to some families of parametric distributions. Typically, $g_{n_0}(\cdot)$ will correspond in the sequel to the Gaussian density. The integer n_0 in this probabilistic model indexes the BB $\mathcal{B}_{n_0}$, and appears as a hyperparameter which must be estimated from the observed data. Note that, for the generality of presentation, the involved PDFs may depend on the basis index, and may thus express non-homogeneous properties of the signal transformations. In the sequel, we propose various approaches depending on the chosen prior models to solve the BB search problem coupled with the underlying signal estimation.

11.3.2 Maximum Likelihood method

Using the independence of the processes $\mathbf{x}_n$ and ϵ_n and the orthonormal property of the decompositions, we can easily obtain the law of $\mathbf{y}_n$. The resulting estimate of n_0 is thus

$$\hat{n}_0 = \arg\max_n (f_n * g_n)(\mathbf{y}_n) = \arg\max_n \int_{\mathbb{R}^K} f_n(\mathbf{u})g_n(\mathbf{y}_n - \mathbf{u})d\mathbf{u} . \qquad (4)$$

Recall that the distribution associated with the Maximum Likelihood (ML) estimate asymptotically minimizes the Kullback-Leibler distance to the true distribution when the BB index n_0 does not belong to the dictionary [Hub67].

An estimate of $\mathbf{x}_{\hat{n}_0}$ can subsequently be obtained (depending on the chosen cost function) by a Maximum *A Posteriori* (MAP) estimate

$$\widehat{\mathbf{x}}_{\hat{n}_0} = \arg\max_{\mathbf{x}_{\hat{n}_0}} g_{\hat{n}_0}(\mathbf{y}_{\hat{n}_0} - \mathbf{x}_{\hat{n}_0}) f_{\hat{n}_0}(\mathbf{x}_{\hat{n}_0}) , \qquad (5)$$

or by an *A Posteriori* Mean estimate

$$\widehat{\mathbf{x}}_{\hat{n}_0} = \frac{1}{p_{\hat{n}_0}(\mathbf{y}_{\hat{n}_0})} \int_{\mathbb{R}^K} \mathbf{u}\, g_{\hat{n}_0}(\mathbf{y}_{\hat{n}_0} - \mathbf{u})\, f_{\hat{n}_0}(\mathbf{u})d\mathbf{u} . \qquad (6)$$

11.3.3   Maximum Generalized Likelihood method

We point out that the previous Maximum Likelihood (ML) approach requires the evaluation of $p(\mathbf{y}_n)$, which may not be available in closed-form
for some prior distributions of interest (e.g. exponential power distributions
proposed by Mallat [Mal89]). An alternative approach then consists of call-
ing upon the Maximum Generalized Likelihood (MGL) method, which is
based on optimizing the joint PDF of $\mathbf{x}_n$ and $\mathbf{y}_n$, i.e.

$$(\hat{n}_0, \widehat{\mathbf{x}}_{\hat{n}_0}) = \arg \max_{n, \mathbf{X}_n} p(\mathbf{y}_n, \mathbf{x}_n).$$

If $\widehat{\mathbf{x}}_n$ denotes the following MAP estimate

$$\begin{aligned} \widehat{\mathbf{x}}_n &= \arg \max_{\mathbf{X}_n} p(\mathbf{y}_n, \mathbf{x}_n) \\ &= \arg \max_{\mathbf{X}_n} g_n(\mathbf{y}_n - \mathbf{x}_n) f_n(\mathbf{x}_n) , \end{aligned} \qquad (7)$$

the resulting estimate of n_0 is given by

$$\hat{n}_0 = \arg \max_n g_n(\mathbf{y}_n - \widehat{\mathbf{x}}_n) f_n(\widehat{\mathbf{x}}_n) . \qquad (8)$$

It should be pointed out that, unlike the ML method, the MGL is not guaranteed to asymptotically provide consistent estimates. It may nevertheless
lead to better estimates for data sets of relatively short size [CI95].

11.3.4   Maximum Generalized Marginal Likelihood method

In the case of prior models involving mixture of distributions with distinc-
tive dominating measures (e.g. Bernoulli-Gaussian distributions presented
in the next section), we remark that the MAP estimate of $\mathbf{x}_n$ in (5) or
(7) will be degenerate and will not take into account the value of the observations. We can then introduce a hidden random vector $\mathbf{q}_n$ which depends on $\mathbf{x}_n$ and account for all singular events as discrete events, so that
the joint probability distribution $p(\mathbf{y}_n, \mathbf{q}_n, \mathbf{x}_n)$ provides non-degenerate
marginal MAP estimates. Consequently, we define $\widehat{\mathbf{q}}_n$ as the following MAP
estimate

$$\begin{aligned} \widehat{\mathbf{q}}_n &= \arg \max_{\mathbf{q}_n} p(\mathbf{y}_n, \mathbf{q}_n) \qquad &(9) \\ &= \arg \max_{\mathbf{q}_n} \int_{\mathbb{R}^K} p(\mathbf{y}_n \mid \mathbf{x}_n) p(\mathbf{x}_n \mid \mathbf{q}_n) P(\mathbf{q}_n) d\mathbf{x}_n. \qquad &(10) \end{aligned}$$

and may then express the Maximum Generalized Marginal Likelihood (MG-
ML) method as

$$\hat{n}_0 = \arg \max_n p(\mathbf{y}_n, \widehat{\mathbf{q}}_n).$$

An estimate of $\mathbf{x}_{\hat{n}_0}$ can subsequently be obtained by a marginal MAP estimate

$$\widehat{\mathbf{x}}_{\hat{n}_0} = \arg \max_{\mathbf{x}_{\hat{n}_0}} p(\mathbf{y}_{\hat{n}_0}, \widehat{\mathbf{q}}_{\hat{n}_0}, \mathbf{x}_{\hat{n}_0}).$$

While this section briefly describes the various techniques one might use to seek a solution to a Bayesian optimization problem, the next section addresses the choice of a prior which will somewhat faithfully reflect the state of nature.

11.4   Prior Models

The choice of a PDF $f_{n_0}(\cdot)$ should be based on the prior information about the original signal $\mathbf{x}$ in some basis $\mathcal{B}_{n_0}$. Intuitively, the signal would be most unambiguously expressed when $\mathcal{B}_{n_0}$ represents the *best matched basis* which can in turn, qualitatively be measured by its parsimony. An adequate prior which would reflect the so-desired concentration of energy, might correspond to a distribution with a certain amount of "spikyness". A large set of choices are possible for such a distribution, an example of which is addressed in the sequel. It must, however, be emphasized that the choice of a "good" prior model is often heuristic within a Bayesian approach and is ultimately validated by the resulting estimation performances. Note that parameters of a given model have to be estimated at each node of the decomposition tree. This problem will be discussed more specifically in Section 5. Care is however required to avoid computationally intensive algorithms when dealing with complex models. In so doing, we choose to reflect the properties of a representation of $\mathbf{x}$ by non-homogeneous Bernoulli-Gaussian (B-G) distributions. The non-homogeneity arizes from the possibility of node-varying prior parameters. Another advantage of this non-homogeneity assumption is that it allows one to introduce i.i.d. models for the wavelet packet coefficients without restricting the processes to be uncorrelated in the time domain. We point out that the B-G distribution is a classical model for representing spiky processes in various application fields such as seismic exploration and non destructive control [CGI96, CCL96]. Note that this mixture model may straightforwardly be extended to more sophisticated Bernoulli-mixture of Gaussians models which may afford a closer approximation of the distribution of the underlying signal in its BB [Lep98]. In the sequel, the notation $\mathbf{x}_{j,m}$ representing the vector of the components of $\mathbf{x}$ in $\mathcal{B}_{j,m}$ will often be used to emphasize the node dependent properties of the distributions.

11.4.1   Bernoulli-Gaussian priors

The model for $\mathbf{x}_{j,m} = (x_{j,m}^1, \ldots, x_{j,m}^{2^{-j}K})^T$ is classically used in tandem with a hidden indicator vector $\mathbf{q}_{j,m} = (q_{j,m}^1, \ldots, q_{j,m}^{2^{-j}K})^T$ of independent binary random variables (also known as latent variables in the statistical literature). More specifically, for $(j, m) \in \mathcal{P}_{n_0}$, $\mathbf{x}_{j,m}$ is an i.i.d. sequence whose conditional densities are

$$p(x_{j,m}^k \mid q_{j,m}^k = 0) = \delta(x_{j,m}^k), \tag{11}$$

$$p(x_{j,m}^k \mid q_{j,m}^k = 1) = \phi(x_{j,m}^k \mid \tau_{j,m}^2), \tag{12}$$

where $\delta(\cdot)$ is the Dirac distribution, and $\phi(\cdot \mid s^2)$ defines throughout the chapter a Gaussian PDF with zero-mean and variance s^2. We further define the mixture parameter $\pi_{j,m} = P(q_{j,m}^k = 1) \in [0, 1]$.
Throughout this development, we have also assumed for simplicity that the noise components are independent Gaussian random variables with zero-mean and variance σ^2. Under these hypotheses, we obtain by using the unitary transform property

$$\mathbb{E}[\epsilon_{j,m} \epsilon_{j,m}^T] = \sigma^2 I_{2^{-j}K},$$

where I_p denotes the $p \times p$ identity matrix.
As a result, we can easily establish that the distribution of $\mathbf{y}_{j,m}$ is a mixture of Gaussian with PDF

$$p(\mathbf{y}_{j,m}) = \prod_{k=1}^{2^{-j}K} \left[(1 - \pi_{j,m})\phi(y_{j,m}^k \mid \sigma^2) + \pi_{j,m}\phi(y_{j,m}^k \mid \tilde{\tau}_{j,m}^2) \right],$$

where, using the independence property of the processes $\mathbf{x}_{j,m}$ and $\epsilon_{j,m}$, $\tilde{\tau}_{j,m}^2 = \sigma^2 + \tau_{j,m}^2$.

11.4.2 BB selection algorithms

The presence of a probability point mass at zero in B-G priors naturally leads us to invoke the MGML approach over the MGL, in light of which we proceed to determine the MAP estimate $\hat{\mathbf{q}}_{j,m}$ of the hidden sequence $\mathbf{q}_{j,m}$ when (j, m) is "guessed" to belong to $\mathcal{P}_{n_0}$. Solving (10) is shown to amount to thresholding the components $y_{j,m}^k$ of $\mathbf{y}_{j,m}$, by noting that

$$
\begin{aligned}
p(\mathbf{y}_{j,m}, \hat{\mathbf{q}}_{j,m}) &= \prod_{k=1}^{2^{-j}K} \int_{-\infty}^{\infty} \phi(y_{j,m}^k - x_{j,m}^k \mid \sigma^2) p(x_{j,m}^k \mid \hat{q}_{j,m}^k) P(\hat{q}_{j,m}^k) dx_{j,m}^k \\
&= \prod_{k=1}^{2^{-j}K} \phi(y_{j,m}^k \mid \hat{\tau}_{k,j,m}^2) P(\hat{q}_{j,m}^k) \tag{13}
\end{aligned}
$$

with

$$\hat{\tau}_{k,j,m}^2 = \begin{cases} \tilde{\tau}_{j,m}^2 & \text{if } \hat{q}_{j,m}^k = 1, \\ \sigma^2 & \text{if } \hat{q}_{j,m}^k = 0, \end{cases} \tag{14}$$

and maximizing the conditional likelihoods in (13). It then follows that

$$\hat{q}_{j,m}^k = \begin{cases} 1 & \text{if } |y_{j,m}^k| \geq \chi_{j,m}, \\ 0 & \text{if } |y_{j,m}^k| < \chi_{j,m}, \end{cases} \qquad (15)$$

for which, after some algebra, the corresponding following threshold value $\chi_{j,m} \geq 0$ results:

$$\chi_{j,m}^2 = \max \left\{ \frac{2\sigma^2 \tilde{\tau}_{j,m}^2}{\tau_{j,m}^2} \ln \left(\frac{\tilde{\tau}_{j,m}(1 - \pi_{j,m})}{\sigma \pi_{j,m}} \right), 0 \right\}. \qquad (16)$$

Note that in contrast with the result in [DJ94b, KP95], this value is independent of the data length K. Interestingly enough, the thresholding procedure may be interpreted in terms of detection of $q_{j,m}^k$. Letting $\alpha(\chi_{j,m})$ and $\beta(\chi_{j,m})$ denote the probabilities of false alarm and correct detection respectively, and considering a single component case, we find

$$\alpha(\chi_{j,m}) = \Pr[|y_{j,m}^k| > \chi_{j,m} \mid x_{j,m}^k = 0]$$
$$\beta(\chi_{j,m}) = \Pr[|y_{j,m}^k| > \chi_{j,m} \mid x_{j,m}^k \neq 0].$$

A Bayesian strategy, as a function of the threshold leads to

$$\chi_{j,m} = \arg\max_{\chi} \beta(\chi) - t\alpha(\chi), \qquad (17)$$

where $t = [(1 - \pi_{j,m})\gamma_0]/[\pi_{j,m}\gamma_1]$ is defined in terms of given costs $(C_{00}, C_{01}, C_{10}, C_{11})$ by $\gamma_0 = C_{10} - C_{00}$ and $\gamma_1 = C_{01} - C_{11}$. Recall that C_{ij} represents the cost of choosing hypothesis i when hypothesis j is true. It can be shown that the solution to this problem is given by

$$\chi_{j,m}^2 = \max \left\{ \frac{2\sigma^2 \tilde{\tau}_{j,m}^2}{\tau_{j,m}^2} \ln \left(\frac{\tilde{\tau}_{j,m}(1 - \pi_{j,m})\gamma_0}{\sigma \pi_{j,m}\gamma_1} \right), 0 \right\}.$$

Note in particular that the 0-1 loss function leading to the MAP estimate for $q_{j,m}^k$, gives $\gamma_0 = \gamma_1 = 1$, which results in the value of the threshold given in (16).

Following the estimation of the indicator process $\mathbf{q}_n$, the BB index $\hat{n}_0$ is subsequently estimated in the MGL approach as the integer n which maximizes the PDF $p(\mathbf{y}_n, \hat{\mathbf{q}}_n)$. Equivalently, the BB partition $\mathcal{P}_{\hat{n}_0}$ results by minimizing the following criterion which is additive with respect to the components of $\mathbf{y}_{j,m}$ noting that $q_{j,m}^k$ only depends on $y_{j,m}^k$,

$$\eta(\mathcal{P}) = - \sum_{(j,m)/I_{j,m} \in \mathcal{P}} \sum_{k=1}^{2^{-j}K} \ln[\phi(y_{j,m}^k \mid \hat{\tau}_{k,j,m}^2) P(\hat{q}_{j,m}^k)], \qquad (18)$$

with $\hat{\tau}_{k,j,m}^2$ given by (14). This criterion clearly preserves a fast tree search structure for both wavelet packets and Malvar's wavelets.

Upon obtaining $\widehat{\mathbf{q}}_{\hat{n}_0}$ and $\hat{n}_0$, we proceed to estimate the vector $\widehat{\mathbf{x}}_{\hat{n}_0}$ of coefficients in the BB as the following marginal MAP criterion

$$\widehat{\mathbf{x}}_{\hat{n}_0} = \arg\max_{\mathbf{x}_{\hat{n}_0}} p(\mathbf{y}_{\hat{n}_0}, \widehat{\mathbf{q}}_{\hat{n}_0}, \mathbf{x}_{\hat{n}_0})$$

where

$$p(\mathbf{y}_{\hat{n}_0}, \widehat{\mathbf{q}}_{\hat{n}_0}, \mathbf{x}_{\hat{n}_0}) = \prod_{(j,m)/I_{j,m}\in\widehat{\mathcal{P}}_{n_0}} \prod_{k=1}^{2^{-j}K} \phi(y_{j,m}^k - x_{j,m}^k \mid \sigma^2) p(x_{j,m}^k \mid \widehat{q}_{j,m}^k) P(\widehat{q}_{j,m}^k).$$

The solution of this optimization results in

$$\widehat{x}_{j,m}^k = \begin{cases} \dfrac{\tau_{j,m}^2}{\sigma^2 + \tau_{j,m}^2} y_{j,m}^k & \text{if } \widehat{q}_{j,m}^k = 1, \\ 0 & \text{if } \widehat{q}_{j,m}^k = 0. \end{cases} \tag{19}$$

Note that this estimate is very reminiscent in form of a (degenerate) Wiener estimate, albeit nonlinear. The above expression shows that the components are weighted according to the prevailing SNR.

As previously mentioned, the ML approach can also be similarly applied. The selection of $\mathcal{P}_{n_0}$ in this case, is also carried out by the minimization of an additive criterion, namely

$$\nu(\mathcal{P}) = - \sum_{(j,m)/I_{j,m}\in\mathcal{P}} \sum_{k=1}^{2^{-j}K} \ln\left[(1 - \pi_{j,m})\phi(y_{j,m}^k \mid \sigma^2) + \pi_{j,m}\phi(y_{j,m}^k \mid \tilde{\tau}_{j,m}^2)\right].$$

$$\tag{20}$$

11.5   Implementation and Parameter Estimation

In describing the statistical optimization techniques in the last section, we assumed all the relevant parameters (usually referred to as hyperparameters) to be available. In realistic cases these have to be estimated for each node (j, m) of the decomposition tree. In this section, we will detail a non-standard form of the Expectation-Maximization (EM) algorithm for the parameter estimation of the Bernoulli-Gaussian priors.

The estimation of the hyperparameters being crucial to our BB optimization procedure, behooves us to devise a simple way of regularizing the whole search which we describe below.

11.5.1   Regularization with pre-determined noise variance

As made clear in the last section, the estimation necessary at each of the nodes, may make the algorithms prohibitively intensive. Moreover, a general implementation of the estimation methods would not take into account the unitary transform property which leads to the invariance of the noise variance in the whole decomposition tree, and may ultimately lead to variable results. In light of this, we assume that the noise level σ is known and constant, and estimate the other parameters. If σ is unknown, we can proceed to determine an estimate $\hat{\sigma}$ at a particular suitable node and take this value as the true variance at the other nodes of the tree. This approach allows us to decrease the computational cost of the algorithms, and provide at the same time a regularization at the unknown nodes where the model is likely to be inappropriate. Different methods may be implemented to determine $\hat{\sigma}$. We can for instance, obtain an estimate of σ in a particular basis (e.g. the wavelet basis) with a particular signal prior, or use standard robust estimators in the highest resolution of the wavelet decomposition like [Vid98].

$$\hat{\sigma} = \frac{\text{Median}|y_{1,1}^k|_{1 \le k \le K/2}}{0.6745}.$$

11.5.2   Regularization of the model order

As stated above, the parameters of the model have to be estimated in practice, at each node (j, m) of the tree, which results in an increasing model order with the decomposition level j. In criteria (18) and (20), we have to add a regularization term as a result. Subsequently, a Minimum Decription Length (MDL) regularization term $3P_{j,m} \log(K)/2$ is applied, where $P_{j,m}$ denotes the number of *unknown* parameters of the considered model.

11.5.3   Incremental algorithms (IA)

To alleviate the computational burden of a full ML estimation procedure, we call upon a non-standard form of the EM algorithm which can be implemented using a set of sufficient statistics [NH98]. Recall that, by introducing the hidden indicator vector $\mathbf{q}_{j,m}$ in our mixture of Gaussians models for the observation $\mathbf{y}_{j,m}$, we are back to the classical missing data problem [DLR77]. The vector $\mathbf{z}_{j,m} = (\mathbf{y}_{j,m}, \mathbf{q}_{j,m})^T$ then has an exponential type distribution and, as a result, admits a vector of sufficient statistics $\mathbf{s}_{j,m}$ which, using the independence property, is written as

$$\mathbf{s}_{j,m} = \sum_{k=1}^{2^{-j}K} \mathbf{s}_{j,m}^k,$$

where $\mathbf{s}_{j,m}^k$ only depends on $\mathbf{z}_{j,m}^k = (y_{j,m}^k, q_{j,m}^k)^T$. Using the Neyman-Fisher factorization theorem [Zac98], we derive

$$
\begin{aligned}
p(\mathbf{z}_{j,m} \mid \boldsymbol{\theta}_{j,m}) &= h(\mathbf{z}_{j,m})p(\mathbf{s}_{j,m} \mid \boldsymbol{\theta}_{j,m}), & (21) \\
&= h(\mathbf{z}_{j,m}) \exp\left(\mathbf{s}_{j,m}^T \mathbf{F}(\boldsymbol{\theta}_{j,m}) + G(\boldsymbol{\theta}_{j,m})\right) & (22)
\end{aligned}
$$

where $h(\mathbf{z}_{j,m})$ does not depend on $\boldsymbol{\theta}_{j,m}$, and the functions $\mathbf{F}$ and G will solely be determined by the distribution $p(\mathbf{z}_{j,m} \mid \boldsymbol{\theta}_{j,m})$. This property provides the expectation (E) step of the EM algorithm at iteration ℓ

$$\mathbb{E}\left[\log p(\mathbf{z}_{j,m} \mid \boldsymbol{\theta}_{j,m}) \mid \boldsymbol{\theta}_{j,m}^{(\ell-1)}, \mathbf{y}_{j,m}\right] = \mathbf{F}(\boldsymbol{\theta}_{j,m})^T \mathbb{E}\left[\mathbf{s}_{j,m} \mid \boldsymbol{\theta}_{j,m}^{(\ell-1)}, \mathbf{y}_{j,m}\right]$$
$$+ G(\boldsymbol{\theta}_{j,m}) + \mathbb{E}\left[\log h(\mathbf{z}_{j,m}) \mid \boldsymbol{\theta}_{j,m}^{(\ell-1)}, \mathbf{y}_{j,m}\right],$$

where we recall that the expectation is taken with respect to $P(\mathbf{q}_{j,m} \mid \boldsymbol{\theta}_{j,m}^{(\ell-1)}, \mathbf{y}_{j,m})$. Noting $\bar{\mathbf{s}}_{j,m}^{(\ell)} = \mathbb{E}\left[\mathbf{s}_{j,m} \mid \boldsymbol{\theta}_{j,m}^{(\ell-1)}, \mathbf{y}_{j,m}\right]$, the maximization (M) step is in turn given by

$$\boldsymbol{\theta}_{j,m}^{(\ell)} = \arg\max_{\boldsymbol{\theta}_{j,m}}[\mathbf{F}(\boldsymbol{\theta}_{j,m})^T \bar{\mathbf{s}}_{j,m}^{(\ell)} + G(\boldsymbol{\theta}_{j,m})], \tag{23}$$

$$= \arg\max_{\boldsymbol{\theta}_{j,m}} p(\bar{\mathbf{s}}_{j,m}^{(\ell)} \mid \boldsymbol{\theta}_{j,m}), \tag{24}$$

which corresponds to maximum likelihood estimates associated with $\bar{\mathbf{s}}_{j,m}$. For a faster convergence of the parameters, we proceed to incrementally implement the E step and perform the M step. This in fact forms the essence of incremental algorithms. Typically, only one component, say $\mathbf{s}_{j,m}^{k_\ell}$, will be cyclically updated to obtain

$$\bar{\mathbf{s}}_{j,m}^{(\ell)} = \bar{\mathbf{s}}_{j,m}^{(\ell-1)} - (\bar{\mathbf{s}}_{j,m}^{k_\ell})^{(\ell-1)} + (\bar{\mathbf{s}}_{j,m}^{k_\ell})^{(\ell)}, \tag{25}$$

where $(\bar{\mathbf{s}}_{j,m}^{k_\ell})^{(\ell)} = \mathbb{E}\left[\mathbf{s}_{j,m}^{k_\ell} \mid \boldsymbol{\theta}_{j,m}^{(\ell-1)}, \mathbf{y}_{j,m}^k\right]$. Then $\boldsymbol{\theta}_{j,m}^{(\ell)}$ is still given by (23). At the end of the iterative procedure, the MAP estimate of $\mathbf{q}_{j,m}^k$ is determined (see Section IV). Although this method can be applied to different priors [Lep98], we describe the implementation of this technique next for the B-G model adopted in this chapter.

The vector of sufficient statistics is here given by

$$\mathbf{s}_{j,m}^k = [q_{j,m}^k, (1 - q_{j,m}^k), q_{j,m}^k (y_{j,m}^k)^2]^T, \tag{26}$$

for the vector of parameters $\boldsymbol{\theta}_{j,m} = (\pi_{j,m}, \tilde{\tau}_{j,m}^2)$. Note in this case that $\mathbf{F}(\boldsymbol{\theta}_{j,m}) = [\log(\pi_{j,m}/\tilde{\tau}_{j,m}), \log(1 - \pi_{j,m}), -1/2\tilde{\tau}_{j,m}^2]^T$ and $G(\boldsymbol{\theta}_{j,m}) = 0$. The E step is then simply derived from

$$\mathbb{E}\left[q_{j,m}^k \mid \boldsymbol{\theta}_{j,m}^{(\ell-1)}, \mathbf{y}_{j,m}\right] = P(q_{j,m}^k = 1 \mid \boldsymbol{\theta}_{j,m}^{(\ell-1)}, y_{j,m}^k)$$

$$= \frac{p(y_{j,m}^k \mid q_{j,m}^k = 1, \boldsymbol{\theta}_{j,m}^{(\ell-1)}) P(q_{j,m}^k = 1 \mid \boldsymbol{\theta}_{j,m}^{(\ell-1)})}{p(y_{j,m}^k \mid \boldsymbol{\theta}_{j,m}^{(\ell-1)})}$$

$$= \frac{\pi_{j,m}^{(\ell-1)} \phi\left(y_{j,m}^k \mid (\tilde{\tau}_{j,m}^2)^{(\ell-1)}\right)}{p(y_{j,m}^k \mid \boldsymbol{\theta}_{j,m}^{(\ell-1)})},$$

where

$$p(y_{j,m}^k \mid \boldsymbol{\theta}_{j,m}^{(\ell-1)}) = \pi_{j,m}^{(\ell-1)} \phi \left(y_{j,m}^k \mid (\tilde{\tau}_{j,m}^2)^{(\ell-1)} \right) + (1 - \pi_{j,m}^{(\ell-1)}) \phi \left(y_{j,m}^k \mid \sigma^2 \right).$$

Given the vector of sufficient statistics $\bar{\mathbf{s}}_{j,m}^{(\ell)}$, it is, upon recalling expression (26), simple to verify that the maximum likelihood estimate of the vector of parameters $\boldsymbol{\theta}_{j,m}$ is

$$\widehat{\pi}_{j,m}^{(\ell)} = \frac{(\bar{\mathbf{s}}_{j,m})_1^{(\ell)}}{2^{-j} K}, \tag{27}$$

$$\widehat{\tilde{\tau}}_{j,m}^{(\ell)} = \max \left\{ \sqrt{\frac{(\bar{\mathbf{s}}_{j,m})_3^{(\ell)}}{(\bar{\mathbf{s}}_{j,m})_1^{(\ell)}}}, \sigma \right\}, \tag{28}$$

where $(\bar{\mathbf{s}}_{j,m})_1^{(\ell)}$ and $(\bar{\mathbf{s}}_{j,m})_3^{(\ell)}$ denote respectively the first and third components of $\bar{\mathbf{s}}_{j,m}^{(\ell)}$ given by (25) with

$$(\bar{\mathbf{s}}_{j,m}^{k_\ell})_1^{(\ell)} = \mathbb{E} \left[q_{j,m}^{k_\ell} \mid \boldsymbol{\theta}_{j,m}^{(\ell-1)}, \mathbf{y}_{j,m} \right],$$

$$(\bar{\mathbf{s}}_{j,m}^{k_\ell})_3^{(\ell)} = (y_{j,m}^{k_\ell})^2 (\bar{\mathbf{s}}_{j,m}^{k_\ell})_1^{(\ell)}.$$

11.6   Simulation results

The goal of this section is to demonstrate the viability and performance of the techniques proposed above for implementing a Bayesian framework. Recall that this framework is particularly useful in the low Signal to Noise Ratio scenarios where even the statistically sound existing methods show some limitation. The Bayesian approach may be viewed as providing the probabilistic prior used to inject some regularization in the solution.

Classical signals from the Stanford database [BD95] along with real-world signals are used to substantiate the proposed methodology in denoising applications. Sample reconstruction ℓ^1 and ℓ^2 error criteria have been computed in order to compare and discuss the performance of the methods for both ML and MGML BB selections (with marginal MAP estimation of the underlying signal in both cases). In what follows, and for each of the examples, we estimate the reconstruction Normalized Mean Absolute Value Error (NMAVE) by

$$\text{NMAVE} = \frac{1}{M \|\mathbf{x}\|_{\ell^1}} \sum_{i=1}^{M} \|\widehat{\mathbf{x}}_i - \mathbf{x}\|_{\ell^1},$$

and the reconstruction Normalized Mean Square Error (NMSE) by

$$\text{NMSE} = \frac{1}{M \|\mathbf{x}\|_{\ell^2}^2} \sum_{i=1}^{M} \|\widehat{\mathbf{x}}_i - \mathbf{x}\|_{\ell^2}^2,$$

where M denotes the number of realizations, $\|\cdot\|_{\ell^p}$ represents the ℓ^p norm of $\mathbb{R}^K$ and $\hat{\mathbf{x}}_i$ is the ith estimate of the signal $\mathbf{x}$. Note that, in all our simulations, we took $M = 100$. For good measure, we also tabulate the performance of an existing BB search algorithm, based on the MDL criterion, which hinges upon the minimization of the following additive criterion

$$\varepsilon(n) = \sum_{k=1}^{K} \min\left[(y_n^k)^2, \chi^2\right],$$

with $\chi = \sigma\sqrt{2\log(K)}$ [KP95]. As we shall see, the MDL method generally tends to be overwhelmed by the noise and, as a result, underfits the signal.

11.6.1   Synthetic signals

• Doppler signal

In this first example, the noise level is set to $\sigma = 0.4$ resulting in a severe SNR of -2.8 dB. Note that 8 tap Daubechies' filters are used for the wavelet packet decomposition with a maximum resolution level of 3. Interestingly, incremental algorithms lead to good performances while the MDL performs poorly (see Table 1) at such low signal to noise ratio. To further substantiate the proposed methods, we provide a relevant comparison of performances in terms of NMSE as a function of the noise deviation σ. This graph clearly shows the interest of Bayesian methods at low SNR. The results are plotted in Fig. 1 for both MDL and IA methods using ML BB selection.

	MDL	IA
NMAVE (ML BB search)	0.90	**0.42**
NMSE (ML BB search)	0.88	**0.22**
NMAVE (MGML BB search)	0.90	0.49
NMSE (MGML BB search)	0.88	0.28

TABLE 1. Reconstruction errors for the Doppler signal.

• Blocks signal

For this second process, we choose $\sigma = 1.5$ which results in a SNR of 2.1 dB. Haar filters were used to perform a 4-level wavelet packet decomposition. Results are stated in Table 2 and clearly exhibit the performance of the proposed methods.

	MDL	IA
NMAVE (ML BB search)	0.28	0.23
NMSE (ML BB search)	0.10	0.065
NMAVE (MGML BB search)	0.28	**0.22**
NMSE (MGML BB search)	0.10	**0.063**

TABLE 2. Reconstruction errors for the Blocks signal.

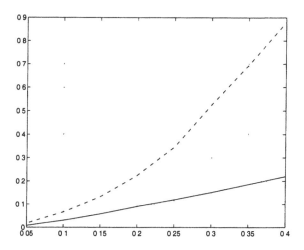

FIGURE 1. Comparison of performances (NMSE) of the MDL method (dashed line) and the IA algorithm (solid line) as a function of σ for the Doppler signal.

• Quadratic Chirp signal

We also provide a denoising example involving local cosine bases. In this case, the noise level is given by $\sigma = 0.6$, which results in a SNR of 1 dB. Table 3 shows that all methods involving prior statistical models provide satisfactory performances.

	MDL	IA
NMAVE (ML BB search)	0.66	**0.53**
NMSE (ML BB search)	0.56	**0.32**
NMAVE (MGML BB search)	0.66	**0.53**
NMSE (MGML BB search)	0.56	0.35

TABLE 3. Reconstruction errors for the Quadratic Chirp signal.

11.6.2   Real-world signals

• Bird sound

We first focus on a recording excerpt corresponding to a bird song. We set $\sigma = 0.5$ and obtain a SNR of 6 dB. We choose 4 tap Daubechies' filters to implement a 4-level wavelet packet decomposition. Results obtained with both ML and MGML BB search are stated in Table 4. For illustration, reconstruction examples are also presented in Fig. 2.

	MDL	IA
NMAVE (ML BB search)	0.47	0.40
NMSE (ML BB search)	0.22	**0.14**
NMAVE (MGML BB search)	0.47	**0.39**
NMSE (MGML BB search)	0.22	**0.14**

TABLE 4. Reconstruction errors for the bird signal.

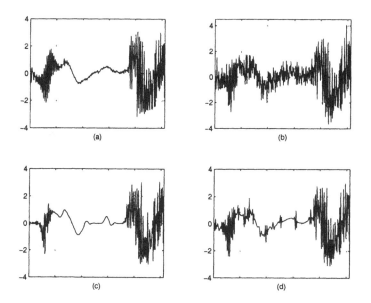

FIGURE 2. (a) original bird sound, (b) noisy signal, (c) and (d) signal reconstructions for the MDL method and the IA algorithm (with MGML BB search) respectively.

• Underwater acoustic signal

To further demonstrate the relevance of our approach in denoising applications, we finally consider an underwater acoustic signal corresponding to a whale sound. The noise variance is set to 0.25 providing a SNR of 6 dB.

Examples of signal reconstruction are presented in Fig. 3, while results are stated in Table 5 and confirm, as in the previous example, the improvement of our method over the MDL approach.

	MDL	IA
NMAVE (ML BB search)	0.64	**0.60**
NMSE (ML BB search)	0.25	0.16
NMAVE (MGML BB search)	0.64	**0.60**
NMSE (MGML BB search)	0.25	**0.15**

TABLE 5. Reconstruction errors for the underwater acoustic signal.

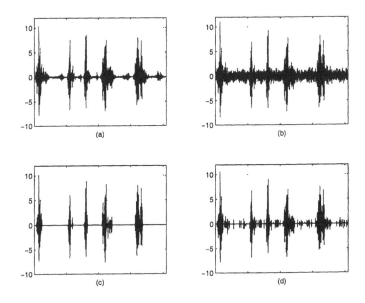

FIGURE 3. (a) original underwater acoustic signal, (b) noisy signal, (c) and (d) signal reconstructions for the MDL method and the IA algorithm (with ML BB search) respectively.

References

[ASS98]   Abramovich, F., Sapatinas, T. & Silverman, B.W. (1998a). Wavelet thresholding via a Bayesian approach. *J. Roy. Statist. Soc. B* **60**, 725-749.

[BD95]   J.B. Buckheit and D.L. Donoho. Wavelab and Reproducible Research. Stanford University, 1995.

[Ber85]   J. Berger. *Statistical Decision Theory and Bayesian Analysis.*
           Springer Verlag, New York, 1985.

[CCL96]   Q. Cheng, R. Chen, and T. Li. Simultaneous Wavelet Estima-
           tion and Deconvolution of Reflection Seismic Signals. *IEEE
           Trans. Geoscience and Rem. Sens.*, 34:377–384, 1996.

[CDV93]   A. Cohen, I. Daubechies, and P. Vial. Wavelets on the inter-
           val and fast wavelet algorithms. *Applied and Computational
           Harmonic Analysis*, 1:54–81, Dec. 1993.

[CGI96]   F. Champagnat, Y. Goussard, and J. Idier. Unsupervised De-
           convolution of Sparse Spike Trains using Stochastic Approxi-
           mation. *IEEE Trans. Signal Processing*, 44:2988–2998, 1996.

[CI95]   F. Champagnat and J. Idier. An alternative to standard max-
           imum likelihood for Gaussian mixtures. In *Proc. IEEE Conf.
           Acoust., Speech, Signal Processing*, pages 2020–2023, Detroit,
           USA, May 9-12 1995.

[CKM97]   H. A. Chipman, E. D. Kolaczyck, and R. E. McCulloch. Adap-
           tive Bayesian Wavelet Shrinkage. *J. Amer. Statist. Assoc.*,
           92:1413–1421, 1997.

[CNB98]   M. S. Crouse, R. D. Nowak, and R. G. Baraniuk. Wavelet-
           Based Statistical Signal Processing Using Hidden Markov
           Models. *IEEE Trans. Signal Processing*, 46:886–902, 1998.

[CPV98]   M. Clyde, G. Parmigiani, and B. Vidakovic. Multiple Shrink-
           age and Subset Selection in Wavelets. *Biometrika*, 85:391–401,
           1998.

[CW92]   R. R. Coifman and M. V. Wickerhauser. Entropy-based algo-
           rithms for best basis selection. *IEEE Trans. Informat. Theory*,
           IT-38:713–718, Mar. 1992.

[DJ94a]   D. L. Donoho and I. M. Johnstone. Ideal denoising in an
           orthogonal basis chosen from a library of bases. *C. R. Acad.
           Sci. Paris*, 319:1317–1322, 1994.

[DJ94b]   D. L. Donoho and I. M. Johnstone. Ideal spatial adaptation
           by wavelet shrinkage. *Biometrika*, 81:425–455, Sept. 1994.

[DLR77]   A. P. Dempster, N. M. Laird, and D. B. Rubin. Maximum
           Likelihood from Incomplete Data via the EM Algorithm. *J.
           R. Statist. Soc. B*, 39:1–38, 1977.

[Hub67]   P. Huber. The behaviour of maximum-likelihood estimates
           under nonstandard conditions. In *Proc. Berkeley Symp. on
           Math. Stat. and Prob.*, volume 1, pages 73–101, 1967.

[HV94]   C. Herley and M. Vetterli. Orthogonal time-varying filter banks and wavelet packets. *IEEE Trans. Signal Processing*, 42:2650–2663, Oct. 1994.

[KMDW95] H. Krim, S. Mallat, D. Donoho, and A. Willsky. Best basis algorithm for signal enhancement. In *Proc. IEEE Conf. Acoust., Speech, Signal Processing*, Detroit, MI, May 1995.

[KP95]   H. Krim and J.-C. Pesquet. On the statistics of best bases criteria. In A. Antoniadis, editor, *Wavelets and statistics*. Lecture Notes in Statistics, Springer Verlag, 1995.

[Lep98]   D. Leporini. *Modélisation Statistique et Paquets d'Ondelettes: Application au Débruitage de Signaux Transitoires d'Acoustique Sous-Marine*. PhD thesis, Université Paris XI, Sept. 1998.

[Mal89]   S. Mallat. A theory for multiresolution signal decomposition: the wavelet representation. *IEEE Trans. Patt. Anal. Mach. Intell.*, PAMI-11:674–693, Jul. 1989.

[NH98]   R. M. Neal and G. E. Hinton. A new view of the EM algorithm that justifies incremental and other variants. In M.I. Jordan, editor, *Learning in Graphical Models*. Kluwer, 1998.

[PKLH96]   J.-C. Pesquet, H. Krim, D. Leporini, and E. Hamman. Bayesian Approach to Best Basis Selection. In *Proc. IEEE Conf. Acoust., Speech, Signal Processing*, pages 2634–2638, Atlanta, USA, May 7-9 1996.

[Sai94]   N. Saito. *Local feature extraction and its applications using a library of bases*. PhD thesis, Yale University, Dec. 1994.

[Vid98]   B. Vidakovic. Nonlinear wavelet shrinkage with Bayes rules and Bayes factors. *J. Amer. Statist. Assoc.*, 93:173–179, 1998.

[Zac81]   S. Zacks. *Parametric Statistical Inference*. International Series in Nonlinear Mathematics, Pergamon Press, New York, 1981.

12

Modeling Dependence in the Wavelet Domain

Marina Vannucci and Fabio Corradi

ABSTRACT Here we present theoretical results on the random wavelet co-efficients covariance structure. We use simple properties of the coefficients to derive a recursive way to compute the within- and across-level covari-ances. We then show the usefulness of those findings in some of the best known applications of wavelets in statistics. Wavelet shrinkage attempts to estimate a function from noisy data. When approaching the problem from a Bayesian point of view, a prior distribution is imposed on the coefficients of the unknown function. We show how it is possible to specify priors that take into account the full correlation among coefficients through a parsi-monious number of hyperparameters. We then concentrate on the wavelet analysis of random processes. Given discrete measurements from a long-memory process, a wavelet transform leads to a set of wavelet coefficients. We show how the statistical properties of the coefficients depend on the process underlying the data and how a Bayesian approach gives rise to a natural way of including in the model information about the process itself.

12.1   Introduction

Here we describe a recursive algorithm proposed by Vannucci and Corradi (1999) to compute within- and across-scale covariances of random wavelet coefficients. The algorithm has an interesting link to the two dimensional discrete wavelet transform, making computations feasible, and it is gener-ally applicable in many problems that involve wavelet coefficients modeling.

Bayesian approaches to wavelet techniques, where wavelet coefficients are considered to be random, seem to be a natural framework for those findings. Vannucci and Corradi (1999) focus on the wavelet shrinkage, a well known application of wavelets to attempt the recovery of a signal from noisy data. Originally proposed by Donoho and Johnstone (1994, 1995, 1998), wavelet shrinkage has been recently considered within a Bayesian framework where a prior distribution is imposed on the wavelet coefficients of the unknown signal. Existing works, Chipman et al. (1997), Clyde et al. (1998) and Abramovich et al. (1998)) assume independent coefficients. Van-nucci and Corradi (1999) adopt an approach suggested by Vidakovic and Müller (1995) to incorporate correlation and employ their recursive covariance structure obtaining a model that allows for a full correlation among coefficients. In addition to the theoretical appeal of totally relaxing the independence assumption, this proposal has the advantage of incorporating knowledge about stochastic relationships among wavelet coefficients. The practical implication of this is a model that depends on a parsimonious number of hyperparameters. Inference is obtained via Markov Chain Monte Carlo methods.

Wavelet analysis of random processes is another field where the recursive covariance structure can be useful. Here variances and covariances of the wavelet coefficients can be computed from the autocovariance function of the process generating the data. Vannucci et al. (1998) focus on a class of discrete long-memory processes. Wavelets, by definition, have the ability to simultaneously localise a process in time and scale domain. This ability results in representing many dense matrices in a sparse form. The de-correlation ability of the wavelet transforms is particularly appealing in the case of measurements from long-memory processes, where, in contrast with the dense long-memory covariance structure of the data, wavelet coefficients will be approximately uncorrelated. Vannucci et al. (1998) discuss the de-correlation property of the wavelet transform and relate the variances of the wavelet coefficients to the characteristic parameters of the process. Inference on the parameters is done using Bayesian methods.

The paper is organized as follows: Section 2 briefly reviews basic concepts about wavelet transforms. Section 3 states the recursive algorithm to compute the wavelet coefficients' covariance structure. Section 4 presents a Bayesian shrinkage model and Section 6 a Bayesian approach to the analysis of long-memory processes.

12.2   Notation

For a nice introduction on wavelets and wavelet transforms we refer readers to the introductory chapter of this book, Vidakovic and Müller (1999). Here we briefly recall the recursive relationships that hold between wavelet and scaling coefficients, given their key role in the derivation of the results we will later introduce. Let $d_{j,k} = \langle f, \psi_{j,k} \rangle = \int f(t)\psi_{j,k}(t)\,dt$ indicate the wavelet coefficients and $c_{j,k} = \langle f, \phi_{j,k} \rangle = \int f(t)\phi_{j,k}(t)\,dt$ the scaling coefficients. Coefficients at scale j can be obtained from scaling coefficients at the finer scale $j+1$ as

$$c_{j,k} = \sum_{m \in \mathbb{Z}} h_{m-2k} c_{j+1,m}, \quad d_{j,k} = \sum_{m \in \mathbb{Z}} g_{m-2k} c_{j+1,m} \tag{1}$$

with filter coefficients h_k and $g_k = (-1)^l h_{l-1}$, see Mallat (1989). The above equations are convolutions followed by a downsampling operation. Using signal processing terminology, let F indicate a linear filter defined by an infinite sequence f_l of coefficients and acting as $(Fa)_k = \sum_{m \in \mathbb{Z}} f_{m-k} a_m$, with a_m an infinite sequence. In this paper we are concerned with filters that have a finite number of nonzero coefficients f_l and issues of convergence do not arise. The filter H, defined by the sequence h_l, is a *low pass* filter capturing low frequency components, while G, defined by the sequence g_l, is a *high pass* filter capturing high frequency components. Let now D_0 indicate the downsampling operator $(D_0 a)_j = a_{2j}$ that chooses every other element of a sequence. With $c^{(j)}$ and $d^{(j)}$ indicating the coefficient vectors at scale j, equations (1) can be expressed as

$$c^{(j)} = H_{j+1} c^{(j+1)}, \quad d^{(j)} = G_{j+1} c^{(j+1)}, \tag{2}$$

where H_{j+1} and G_{j+1} are the linear functions corresponding to the application of the filters $D_0 H$ and $D_0 G$. Index $j+1$ points out that the dimensions of the matrices change with the scale, due to the downsampling operation. Equations (1) are used in wavelet theory to derive a fast algorithm for decomposing a function into a set of wavelet coefficients, known as *Discrete Wavelet Transform* (DWT). The algorithm for the inverse construction is called *Inverse Wavelet Transform* (IWT). See Strang (1989) for a detailed exposition of these algorithms.

12.3   Covariances of Random Coefficients

The results of this section refer to the correlation structure of random wavelet coefficients and can be used in a variety of wavelet applications. Later we will illustrate their usefulness in a Bayesian approach to wavelet shrinkage and in the wavelet analysis of long-memory processes.

12.3.1   Recursive Structure

In Vannucci and Corradi (1999) a recursive algorithm is developed to calculate the *within-* and *across-scale* coefficients covariances. To clarify the terminology, we call within-scale the covariances between coefficients that belong to a same scale and across-scales those between coefficients that belong to different scales. From relations (1) it is straightforward to prove that

Proposition 12.3.1 *Given a wavelet basis in $L^2(\mathbb{R})$, the following results hold*

$$i) \quad \mathrm{cov}[d_{j,k}, d_{j',k'}] = \sum_m \sum_n g_{m-2k} g_{n-2k'} \mathrm{cov}[c_{j+1,m}, c_{j'+1,n}]$$

$$ii) \quad \mathrm{cov}[c_{j,k}, c_{j',k'}] = \sum_m \sum_n h_{m-2k} h_{n-2k'} \mathrm{cov}[c_{j+1,m}, c_{j'+1,n}]$$

$$iii) \ \text{cov}[c_{j,k}, d_{j',k'}] = \sum_m \sum_n h_{m-2k} g_{n-2k'} \text{cov}[c_{j+1,m}, c_{j'+1,n}]$$

with j, j' and k, k' integers.

Assume now that the within-scale variance-covariance matrix of scaling coefficients $c^{(j+1)}$ at scale $(j + 1)$ is known and indicate that matrix as $CC^{\{j+1,j+1\}}$. We state the following results in filter notation. The within-scale covariances at the coarser scale j can be easily computed as

$$DD^{\{j,j\}} = G_{j+1}[CC^{\{j+1,j+1\}}]G^T_{j+1} \tag{3}$$

$$CC^{\{j,j\}} = H_{j+1}[CC^{\{j+1,j+1\}}]H^T_{j+1} \tag{4}$$

$$CD^{\{j,j\}} = G_{j+1}[CC^{\{j+1,j+1\}}]H^T_{j+1} \tag{5}$$

where $DD^{\{j,j\}}$ indicates the within-scale variance-covariance matrix of wavelet coefficients at scale j and $CD^{\{j,j\}}$ the within-scale variance-covariance matrix of scaling and wavelet coefficients at scale j. Moreover, the across-scales covariances between scales $j - 1$ and j are

$$CD^{\{j-1,j\}} = H_j H_{j+1}[CC^{\{j+1,j+1\}}]G^T_{j+1} \tag{6}$$

and

$$DD^{\{j-1,j\}} = G_j H_{j+1}[CC^{\{j+1,j+1\}}]G^T_{j+1}, \tag{7}$$

with $CD^{\{j-1,j\}}$ the across-scales variance-covariance matrix of scaling co-efficients at scale $j - 1$ and wavelet coefficients at scale j, and $DD^{\{j-1,j\}}$ the across-scales variance-covariance matrix of wavelet coefficients at scales $j - 1$ and j. The above formulae can be applied to the matrix $CC^{\{j,j\}}$ to get the within-scale covariances at the coarser scale $j - 1$ and the across-scales covariances between scales $j - 2$ and $j - 1$, and so on until a desired scale. Figure 1 shows the algorithm for the first two scales j and $j - 1$. The resultant matrix is symmetric.

12.3.2   Link to the DWT2

To further understand the described algorithm, consider the *two dimensional discrete wavelet transform* (DWT2). Given C, a $2^{j+1} \times 2^{j+1}$ matrix of pixels, a wavelet decomposition of the matrix can be calculated; this results in first applying the linear filters to the rows of the matrix C, obtaining two matrices $H_{j+1}[C]$ and $G_{j+1}[C]$, and then to the columns of $H_{j+1}[C]$ and $G_{j+1}[C]$, obtaining four matrices $H_{j+1}[C]H^T_{j+1}$, $H_{j+1}[C]G^T_{j+1}$, $G_{j+1}[C]H^T_{j+1}$ and $G_{j+1}[C]G^T_{j+1}$ of dimensions $2^j \times 2^j$. This procedure can be repeated

$$CC^{\{j+1,j+1\}}$$
$$\downarrow$$

$CC^{\{j-1,j-1\}} =$ $H_j[CC^{\{j,j\}}]H_j^T$	$DC^{\{j-1,j-1\}} =$ $H_j[CC^{\{j,j\}}]G_j^T$	$CD^{\{j-1,j\}} =$ $H_j H_{j+1}[CC^{\{j+1,j+1\}}]G_{j+1}^T$
$[DC^{\{j-1,j-1\}}]^T$	$DD^{\{j-1,j-1\}} =$ $G_j[CC^{\{j,j\}}]G_j^T$	$DD^{\{j-1,j\}} =$ $G_j H_{j+1}[CC^{\{j+1,j+1\}}]G_{j+1}^T$
$[CD^{\{j-1,j\}}]^T$	$[DD^{\{j-1,j\}}]^T$	$DD^{\{j,j\}} =$ $G_{j+1}[CC^{\{j+1,j+1\}}]G_{j+1}^T$

FIGURE 1. *Left side:* Variance-covariance matrix of the vector $c^{(j+1)}$. *Right side:* Variance-covariance matrix of $(c^{(j-1)}, d^{(j-1)}, d^{(j)})$ obtained from $CC^{\{j+1,j+1\}}$.

with the matrix $B = H_{j+1}[C]H_{j+1}^T$ (see Figure 2), and so on until a desired scale.

The recursive algorithm of Vannucci and Corradi has an interesting link to the DWT2. A comparison of Figures 1 and 2 shows that, having applied the two-dimensional discrete wavelet transform at the matrix $CC^{\{j+1,j+1\}}$, the diagonal blocks will correspond to the within-scale variance-covariance matrices; moreover, the across-scale variance-covariance matrices will be obtained by suitably applying the one-dimensional DWT to the rows of the non diagonal blocks. Notice that, since $CC^{\{j+1,j+1\}}$ is a symmetric matrix, the matrices $G_{j+1}[CC^{\{j+1,j+1\}}]H_{j+1}^T$ and $H_{j+1}[CC^{\{j+1,j+1\}}]G_{j+1}^T$ are transposes of each other. This link to the DWT2 makes the implementation of the recursive algorithm extremely simple in any of the available wavelet packages.

12.4   Bayesian Wavelet Shrinkage

Let $y_1, \ldots, y_n$, $n = 2^J$, be a sequence of observations modeled as

$$y_i = f(t_i) + \epsilon_i, \quad i = 1, \ldots, n \tag{8}$$

where f is a function to be estimated, $t_i = i/n$ are equally spaced points and ϵ_i are i.i.d. normal random variables with zero mean and variance σ^2. Wavelet shrinkage was originally proposed by Donoho and Johnstone (1994, 1995, 1998) as a technique consisting of three steps: Firstly, the DWT is applied to the data y_i obtaining a vector $\tilde{d}$ of empirical wavelet coefficients.

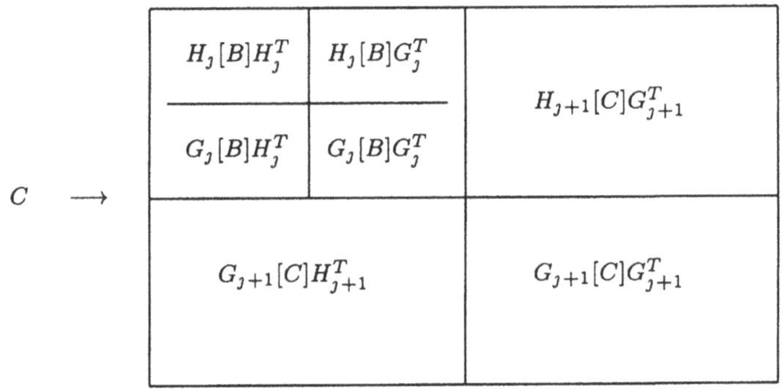

$$
\begin{array}{|c|c|c|}
\hline
H_J[B]H_J^T & H_J[B]G_J^T & \multirow{2}{*}{$H_{J+1}[C]G_{J+1}^T$} \\
\cline{1-2}
G_J[B]H_J^T & G_J[B]G_J^T & \\
\hline
\multicolumn{2}{|c|}{G_{J+1}[C]H_{J+1}^T} & G_{J+1}[C]G_{J+1}^T \\
\hline
\end{array}
$$

$$C \longrightarrow$$

FIGURE 2. Two-dimensional Discrete Wavelet Transform $(B = H_{J+1}[C]H_{J+1}^T)$.

Secondly the noise component is removed by shrinking and/or thresholding the coefficients. Finally, the IWT is applied leading to an estimate of the unknown function. Bayesian approaches to the wavelet shrinkage have recently attracted attention in the literature. Since the DWT is linear and orthogonal, $\tilde{d}$ can still be modeled as

$$\tilde{d} = d + \tilde{\epsilon}, \qquad (9)$$

where d is the vector of wavelet coefficients of the unknown function f and $\tilde{\epsilon}$ is a Gaussian white noise with mean zero and variance-covariance matrix $\sigma^2 I$. Prior distributions can now be chosen for d and σ^2 and the shrinkage step becomes then the result of deriving a Bayes rule from the posterior distribution of d. Vidakovic (1998), Chipman et al. (1997), Clyde et al. (1998) and Abramovich et al. (1998) explore different ways to specify the priors. All these contributions share the assumption that the components of d are conditionally independent.

Vannucci and Corradi (1999) develop algorithms for Bayesian wavelet shrinkage that do not rely on the assumption of independence. In addition to the theoretical appeal of totally relaxing the independence assumption, their proposal has the advantage of incorporating knowledge about stochastic relationships among wavelet coefficients. The practical implication of this is a model that depends on a parsimonious number of hyperparameters. Vannucci and Corradi build their recursive algorithm into a model proposed by Vidakovic and Müller (1995) in the context of density estimation and applied by Vannucci (1996) to denoise data. This model, $V\&M$ in the sequel, represents the first attempt to take into account the coefficients' correlation structure.

12.4.1   The V&M Model

According to equation (9), uncertainty about the empirical coefficients is described by

$$\tilde{d}|d, \sigma^2 \sim \mathcal{N}(d, \sigma^2 I). \tag{10}$$

A closed form learning procedure about the d's can be achieved (see, for example, O'Hagan (1994)) assuming

$$d, \sigma^2 \sim \mathcal{NIG}_{n+1}(\alpha, \delta, m, \Sigma) \tag{11}$$

where $\mathcal{NIG}_{n+1}$ stands for the multivariate normal-inverse gamma distribution with dimension $n + 1$. The model requires the specification of the hyperparameters α, δ, m and Σ. Since the marginal prior distribution of σ^2 is an inverse gamma $\mathcal{IG}(\alpha/2, \delta/2)$, the parameters α and δ can be used to specify beliefs about the noise scale, considering that $E[\sigma^2] = \alpha/(\delta - 2)$ and $Var(\sigma^2) = 2\alpha^2/((\delta - 1)(\delta - 2)^2)$. Moreover, for a given σ^2, the prior conditional distribution of d given σ^2 is a multivariate normal with mean vector m and covariance matrix $\sigma^2\Sigma$. The posterior distribution for d and σ^2 is

$$d, \sigma^2|\tilde{d} \sim \mathcal{NIG}_{n+1}(\alpha^\star, \delta^\star, m^\star, \Sigma^\star), \tag{12}$$

with updated parameters

$$\Sigma^\star = (I + \Sigma^{-1})^{-1}, \quad m^\star = \Sigma^\star(\tilde{d} + \Sigma^{-1}m) \tag{13}$$

and $\alpha^\star = \alpha + m^T\Sigma^{-1}m + \tilde{d}^T\tilde{d} + (m^\star)'(\Sigma^\star)^{-1}m^\star$, $\delta^\star = \delta + n$. The adoption of a squared error loss function leads to the posterior mean $m^\star = \Sigma^\star\tilde{d}$ as a Bayes estimator of d. Thus, the IWT can be applied to $m^\star$ to get the function estimate. Notice that this estimation procedure does not require the specification of α and δ.

12.4.2   Recursive Specification of Σ

Vannucci and Corradi (1999) proposed a way for specifying Σ that takes into account the correlation structure of the wavelet coefficients. Let us refer to their proposal as the $V\&C$ model. The main idea of this model is to use the recursive algorithm described in Section 3 to derive the matrix Σ from a matrix $\Sigma_{J,\phi}$ representing the covariance matrix of scaling coefficients at level J (i.e. the covariance matrix $CC^{\{J,J\}}$ in the notation of Section 3) of the unknown function f underlying the data. The specification of $\Sigma_{J,\phi}$ should be based on knowledge about f. Vannucci and Corradi (1999) describe uncertainty about f by assuming f to be a realization of a quite general stochastic process $X(t)$ having finite (constant) mean $E[X(t)] = c$ (that is, $c = 0$ without loss of generality) and finite $E[X(t)X(s)]$. They then prove that, using Daubechies wavelets (see Daubechies (1992)), within- and across-scale covariance matrices of wavelet and scaling coefficients have zero entries outside diagonal bands that depend on the number of vanishing moments of the wavelet family.

Proposition 12.4.1 *(Vannucci and Corradi 1999)*
Consider the Daubechies minimum phase wavelets. Then, at fixed integer scales j and j' with $j' - j = l$, l positive integer, scaling coefficients $c_{j,k}$ and $c_{j',k'}$ are uncorrelated for integer k and k' such that

$$(k' - 2^l k) > 2^l (2N - 1) \quad or \quad (k' - 2^l k) < 1 - 2N$$

and wavelet coefficients $d_{j,k}$ and $d_{j',k'}$ are uncorrelated for k and k' such that

$$(k' - 2^l k) > (1 + 2^l)N - 1 \quad or \quad (k' - 2^l k) < 2^l - (1 + 2^l)N$$

where N is the number of vanishing moments of the wavelet family.

Using this result, $\Sigma_{J,\phi}$, viewed as the covariance matrix of scaling coefficients at the level J, can be specified as diagonal. Within the diagonal band, one may specify a covariance structure that decreases in inverse proportion to the distance between coefficients. Decay properties of the coefficients' correlation are now well known for a large variety of stationary and non-stationary stochastic processes. Thus, in the $V\&C$ model $\Sigma_{J,\phi}$ is specified as $\sigma_{k,k'} = \lambda \rho^{|k-k'|}$ for $|k - k'| < 2N - 1$ and $\sigma_{k,k'} = 0$ otherwise. Notice that, because of the recursive structure, since $\Sigma_{J,\phi}$ is specified as a band matrix, one has within- and across-scale band covariance matrices at each scale. Later in Section 5, when dealing with random processes with a known correlation structure, we will see a different way to specify $\Sigma_{J,\phi}$.

The specification of Σ in the $V\&C$ model incorporates knowledge about the wavelet coefficients correlation structure; it allows within- and across-scale correlation modeling; it leads to a remarkable reduction of the number of the parameters implied in the model. The matrix $\Sigma_{J,\phi}$ (and hence Σ) depends in fact only on the two parameters λ and ρ, and based on the construction above one can write $\Sigma(\lambda, \rho) = \lambda \Sigma(\rho)$. The parameter λ is a smoothing parameter. Smaller values of λ imply a more precise prior, that is greater shrinkage of the wavelet coefficients towards the prior mean m. Suitable values for ρ are those that imply a positive definite matrix $\Sigma_{J,\phi}$, that is a positive minimum eigenvalue. See Vannucci and Corradi (1999) for a discussion and some numerical results.

The recursive specification of Σ requires choosing λ and ρ. Two different methods can be used. One uses an empirical Bayes approach searching over a grid for the values that minimize a general measure of discrepancy. A suitable score, that measures the goodness-of-fit of the wavelet estimator, is the mean squared error $\frac{1}{n} \sum_{i=1}^{n} (\hat{f}_i - f_i)^2$. A more realistic method, that learns about λ and ρ using a Bayesian hierarchical model, requires the use of

Markov chain Monte Carlo methods and will be briefly described in the next section. See Vannucci and Corradi (1999) for illustration of these methods on the simulated signals *HeaviSine*, *Blocks*, *Bumps* and *Doppler* of Donoho and Johnstone (1994, 1995) and for visual and numerical comparisons with the *V&M* model as originally proposed by Vidakovic and Müller (1995). Allowing for across-scale correlation among coefficients seems to improve reconstruction of the signals.

12.4.3 Learning about λ and ρ

Because results can be sensitive to the choice of λ and ρ, it can be attractive to include them in the inferential process. A possible solution is to specify a third scale in the *V&C* model. Empirical coefficients $\tilde{d}$ are still modeled through (10) with the prior distribution on d and σ^2 expressed in the conjugate form (11). Then, λ and ρ are also random and can be assumed independent and a priori distributed as follows:

$$\lambda \sim \mathcal{IG}(p/2, q/2), \tag{14}$$

and

$$p(\rho) \propto (C - \rho)^{r_1 - 1}(C + \rho)^{r_2 - 1}, \quad |\rho| < C \tag{15}$$

such that $(C - \rho)/2C$ is proportional to a Beta distribution with parameters r_1 and r_2. The constant C takes into account the constraints on the support of ρ. See Vannucci and Corradi (1999) for a discussion. The complex structure of $\Sigma(\lambda, \rho)$ does not allow for inference in closed form. A sample from the posterior distribution of the parameters can be obtained by a Markov chain Monte Carlo method (see, for example, Tierney (1994) and Smith and Roberts (1993)). The full conditional distributions of d, σ^2 and λ can be easily computed, while it is harder to specify the full conditional distribution of ρ. Consequently, the chain is simulated combining in a cycle Gibbs steps for the parameters d, σ^2 and λ with a Metropolis step for ρ. More precisely, given starting values for σ^2, λ and ρ , parameters can be sampled in the following order. The vector d from

$$d|\tilde{d}, \sigma^2, \lambda, \rho \sim \mathcal{N}(m^\star, \sigma^2 \Sigma^\star) \tag{16}$$

with $\Sigma^\star = (I + \lambda^{-1}(\Sigma(\rho))^{-1})^{-1}$ and $m^\star = \Sigma^\star(\tilde{d} + \lambda^{-1}(\Sigma(\rho))^{-1}m)$. This is done in a single step to exploit the correlation structure and improve the speed of convergence. The variance noise σ^2 is simulated from

$$\sigma^2|\tilde{d}, d, \lambda, \rho \sim \mathcal{IG}(\alpha^\star/2, \delta^\star/2) \tag{17}$$

with $\alpha^\star = \alpha + (d-m)^T\lambda^{-1}(\Sigma(\rho))^{-1}(d-m) + (\tilde{d}-d)^T(\tilde{d}-d)$ and $\delta^\star = \delta + 2n$. The parameter λ from

$$\lambda|d, \sigma^2, \rho \sim \mathcal{IG}(p^\star/2, q^\star/2) \tag{18}$$

with $p^\star = p + [(d - m)^T(\Sigma(\rho))^{-1}(d - m)]/\sigma^2$ and $q^\star = q + n$. Finally, choosing $\mathcal{N}(\hat{\rho}|\rho^{(j)}, \sigma_\rho^2)$, $\hat{\rho} < |C|$ as a proposal distribution, ρ is simulated by

1) sampling $\hat{\rho}$ from $\mathcal{N}(\hat{\rho}|\rho^{(j)}, \sigma_\rho^2)$, $\hat{\rho} < |C|$, with $\rho^{(j)}$ the sample value generated from the previous cycle;

2) computing

$$a = \min\left[1, p(\hat{\rho}|d, \sigma^2, \lambda)/p(\rho^{(j)}|d, \sigma^2, \lambda)\right]; \tag{19}$$

3) accepting $\hat{\rho}$ if $0 < U(0, 1) < a$.

After running the transient phase of the chain, the mean vector of d can be estimated by averaging over the simulated samples of the d's. Finally, the IWT can be applied to obtain the function estimate.

12.5 Wavelet analysis of long-memory processes

We now turn to the wavelet analysis of random processes. Let us assume that $X = (X_1, \ldots, X_n), n = 2^J$, is a realization from a random process $\{X_t, t = 0, \pm 1, \ldots\}$ with finite mean $E[X(t)] = c$ (that is, $c = 0$ without loss of generality). When we apply the DWT to X and get a vector Z of wavelet coefficients, we implicitly treat the $n = 2^J$ data as a vector of scaling coefficients at level J. If the data were generated from a random process with autocovariance function $\gamma(\tau)$, we can write the variance-covariance matrix of the vector X of observations as

$$\Sigma_X(i, j) = [\gamma(|i - j|)] \tag{20}$$

and use the algorithm of Vannucci and Corradi (1999) to compute the variance-covariance matrix Σ_Z of Z from Σ_X, as described here in Section 3.

Vannucci et al. (1998) concentrate on long-memory processes. Long-memory processes have been used to model a number of financial and economic time series, including stock market prices (Lo (1991)) and exchange rates (Cheung (1993)). Wavelets are strongly connected to long-memory processes in that the same shapes repeat at different orders of magnitude. Wavelets, by definition, have the ability to simultaneously localise a process in time and scale domain. This ability results in representing many dense matrices in a sparse form. When dealing with long-memory processes, the dense long-memory covariance structure would make the exact likelihood difficult to handle. Wavelets, de-correlating the data, become a useful tool to simplify the likelihood.

12.5.1 Long-memory Processes

A long-memory process is a process for which the dependence between distant observations is not negligible (though small). Long-memory processes are known as $1/f$-type processes, given the fact that their power spectra appear as straight lines on log-frequency/log-power scales over many octaves of frequency. The autocorrelation function behaves asymptotically as

$$\rho(\tau) \sim \tau^{2h-1} \text{ as } \tau \to +\infty \tag{21}$$

for $|h| < 0.5$. Long-memory processes are *self-similar*, meaning that X_{ct} and $c^H X_t$, $c > 0$, with $H = 2h - 1$, have the same distribution. In other words, the statistical properties of the process do not change with of the resolution of $\{X_t\}$. For a nice review of long-memory processes see Beran (1992).

Vannucci et al. (1998) concentrate on a specific class of discrete-time long-memory processes, known as fractionally differenced processes (see Hosking (1981) and Granger and Joyeux (1980) for a definition) that have autocorrelation and autocovariance functions defined as

$$\rho(0) = 1, \quad \rho(\tau) = \frac{\Gamma(\tau + h)\Gamma(1 - h)}{\Gamma(\tau - h + 1)\Gamma(h)}, \quad \tau \geq 1 \tag{22}$$

$$\gamma_0 = \sigma^2 \frac{\Gamma(1 - 2h)}{\Gamma^2(1 - h)} \tag{23}$$

with $\Gamma(\cdot)$ the Gamma function. These processes are long memory for $0 < h < 1/2$.

Figure 3 shows the variance-covariance matrix of wavelet coefficients when $\gamma(\tau)$ is given by (22) multiplied by (23) with $h = 0.2$ and $\sigma^2 = 0.9$. The highest gray-scale values of the images correspond to the largest entries of the matrices. Scales are graphed from coarser to finer. The two plots correspond to wavelet families with different numbers of vanishing moments (Haar wavelets and Daubechies wavelets with 7 vanishing moments). Notice that there is essentially no correlation among coefficients at the same scale. Some correlation between scales is present, as shown by the extra-diagonal gray lines, but it dramatically decreases when using wavelets with higher numbers of vanishing moments. The whitening filter becomes more and more effective when using more regular wavelets.

12.5.2 A Bayesian Approach

Vannucci et al. (1998) propose a Bayesian approach to the wavelet analysis of long-memory processes that leads to Bayesian estimates of the differencing parameter h and the nuisance parameter σ^2. If the long-memory process generating the data is Gaussian with zero mean, wavelet coefficients

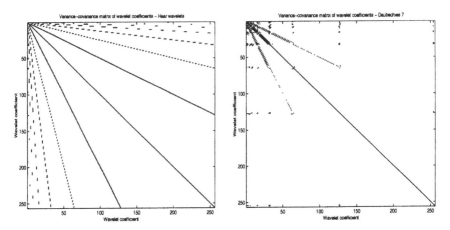

FIGURE 3. Wavelet coefficients' variance-covariance matrix for a fractionally differenced process and Haar wavelets (left panel) and Daubechies 7 (right panel).

will also be zero mean Gaussian. As shown in the previous section, wavelet transforms de-correlate the data, i.e. wavelet coefficients are approximately uncorrelated, and one loses relatively little in writing the likelihood as

$$[Z_i|h, \sigma^2] \sim N(0, \sigma^2_{Z_i})$$ (24)

independently for $i = 1, \ldots, n$. Variances $\sigma^2_{Z_i}$ are obtained from the auto-covariance function of the process generating the data, as previously described. When the process is a fractionally differenced one, those variances will depend on the differencing parameter h and on the nuisance parameter σ^2. They can be written as σ^2 times an expression ($\sigma^2_{Z_i}(h)$ in the sequel) depending only on h. Inference on h and σ^2 can be now carried out. One choice is to assume them independent and a priori distributed as $\sigma^2 \sim IG(\alpha/2, \beta/2)$ and $2h \sim Beta(\eta, \nu)$. The recursive way in which variances of wavelet coefficients are computed, although computationally simple, leads to a complex structure for $\sigma^2_{Z_i}(h)$ and does not allow inference in closed form. A sample from the posterior distribution of the parameters can be obtained by a Markov chain Monte Carlo approach. A chain is simulated by combining in a cycle a Gibbs step for the parameter σ^2 with a Metropolis step for h. After a burn-in period, h and σ^2 can be estimated by averaging over the simulated samples. Wavelet coefficients-based maximum likelihood estimates of h and σ^2 have been proposed by McCoy and Walden (1996).

Vannucci et al. (1998) perform a comprehensive simulation study. To assess performances, they compute Monte Carlo mean squared error (MSE) and bias of the estimates by simulating a number of time series using $\rho(\tau)$ as in (22) with h fixed and repeat the procedure for different values of h to check robustness of the estimates to parameter values. Their results show a remarkable robustness to the different values of the parameter. The small values of the sample bias confirm the usefulness of wavelet methods in long-memory analysis.

12.6   Concluding Remarks

Here we have presented a recursive algorithm to calculate within- and across-scale covariances of random wavelet coefficients. The proposed algorithm can be a useful tool in any problem that involves wavelet coefficients modelling. We have employed it in a Bayesian approach to the wavelet shrinkage, motivating a parsimonious model specification that depends only on two parameters, and in the wavelet analysis of long-memory processes, deriving wavelet coefficients' variances-covariances from the autocorrelation structure of the process and proposing Bayesian estimates of characteristic parameters.

Different wavelet methods are addressed by Kovac and Silverman (1998), who independently explore the recursive way of computing variances and covariances of wavelet coefficients. They concentrate only on variances and within-scale covariances and investigate the use of the algorithm in wavelet regression methods with irregularly spaced data, regularly spaced data sets of arbitrary size and correlated data.

Acknowledgments: We thank Peter Müller and Brani Vidakovic for inviting us to write this chapter.

References

Abramovich, F., Sapatinas, T., and Silverman, B. (1998), "Wavelet Thresholding via a Bayesian Approach," *J. Roy. Statist. Soc. B* , 60(4), 725–749.

Beran, J. (1992), "Statistical Methods for Data with long-memory Dependence." *Statistical Science*, 7, 404–427.

Cheung, Y. (1993), "Long memory in foreign-exchange rates," *Journal of Business and Economic Statistics*, 11, 93–101.

Chipman, H., Kolaczyk, E., and McCulloch, R. (1997), "Adaptive Bayesian Wavelet Shrinkage," *J. Am. Stat. Ass.*, 92, 1413–1421.

Clyde, M., Parmigiani, G., and Vidakovic, B. (1998), "Multiple Shrinkage and Subset Selection in Wavelets," *Biometrika*, 85(2), 391–402.

Daubechies, I. (1992), *Ten Lectures on Wavelets*, volume 61, SIAM, CBMS-NSF Conference Series.

Donoho, D. and Johnstone, I. (1994), "Ideal Spatial Adaption via Wavelet Shrinkage," *Biometrika*, 81, 425–455.

— (1995), "Adapting to Unknown Smoothness by Wavelet Shrinkage," *J. Am. Stat. Ass.*, 90, 1200–1224.

— (1998), "Minimax Estimation via Wavelet Shrinkage," *Ann. Stat.*, 26(3), 879–921.

Granger, C. and Joyeux, R. (1980), "An introduction to long memory time series models and fractional differencing," *J. Time Series Analysis*, 1, 15–29.

Hosking, J. (1981), "Fractional differencing," *Biometrika*, 68, 165–176.

Kovac, A. and Silverman, B. (1998), "Extending the scope of wavelet regression methods by coefficient-dependent thresholding," Technical Report 5/1998, University of Dortmund, Germany.

Lo, A. (1991), "Long term memory in stock market prices," *Econometrica*, 59, 1279–1313.

Mallat, S. (1989), "Multiresolution Approximations and Wavelet Orthonormal Bases of $L_2(R)$," *Transactions of the American Mathematical Society*, 315(1), 69–87.

McCoy, E. and Walden, A. (1996), "Wavelet Analysis and Synthesis of Stationary Long-Memory Processes," *J. Computational and Graphical Statistics*, 5(1), 26–56.

O'Hagan, A. (1994), *Kendall's Advanced Theory of Statistics*, volume 2B, Cambridge: Cambridge University Press.

Smith, A. and Roberts, G. (1993), "Bayesian computation via the Gibbs sampler and related Markov chain Monte Carlo methods," *J. Roy. Statist. Soc. B* , 55, 3–23.

Strang, G. (1989), "Wavelets and Dilation Equations: a Brief Introduction," *SIAM Review*, 31(4), 614–627.

Tierney, L. (1994), "Markov chains for exploring posterior distributions," *Ann. Stat.*, 22, 1701–1728.

Vannucci, M. (1996), *On the Application of Wavelets in Statistics*, Ph.D. thesis, Dipartimento di Statistica G.Parenti, University of Florence, Italy, in italian.

Vannucci, M., Brown, P., and Fearn, T. (1998), "Wavelet analysis of long-memory processes," Technical Report IMS/98-22, University of Kent at Canterbury.

Vannucci, M. and Corradi, F. (1999), "Covariance Structure of Wavelet Coefficients: Theory and Models in a Bayesian Perspective." *J. Roy. Statist. Soc. B* , to appear.

Vidakovic, B. (1998), "Nonlinear Wavelet Shrinkage with Bayes Rules and Bayes Factor," *J. Am. Stat. Ass.*, 93(441), 173–179.

Vidakovic, B. and Müller, P. (1995), "Wavelet Shrinkage with Affine Bayes Rules with Applications." Technical Report DP 95-34, ISDS, Duke University.

Vidakovic, B. and Müller, P. (1999), "An Introduction to Wavelets," in *Bayesian Inference in Wavelet based Models*, eds. P. Müller and B. Vidakovic. New York: Springer Verlag.

13

MCMC Methods in Wavelet Shrinkage: Non-Equally Spaced Regression, Density and Spectral Density Estimation

Peter Müller and Brani Vidakovic

ABSTRACT We consider posterior inference in wavelet based models for non-parametric regression with unequally spaced data, density estimation and spectral density estimation. The common theme in all three applications is the lack of posterior independence for the wavelet coefficients d_{jk}. In contrast, most commonly considered applications of wavelet decompositions in Statistics are based on a setup which implies *a posteriori* independent coefficients, essentially reducing the inference problem to a series of univariate problems. This is generally true for regression with equally spaced data, image reconstruction, density estimation based on smoothing the empirical distribution, time series applications and deconvolution problems.

We propose a hierarchical mixture model as prior probability model on the wavelet coefficients. The model includes a level-dependent positive prior probability mass at zero, i.e., for vanishing coefficients. This implements wavelet coefficient thresholding as a formal Bayes rule. For non-zero coefficients we introduce shrinkage by assuming normal priors. Allowing different prior variance at each level of detail we obtain level-dependent shrinkage for non-zero coefficients.

We implement inference in all three proposed models by a Markov chain Monte Carlo scheme which requires only minor modifications for the different applications. Allowing zero coefficients requires simulation over variable dimension parameter space (Green, 1995). We use a pseudo-prior mechanism (Carlin and Chib, 1995) to achieve this.

13.1 Introduction

Many statistical inference problems can be thought of as estimating a random function $f(x)$, conditional on data generated from a sampling model which in some form involves f. In particular, this view is natural for non-linear regression, density estimation and spectral density estimation where f has the interpretation of the unknown mean function, probability density function, and spectral density, respectively. Bayesian inference in these problems requires a prior probability model for the unknown function f. Sometimes restriction to a low dimensional parametric model for f is not desirable. Common reasons are, for example, that there is no obvious parametric form for a non-linear regression function; or little is known about the shape of the unknown density in a density estimation. This calls for non-parametric methods, i.e., parameterizations of the unknown function f with infinitely many parameters, allowing a priori a wide range of possible functions. Wavelet bases provide one possibility of formalizing this.

Wavelet decomposition allows representation of any square integrable function $f(x)$ as

$$f(x) = \sum_{k \in Z} c_{J_0 k} \phi_{J_0 k}(x) + \sum_{j \geq J_0} \sum_{k \in Z} d_{jk} \psi_{jk}(x), \qquad (1)$$

i.e., a parameterization of f in terms of the wavelet coefficients $\theta = (c_{J_0 k}, d_{jk}, j \geq J_0; \ k \in Z)$. Here $\psi_{jk}(x) = 2^{j/2} \psi(2^j x - k)$, and $\phi_{jk}(x) = 2^{j/2} \phi(2^j x - k)$ are wavelets and scaling functions at level of detail j and shift k. When we want to emphasize the dependence of $f(x)$ on the wavelet coefficients we will write $f_\theta(x)$. Without loss of generality we will in the following discussion assume $J_0 = 0$. See Vidakovic and Müller (1999) and Marron (1999) for a discussion of basic facts related to wavelet representations.

Perhaps the most commonly used application of (1) in statistical modeling is to non-linear regression where $f(x)$ represents the unknown mean response $E(y|x)$ for an observation with covariate x. Chipman, Kolaczyk and McCulloch (1997), Clyde, Parmigiani and Vidakovic (1998), and Vidakovic (1998) discuss Bayesian inference in such models assuming equally spaced data, i.e., covariate values x_i are on a regular grid. For equally spaced data the discrete wavelet transformation is orthogonal. Together with assuming independent measurement errors and a priori independent wavelet coefficients this leads to posterior independence of the d_{jk}. Thus the problem essentially reduces to a series of univariate problems, one for each wavelet coefficient. See the chapter by Yau and Kohn (1999) in this volume for a review. Generalizations of wavelet techniques to non-equidistant (NES) design impose additional conceptual and computational burdens. A reasonable approximation is to bin observations in equally spaced bins and proceed as in the equally spaced case. If only few observations are missing to complete an equally spaced grid, treating these few as missing data leads to efficient implementations (Antoniadis, Gregoire and McKeague 1994; Cai and Brown, 1997; Hall and Turlach, 1997; Sardy et al., 1997). Alternatively, we will propose in this chapter an approach which does not rely on posterior independence.

Density estimation is concerned with the problem of estimating an unknown probability density $f(x)$ from a sample $x_i \sim f(x)$, $i = 1, \ldots, n$. Donoho et al. (1996) propose to consider a wavelet decomposition of $f(x) =$

$f_\theta(x)$ and estimate the coefficients d_{jk} by applying thresholding and shrinkage to the empirical coefficients $\hat{d}_{jk} = \frac{1}{n} \sum \psi_{jk}(x_i)$ and $\hat{c}_{jk} = \frac{1}{n} \sum \phi_{jk}(x_i)$. This is justified by the fact that as coefficients of an orthonormal basis the wavelet coefficients d_{jk} are the inner products of $f(x)$ and $\psi_{jk}(x)$. The empirical coefficients $\hat{d}_{jk}$ and $\hat{c}_{jk}$ are simply method-of-moments estimators. Similarly, Delyon and Juditsky (1995), Hall and Patil (1995), Pinheiro and Vidakovic (1997) and Vannucci (1995) considered wavelet based density estimation from a classical and data analytic perspective. Instead, in this chapter we propose density estimation as formal posterior inference on the wavelet coefficients, using the exact likelihood $x_i \sim f_\theta(x)$. Using the correct likelihood instead of smoothing empirical coefficients comes at the cost of loosing posterior independence for the d_{jk}.

Wavelet based methods have been applied to the problem of spectral density estimation by Lumeau et al. (1992), Gao (1997) and Moulin (1994). The periodogram $I(\omega)$ provides an unbiased estimate of the spectral density $f(\omega)$ for a weakly stationary time series. But because of sampling variances which are not decreasing with sample size it is important to introduce some notion of smoothness for the unknown spectral density. Popular methods for spectral density estimation achieve this by data analytic smoothing of the raw periodogram. In this chapter we propose an approach based on a wavelet representation (1) of the spectral density $f_\theta(\omega)$ and posterior inference on the unknown coefficients d_{jk}. Again, the problem does not fit into the usual conjugate *a posteriori* independent framework.

For all three problems we introduce a hierarchical mixture prior model $p(\theta)$ on the wavelet coefficients d_{jk}. The model allows for posterior thresholding, i.e., vanishing coefficients $d_{jk} = 0$, and posterior shrinkage of non-zero coefficients. The prior model includes separate variance parameters at different level of detail j, implying differential shrinkage for different j. We discuss a Markov chain Monte Carlo posterior simulation scheme which implements inference in all three applications.

13.2   The Prior Model

We put a prior probability model on the random function $f_\theta(x)$ in (1) by defining a probability model for the coefficients d_{jk} and c_{0k}. To allow wavelet coefficient thresholding the model needs to include a positive point-mass at zero. Thus we use as a marginal prior distribution for each d_{jk} a mixture model with a point-mass at zero and a continuous distribution for non-zero values. A convenient notation for such a mixture is to introduce indicators $s_{jk} \in \{0, 1\}$, and replace the coefficients d_{jk} in (1) by the product

$s_{jk} \cdot d_{jk}$, i.e., let d_{jk} denote the coefficient if it is non-zero only.

$$f(x) = \sum_{k \in Z} c_{0k} \phi_{0k}(x) + \sum_{j \geq 0} \sum_{k \in Z} s_{jk} d_{jk} \, \psi_{jk}(x), \qquad (2)$$

An important rationale for using wavelet bases in place of alternative function bases is the fact that wavelet decompositions allow parsimonious representations, i.e., using only few non-zero coefficients allows close approximation of many functions. Thus our model $p(\theta)$ puts progressively smaller prior probability π_j at $d_{jk} = 1$ for higher levels of details.

$$\pi_j = Pr(s_{jk} = 1) = \alpha^j. \qquad (3)$$

Given d_{jk} is not vanishing, we assume a normal prior with level-dependent variance

$$p(d_{jk}|s_{jk} = 1) = N(0, \tau r_j) \text{ with } r_j = 2^{-j}. \qquad (4)$$

The scale factor r_j compensates for the factor $2^{j/2}$ in the definition of $\psi_{jk}(x)$. In many discussions of wavelet shrinkage and thresholding this extra scale factor r_j does not appear (for example, Vidakovic 1998). This is because instead of explicitly modeling π_j, decreasing prior probabilities for non-zero (or large) wavelet coefficients at higher level of detail are implemented implicitly in the following sense. Shrinkage rules can be thought of as posterior mean functions. For example, in a regression problem a shrinkage rule gives the posterior mean for d_{jk} as a function of the corresponding coefficient $\hat{d}_{jk}$ in the discrete wavelet transform of the data (the empirical wavelet coefficient). The typical nonlinear shrinkage rule function (compare Vidakovic 1998, Figure 4) shrinks small coefficients significantly stronger than large values, and because of the scaling factor $2^{j/2}$ in the definition of ψ_{jk}, the (true) fine detail coefficients tend to be small, i.e., are *a posteriori* shrunk stronger than low level coefficients. Rescaling with a factor like r_j would shrink all coefficients equally and defy this. This explains why no factor like r_j is used in the usual thresholding rules. In contrast, we explicitly specify geometrically decreasing prior probabilities π_j for non-zero coefficient. Together, the scaling factors r_j and the probabilities $\pi_j = \alpha^j$ specify prior information on the rate of decay in the magnitudes of the wavelet coefficients and thus indirectly on the smoothness of f. More general choices for π_j and r_j are possible. Abramovich, Sapatinas and Silverman (1998) adopt $\pi_j = \min[1, C_2(1/2)^{\beta j}]$ and $\tau_j^2 = C_1(1/2)^{\alpha j}$, with C_1 and C_2 determined in an empirical Bayes fashion. Chipman, Kolaczyk and McCulloch (1997) take $\pi_j \propto f_j(1/2)^j$ where f_j is the fraction of empirical wavelet coefficients greater than a certain cutoff.

We complete the model with a prior on c_{0k} and hyperpriors for α and τ:

$$c_{0k} \sim N(0, \tau r_0), \quad \alpha \sim Beta(a, b), \quad 1/\tau \sim Ga(a_\tau, b_\tau). \qquad (5)$$

Equations (3) – (5) together define the prior probability model

$$p(\alpha, \tau, c_{0k}, s_{jk}, d_{jk}) =$$
$$p(\alpha, \tau) \cdot \prod_k p(c_{0k}|\tau) \cdot \prod_{j,k} p(s_{jk}|\alpha) \cdot \prod_{s_{jk}=1} p(d_{jk}|s_{jk} = 1, \tau),$$

where the indices for d_{jk} include only the coefficients with $s_{jk} = 1$.

13.3 Estimation

The applications related to regression, density and spectral density estimation require entirely different likelihood specifications. Still, implementation of posterior simulation is very similar in all three cases. In this section we discuss issues which are common to all three applications.

13.3.1 Pseudo Prior

To implement posterior inference in the proposed models we use Markov chain Monte Carlo (MCMC) simulation. See, for example, Tierney (1994) or Smith and Roberts (1993) for a discussion of MCMC posterior simulation.

The proposed prior probability model for the wavelet coefficients includes a variable dimension parameter vector. Depending on the indicators s_{jk} a variable number of coefficients d_{jk} are included. Green (1995) and Carlin and Chib (1995) proposed MCMC variations to address such variable dimension parameter problems. These methods are known as reversible jump MCMC and pseudo prior MCMC, respectively. Both introduce a mechanism to propose a value for additional parameters if in the course of the simulation an augmentation of the parameter vector is proposed. In the following discussion we will follow Carlin and Chib (1995), but note that there is an obvious corresponding reversible jump algorithm. In fact, Dellaportas, Forster and Ntzoufras (1997) show how a pseudo prior algorithm can always be converted to a reversible jump algorithm, and vice versa.

Carlin and Chib (1995) propose to artificially augment the probability model with priors ("pseudo priors") on parameters which are not currently included in the model ("latent parameters"). In the context of model (3) – (5) this requires to add a pseudo prior $h(d_{jk}) = (d_{jk}|s_{jk} = 0)$ for those coefficients d_{jk} which are currently not included in the likelihood. We will discuss below a specific choice for h. The pseudo prior augments the probability model to

$$p(\alpha, \tau) \cdot \prod_k p(c_{0k}|\tau) \cdot \prod_{jk} p(s_{jk}|\alpha) \cdot \prod_{s_{jk}=1} p(d_{jk}|s_{jk} = 1, \tau) \cdot \prod_{s_{jk}=0} \underbrace{p(d_{jk}|s_{jk} = 0)}_{h(d_{jk})}.$$

In the artificially augmented probability model all parameters are always included. The problem with the varying dimension parameter vector is removed. In choosing the pseudo prior it is desirable to specify h such that we generate values for the latent parameters which favor frequent transitions between the different subspaces, i.e., frequent changes of s_{jk}. This is achieved by choosing for $h(d_{jk})$ an approximation of $p(d_{jk}|s_{jk} = 1, data)$. In the current implementation we choose

$$p(d_{jk}|s_{jk} = 0) = N(\hat{d}_{jk}, \hat{\sigma}_{jk}), \tag{6}$$

with $\hat{d}_{jk}$ a rough preliminary estimate of d_{jk}, for example the corresponding coefficient in a wavelet decomposition of an initial estimate $\hat{f}(x)$ for the unknown function; and $\hat{\sigma}_{jk}$ is some initial guess of the marginal posterior variance of d_{jk}, for example τr_j, using the prior mean $1/\tau = a_\tau/b_\tau$. The choice of $\hat{f}(x)$ depends on the application. See Section 13.3.3 for details. In all three applications, after an initial burn-in period of T_0 iterations we replace $\hat{d}_{jk}$ and $\hat{\sigma}_{jk}$ by the ergodic mean and variance of the imputed values for d_{jk} over the initial burn-in period.

A good choice of the pseudo prior h is important for an efficient implementation of the MCMC simulation. Still, note that any choice, subject to some technical constraints only, would result in a MCMC scheme with exactly the desired posterior distribution as asymptotic distribution. Choice of the pseudo prior alters the simulated Markov chain, but not the asymptotic distribution.

13.3.2 An MCMC Posterior Simulation Scheme

We describe an MCMC simulation scheme to implement posterior inference in model (3) – (5). Starting with some initial values for s_{jk}, d_{jk}, c_{0k}, α and τ we proceed by updating each of these parameters, one at a time. See the next section for comments on implementation details and initial values. In the following discussion we will write $\theta = (s_{jk}, d_{jk}, c_{0k}, \alpha, \tau)$ to denote the parameter vector, Y to denote the observed data, and $p(Y|\theta)$ to generically indicate the likelihood. The specific form of $p(Y|\theta)$ depends on the application, and will be noted later. We will denote with $\theta(-x)$ the parameter vector without parameter x.

1a. To update d_{jk} for latent coefficients, i.e., coefficients with $s_{jk} = 0$, generate $d_{jk} \sim h(d_{jk})$. Note that by definition of the pseudo prior $h(d_{jk}) = p[d_{jk}|Y, \theta(-d_{jk})]$ is the complete conditional posterior distribution for d_{jk}.

1b. Updating d_{jk} when $s_{jk} = 1$, is done by a Metropolis-Hastings step. Generate a proposal $\tilde{d} \sim g_d(\tilde{d}_{jk}|d_{jk})$. Use, for example, $g_d(\tilde{d}_{jk}|d_{jk}) = N(d_{jk}, 0.25\,\hat{\sigma}_{jk})$, where $\hat{\sigma}_{jk}$ is some rough estimate of the posterior standard deviation of d_{jk}. Compute

$$a(d_{jk}, \tilde{d}_{jk}) = \min \left\{ 1, p(Y|\tilde{\theta})p(\tilde{d}_{jk})/[p(Y|\theta)p(d_{jk})] \right\},$$

where $\tilde{\theta}$ is the parameter vector θ with d_{jk} replaced by $\tilde{d}$, and $p(d_{jk})$ is the p.d.f. of the normal prior distribution given in (4). With probability $a(d_{jk}, \tilde{d}_{jk})$ replace d_{jk} by $\tilde{d}_{jk}$; otherwise keep d_{jk} unchanged.

2. To update the indicators s_{jk} we replace each s_{jk} by a draw from the complete conditional posterior distribution $p[s_{jk}|Y, \theta(-s_{jk})]$. Denote with θ_0 and θ_1 the parameter vector with s_{jk} replaced by 0 and 1, respectively. Compute $p_0 = p(Y|\theta_0) \cdot (1 - \alpha^j)h(d_{jk})$ and $p_1 = p(Y|\theta_1) \cdot \alpha^j p(d_{jk}|s_{jk} = 1)$. With probability $p_1/(p_0 + p_1)$ set $s_{jk} = 1$; otherwise $s_{jk} = 0$.

3. To update c_{00}, generate a proposal $\tilde{c}_{00} \sim g_c(\tilde{c}_{00}|c_{00})$. Use, for example, $g_c(\tilde{c}_{00}|c_{00}) = N(c_{00}, 0.25 \hat{\rho}_{00})$, where $\hat{\rho}_{00}$ is some rough estimate of the posterior standard deviation of c_{00}. Analogously to step 1b, compute an acceptance probability a and replace c_{00} with probability a.

   If included in the model, c_{0k}, $k \neq 0$, is updated similarly. In our implementation we only have c_{0k}, $k = 0$, because of the constraint to $[0, 1]$ discussed in Section 13.3.3.

4. Generate $\tilde{\alpha} \sim g_\alpha(\tilde{\alpha}|\alpha) = N(\alpha, 0.1^2)$ and compute

$$a(\alpha, \tilde{\alpha}) = \min \left\{ 1, \prod_{jk} [\tilde{\alpha}^j/\alpha^j]^{s_{jk}} [(1 - \tilde{\alpha}^j)/(1 - \alpha^j)]^{1 - s_{jk}} \right\}.$$

   With probability $a(\alpha, \tilde{\alpha})$ replace α by $\tilde{\alpha}$. Otherwise keep α unchanged. See below for comments about $g_\alpha(\cdot)$.

5. Update τ. Resample τ from the complete inverse Gamma conditional posterior $p(\tau|d, Y)$, where $d = \{d_{jk} : s_{jk} = 1\}$. I.e., when resampling τ we marginalize over all latent d_{jk}.

Steps 1a, 2 and 5 are Gibbs sampling steps, i.e., the parameters are replaced by draws from the complete conditional posterior distributions. Steps 1b, 3 and 4 are Metropolis-Hastings steps (Tierney 1994) with random walk proposal distributions. Theoretically, the probing distributions $g_d(\cdot)$, $g_c(\cdot)$ and $g_\alpha(\cdot)$ can be any distributions which are symmetric in the arguments, i.e., $g_d(\tilde{d}|d) = g_d(d|\tilde{d})$, etc. For a practical implementation, g should be chosen such that the acceptance probabilities a are neither close to zero, nor close to one. In the implementations which we used in this paper, we used $g_d(\tilde{d}_{jk}|d_{jk}) = N(d_{jk}|0.25 \hat{\sigma}_{jk})$ with $\hat{\sigma}_{jk}$ as described for the variance of the pseudo prior (6). Similarly for the scale parameter $\hat{\rho}_{00}$ of the probing distribution $g_c(\tilde{c}_{00}|c_{00})$ in Step 3. As probing distribution $g_\alpha(\tilde{\alpha}|\alpha)$ for α we used a normal distribution centered at the current value α with a standard deviation of 0.1. Also, during an initial burn-in of the first T_0 iterations we keep $s_{jk} = 1$ and do not update s_{jk}.

13.3.3   Initialization and Implementation Details

Initialization. In all three applications we compute an initial estimate $\hat{f}(x)$ of the unknown function $f(x)$ using exploratory data techniques. This estimate $\hat{f}(x)$ is used to fix the pseudo prior, as described in Section 13.3.1, and to initialize the coefficients d_{jk}. For the regression application we use a smoothing spline (compare Figures 1 and 3). For the density estimation problem we use the square root of a kernel density estimate (compare Figure 13.4). For the spectral density estimation $\hat{f}(\omega)$ is the smoothed raw periodogram, using a sequence of modified Daniell smoothers (see Figure 7). Note that a good initialization, and a good choice of the pseudo prior change the initial state and possibly the transition probabilities in the simulated Markov chain, but do not alter the asymptotic distribution. Any initialization, and any pseudo prior, subject only to some technical constraints, will result in the desired posterior distribution as ergodic distribution.

As usual in applications of wavelet decomposition, for a practical implementation we equate the scaling coefficients at some high level of detail, say $J^* = 10$, with the random function evaluated on a grid, i.e.,

$$c_{J^* k} \equiv f_\theta \left(k/2^{J^*} \right). \tag{7}$$

This is justified by the fact that $\phi_{J^* k}(x)$ integrates to 1, and has for high enough J^* very narrow effective support, i.e., can to the extent of the plotter resolution be thought of as constant over the interval $[k/2^{J^*}, (k+1)/2^{J^*}]$. Note that the relevant constant would of course be $\gamma = 2^{J^*/2}$, i.e., we really should include a factor γ in (7). However, as a constant factor across all k we can ignore it without affecting the final inference.

Likelihood Evaluation. In all three applications the likelihood function requires evaluation of $f_\theta(x)$. A computationally very efficient way of computing $f_\theta(x)$ for a given vector of wavelet coefficients is the pyramid algorithm for wavelet decomposition and reconstruction. See, for example, Vidakovic and Müller (1999) in this volume for an explanation of the pyramid scheme.

Constraint to $[0,1]$. For reasons of programming efficiency we constrain the data in all three applications essentially to the interval $[0,1]$. Additionally, in the application to regression and spectral density estimation we augment the data by a symmetric mirror image to avoid boundary problems. Specifically, for the regression problem we constrain the covariates x to the interval $[0, 0.5]$, rescaling if necessary. Then we augment the data $(x_i, y_i; \ i = 1, \dots, n)$ with a symmetric mirror image $(x_{n+i} = 1 - x_i, y_{n+i} = y_i; \ i = 1, \dots, n)$. In the spectral density estimation the Fourier frequencies ν_t are constrained to $[0, 0.5]$. Again, we augment the original data

$(\nu_l, y_l; \ l = 0, \ldots, [T/2])$ by a symmetric mirror image. Here y_l is the log periodogram as defined in Section 13.6, and ν_l are the Fourier frequencies. For the density estimation problem we constrain x to $0.1 \leq x \leq 0.9$, rescaling the data if necessary. In all three applications, after these modifications we extend the support of $f(x)$ to the entire real line by defining $f(x + k) = f(x)$, for $x \in [0,1]$ and $k \in Z$.

Hyperparameters. In all three application we fix the hyperparameters $a = 10$, $b = 20$ and $a_\tau = 1$. The scale parameter b_τ is fixed at $b_\tau = 1$ for the density estimation and spectral density estimation, $b_\tau = 0.01$ for Example 1, and $b_\tau = 100$ for Example 2. Also, in the prior we constrain $\alpha < 0.7$. This is necessary for the density estimation to avoid unbounded likelihood values. The maximum level of detail was chosen as $J = 5$.

Simulation Length. We discarded the first $T_0 = 100$ iterations as burn-in period, and then simulated $T = 5000$ iterations of steps 1 through 5. For each j, k we repeated Step 1b three times. Also, in Step 1b we used two different probing distributions. Every third time we used in place of $g_d(\tilde{d}|d)$ an alternative probing distribution $g_2(\tilde{d}) = N(\hat{d}_{jk}, \hat{\sigma}_{jk})$, where $\hat{d}_{jk}$ is the approximation to the posterior mean $E(d_{jk}|y)$ based on the first T_0 burn-in iterations. Note that g_2 is independent of the currently imputed value d_{jk}. Tierney (1994) refers to such proposals as "independence chain" proposals. The relevant acceptance probability is $a(d_{jk}, \tilde{d}_{jk}) =$ $\min \left\{ 1, p(Y|\tilde{\theta})p(\tilde{d}_{jk})g_2(d_{jk})/[p(Y|\theta)p(d_{jk})g_2(\tilde{d}_{jk})] \right\}.$

13.4 Non-equally Spaced Regression

Consider a non-linear regression model $y_i = f_\theta(x_i) + \epsilon_i$, $i = 1, \ldots, n$, with independent normal error $\epsilon_i \sim N(0, \sigma^2)$. To define a prior probability model on the unknown mean function $f_\theta(\cdot)$ we parameterize f with it's wavelet decomposition and use the model (3) – (5), together with an additional hyperprior $1/\sigma^2 \sim Ga(a_s/2, b_s/2)$. Let $N(x; m, s)$ indicate the normal p.d.f. with moments (m, s), evaluated at x. The likelihood $p(Y|\theta)$ is the usual regression likelihood

$$p(Y|\theta) = \prod N[y_i; f_\theta(x_i), \sigma^2].$$

Using the discussed MCMC scheme we obtain posterior estimates for all wavelet coefficients d_{jk}, and thus for $f_\theta(x)$. Examples 1 and 2 illustrate this. If the data were equally spaced, then the MCMC would significantly simplify because of the orthonormality of the wavelet transformation which implies that the coefficients d_{jk} are *a posteriori* independent given α and τ. For general, non-equally spaced data this is not necessarily the case. Still, practical convergence of the MCMC simulation is very fast. Probably this is because of near posterior independence of the d_{jk}. In Examples 1 and

2, after the first 100 passes through the scheme given in Section 13.3.2 the estimated posterior mean $\bar{f}(x) = E[f_\theta(x)|Y]$ changes only little.

Example 1. Donoho and Johnstone (1994) consider a battery of test functions to evaluate performance of wavelet shrinkage methods. One of them is the Doppler function $f(x) = \sqrt{x(1-x)} \sin[(2.1\pi)/(x+0.05)]$, for $0 \le x \le 1$. We generated $n = 100$ observations with $y_i = f(x_i) + \epsilon_i$, with noise $\epsilon_i \sim N(0, 0.05^2)$ and unequally spaced x_i. The simulated data, together with the estimated mean function $\bar{f}(x) = E[f_\theta(x)|Y]$ is shown in Figure 1. Figure 2 shows the posterior distributions for some of the wavelet coefficients d_{jk}.

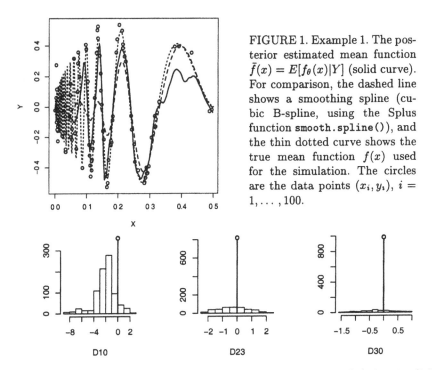

FIGURE 1. Example 1. The posterior estimated mean function $\bar{f}(x) = E[f_\theta(x)|Y]$ (solid curve). For comparison, the dashed line shows a smoothing spline (cubic B-spline, using the Splus function `smooth.spline()`), and the thin dotted curve shows the true mean function $f(x)$ used for the simulation. The circles are the data points (x_i, y_i), $i = 1, \ldots, 100$.

FIGURE 2. Example 1. Marginal posterior distribution $p(d_{jk}|Y)$ for d_{10} (left panel), d_{23} (center panel), and d_{30} (right panel). The histograms show the continuous part $p(d_{jk}|Y, s_{jk} \ne 0)$. The point mass at zero shows the posterior probability $P(s_{jk} = 0|Y)$, i.e., the point mass at zero for the wavelet coefficient $s_{jk} \cdot d_{jk}$. The scale on the y-axis are absolute frequencies.

Example 2. Azzalini and Bowman (1990) analyzed a data set concerning eruptions of the Old Faithful geyser in Yellowstone National Park in Wyoming. The data set records eruption durations and intervals between subsequent eruptions, collected continuously from August 1st until August 15th, 1985. Of the original 299 observations we removed 78 observations which were taken at night and only recorded durations as "short", "medium" or "long". Figure 13.4 plots intervals between eruptions y_i versus duration of the preceding eruption x_i and shows a posterior estimate for the mean response curve $E[y|x, Y]$ as a function of x.

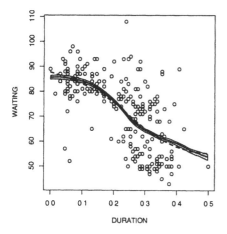

FIGURE 3. Example 2. The posterior estimated mean function $\bar{f}(x) = E[f_\theta(x)|Y]$ (solid curve). The narrow gray shaded band show pointwise 50% HPD intervals for $f(x)$. For comparison, the dashed line shows a smoothing spline (cubic B-spline, using the Splus function `smooth.spline()`). The circles are the data points (x_i, y_i), $i = 1, \ldots, n$.

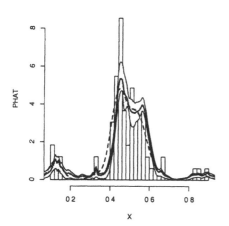

FIGURE 4. Example 3. The figure shows the estimated p.d.f. $\hat{p}(x) = \int f_\theta^2(x)/K dp(\theta|y)$. For comparison, the dashed line plots a conventional kernel density estimate for the same data. The grey-shaded band shows pointwise 50% HPD (highest posterior density) regions. The histogram indicates the data.

13.5   Density Estimation

Assume x_i, $i = 1, \ldots, n$, are independent draws from an unknown p.d.f. $p(x)$. We consider a wavelet decomposition of the (unnormalized) square root p.d.f., i.e., $p(x) = [f_\theta(x)]^2/K$, where $K = \int [f_\theta(x)]^2 dx = \sum c_{0k}^2 + \sum s_{jk} d_{jk}^2$. Without the square root transformation the non-negativity constraint for the p.d.f. would complicate inference and the evaluation of K would become an analytically intractable integration problem. The relevant likelihood becomes

$$p(Y|\theta) = \prod_{i=1}^{n} [f_\theta(x_i)]^2 / K^n.$$

Again we use the MCMC scheme from Section 13.3.2 to obtain posterior estimates $\bar{f}(x) = E[f_\theta(x)|Y]$. Example 3 reports an example discussed in Müller and Vidakovic (1998).

Example 3. We illustrate density estimation with the galaxy data set from Roeder (1990). The data is rescaled to the interval $[0, 1]$.

13.6   Spectral Density Estimation

In Vidakovic and Müller (1999) we consider the problem of estimating the spectral density

$$p(\omega) = 1/(2\pi) \sum_{t=-\infty}^{\infty} \gamma(t) \exp(-it\omega)$$

for a weakly stationary time series. Here $\gamma(t)$ is the auto-covariance function. Assume we observe data x_t, $t = 0, \ldots, T - 1$. Let $\omega_l = 2\pi l/T$, $l = 0, 1, \ldots, [T/2]$ denote the Fourier frequencies. From the data we can compute the periodogram

$$I(\omega_j) = 1/(2\pi T) \left| \sum_{t=0}^{T-1} x_t \exp(-it\omega_i) \right|^2$$

which allows inference about the unknown spectral density. Let $y_l = \log I(\omega_l) + \gamma$, where $\gamma = 0.5771$ is the Euler-gamma constant, and let $f(\omega) = \log p(\omega)$ denote the log spectral density. Then $y_l = f(\omega_l) + \epsilon_l$, with ϵ_l approximately independently distributed with p.d.f.

$$g(\epsilon) = \delta \exp[\epsilon - \delta \exp(\epsilon)], \ \delta = e^{-\gamma}$$

(Wahba, 1980). Figure 5 shows the density $g(\epsilon)$.

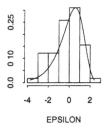

FIGURE 5. Example 4.
The histogram shows the empirical distribution of $\epsilon_t = [\log I(\omega_l) + \gamma] - \log \hat{p}(\omega_l)$. Superimposed as solid line is the p.d.f. $g(\epsilon) = \delta \ \exp[\epsilon - \delta \ \exp(\epsilon)]$.

See Vidakovic and Müller (1999) for more discussion of the spectral density estimation problem.

We parameterize the unknown function $f(\omega)$ by the wavelet decomposition $f_\theta(\omega)$ and use the prior probability model (3) – (5). Formally, we face a very similar inference problem as in the regression and density estimation problems described earlier. The only change is a different likelihood

$$p(Y|\theta) = \prod_{l=0}^{[T/2]} g[y_l - f_\theta(\omega_l)].$$

We proceed as before, and simulate the MCMC scheme described in Section 13.3.2. The desired spectral density estimate is obtained as posterior mean $\bar{f}(\omega) = E[f_\theta(\omega)|Y]$.

Example 4. We illustrate the model by analyzing data giving the annual number of lynx trappings in the Mackenzie River District of North-West Canada for the period 1821 to 1934 (Priestley 1988).

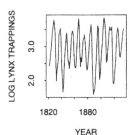

FIGURE 6. Example 4.
Lynx trappings in the Mackenzie River District of North-West Canada for the period 1821 to 1934. The data are transformed by logarithms.

Figure 13.6 shows the time series, on a log scale. Figure 7 plots the estimated spectral density.

Priestley (1988) observes an asymmetry in the behavior of the series. The time spent on the rising side of a period (i.e., rising from "trough" to "peak") seems slightly longer than the time spent on the falling side (i.e. from "peak" to "trough"). The secondary mode in Figure 7 confirms this observation of asymmetry.

13.7   Conclusion

We proposed a hierarchical mixture model for wavelet coefficients and an MCMC scheme using pseudo priors to implement inference in problems without posterior independence, including non-equally spaced regression, density and spectral density estimation. For regression and spectral density estimation convergence is very fast. The additional effort required to use the correct likelihood instead of an approximation guaranteeing *a posteriori* independence is minimal. For density estimation, however, posterior dependence can be considerable and practical convergence requires long

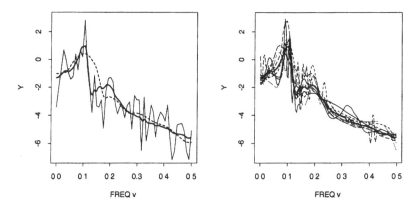

FIGURE 7. Example 4. The left panel shows the raw periodogram, on a log scale and with the Euler-gamma constant added, $y_t = \log I(\omega_t) + \gamma$ (thin line); the posterior estimate $E[\log p(\omega)|Y] = E[f_\theta(\omega)|Y]$ (thick solid line), and for comparison a spectral density estimate using repeated smoothing of the raw periodogram with a sequence of modified Daniell smoothers (dashed line). The right panel shows posterior draws of $f_\theta \sim p(f_\theta|Y)$ as thin lines around the posterior mean curve.

simulation runs. We only recommend using it if the full posterior probability model is important in a given application.

The proposed MCMC scheme is very general. Although the three discussed applications require entirely different likelihood functions, only minor modifications were needed in the program. Note that the discussion in Section 13.3.3 did not hinge upon any assumptions on the likelihood beyond assuming that it can be pointwise evaluated. Thus the proposed model (3) – (5) and posterior simulation could be used for any other problem where lack of posterior independence hinders other schemes. For example, estimation of hazard rate functions as discussed in Anoniadis, Grégoire and Nason (1997) could be considered.

References

Abramovich, F., Sapatinas, T. and Silverman, B.W. (1998). Wavelet thresholding via a Bayesian approach, *Journal of the Royal Statistical Society B*, 60(3), 725–749.

Antoniadis, A.,Grégoire, G. and McKeague, I. (1994), Wavelet methods for curve estimation, *Journal of the American Statistical Association*, 89, 1340–1353.

Antoniadis, A., Grégoire, G, and Nason, G.P. (1997), Density and hazard rate estimation for right censored data using wavelet methods, *J. Roy. Statist. Soc.*, Series B, to appear.

Azzalini, A. and Bowman, A. W. (1990). A look at some data on the Old Faithful geyser. *Applied Statistics* 39, 357–365.

Cai, T. and Brown, L. (1997) Wavelet Shrinkage for Nonequispaced Samples, Technical Report, Purdue University.

Carlin, B.P. and Chib, S. (1995), Bayesian model choice via Markov chain Monte Carlo, *Journal of the Royal Statistical Society, Series B*, 57, 473-484.

Chipman, H., Kolaczyk, E., and McCulloch, R. (1997), Adaptive Bayesian Wavelet Shrinkage, *J. of the American Statistical Association*, 92, 440.

Clyde, M., Parmigiani, G., and Vidakovic, B. (1998), Multiple Shrinkage and Subset Selection in Wavelets, *Biometrika*, 85, 391–402.

Dellaportas, P., Forster, J.J., and Ntzoufras, I. (1997), On Bayesian Model and variable selection using MCMC, Technical Report, Athens University of Economics and Business.

Delyon, B. and Juditsky, A. (1993), Wavelet Estimators, Global Error Measures: Revisited, *Publication interne 782, IRISA-INRIA.*

Delyon, B. and Juditsky, A. (1995), Estimating Wavelet Coefficients, in *Wavelets and Statistics*, A. Antoniadis and G. Oppenheim (eds.), pp. 151–168, New York: Springer-Verlag.

Donoho, D.L. and Johnstone, I.M. (1994), Ideal spatial adaptation by wavelet shrinkage, *Biometrika*, 81 (3), 425–455.

Donoho, D., Johnstone, I., Kerkyacharian, G. and Picard, D. (1995), Wavelet Shrinkage: Asymptopia? (with discussion), *Journal of the Royal Statistical Society B*, 57, 301–369.

Donoho, D., Johnstone, I., Kerkyacharian, G. and Picard, D. (1996), Density Estimation by Wavelet Thresholding, *Annals of Statistics*, 24, 508–539.

Gao, H.-Y. (1997), Choice of thresholds for wavelet shrinkage estimate of the spectrum, *Journal of Time Series Analysis*, 18(3).

Green, P. (1995). Reversible jump Markov chain Monte Carlo computation and Bayesian model determination. *Biometrika*, 82, 711-732.

Hall, P. and Patil, P. (1995). Formulae for mean integrated squared error of nonlinear wavelet-based density estimators. *The Annals of Statistics*, 23, 905–928.

Hall, P. and Turlach, B.A. (1997), Interpolation methods for nonlinear wavelet regression with irregularly spaced design, *The Annals of Statistics*, 25, 1912–1925.

Lumeau, B., Pesquet, J.C., Bercher, J.F. and Louveau, L. (1992), Optimization of bias-variance trade-off in nonparametric specral analysis by decomposition into wavelet packets. In *Progress in Wavelet Analysis and Applications*, Y. Meyer and S. Roques (eds.), Tolouse: Editions Frontieres.

Marron, S. (1999), Spectral view of wavelets and nonlinear regression, in *Bayesian Inference in Wavelet Based Models*, P. Müller and B. Vidakovic (eds.), chapter 2, New York: Springer-Verlag..

Moulin, P. (1994), Wavelet thresholding techniques for power spectrum estimation, *IEEE Transactions on Signal Processing*, 42(11), 3126-3136.

Müller, P., and Vidakovic, B. (1999). "Bayesian Inference with Wavelets: Density Estimation," *Journal of Computational and Graphical Statistics*, 7, 456-468.

Pinheiro, A. and Vidakovic, B. (1997), Estimating the square root of a density via compactly supported wavelets, *Computational Statistics & Data Analysis*, 25, 399–415.

Priestley, M.B. (1988), *Non-linear and Non-stationary Time Series Analysis* Academic Press, London.

Roeder, K. (1990), "Density estimation with confidence sets exemplified by superclusters and voids in the galaxies," *Journal of the American Statistical Association*, 85, 617–624.

Sardy, S., Percival, D., Bruce, A., Gao, H.-Y., and Stuetzle, W. (1997), Wavelet DeNoising for Unequally Spaced Data, Technical Report, StatSci Division of MathSoft, Inc.

Smith, A.F.M. and Roberts, G.O. (1993), Bayesian computation via the Gibbs sampler and related Markov chain Monte Carlo methods (with discussion), *Journal of the Royal Statistical Society B*, 55, 3–23.

Tierney, L. (1994), Markov chains for exploring posterior distributions, *Annals of Statistics*, 22, 1701–1728.

Vannucci, M. (1995), "Nonparameteric Density Estimation Using Wavelets," Discussion Paper 95-27, Duke University.

Vidakovic, B. (1998), Nonlinear wavelet shrinkage with Bayes rules and Bayes Factors, *Journal of the American Statistical Association*, 93, 173–179.

Vidakovic, B. and Müller, P. (1999), Introduction to Wavelets, in *Bayesian Inference in Wavelet Based Models*, P. Müller and B. Vidakovic (eds.), chapter 1, New York: Springer-Verlag..

Vidakovic, B. and Müller, P. (1999), Bayesian wavelet estiamtion of a spectral density, Technical Report, Duke University.

Wahba, G. (1980), Automatic smoothing of the log periodogram, *Journal of the American Statistical Association*, 75, 369, 122-132.

Yau, P. and Kohn, R. (1999), Wavelet nonparametric regression using basis averaging, in *Bayesian Inference in Wavelet Based Models*, P. Müller and B. Vidakovic (eds.), chapter 7, New York: Springer-Verlag..

14

Empirical Bayesian Spatial Prediction Using Wavelets

Hsin-Cheng Huang and Noel Cressie

ABSTRACT Wavelet shrinkage methods, introduced by Donoho and Johnstone (1994, 1995, 1998) and Donoho *et al.* (1995), are a powerful way to carry out signal denoising, especially when the underlying signal has a sparse wavelet representation. Wavelet shrinkage based on the Bayesian approach involves specifying a prior distribution for the wavelet coefficients. In this chapter, we consider a Gaussian prior with *nonzero* means for wavelet coefficients, which is different from other priors used in the literature. An empirical Bayes approach is taken by estimating the mean parameters using Q-Q plots, and the hyperparameters of the prior covariance are estimated by a pseudo maximum likelihood method. A simulation study shows that our empirical Bayesian spatial prediction approach outperforms the well known VisuShrink and SureShrink methods for recovering a wide variety of signals.

14.1 Introduction

Wavelets in $\mathbb{R}$ are functions with varying scales and locations obtained by dilating and translating a basic function (mother wavelet) ψ that has (almost) bounded support. For certain functions $\psi \in L^2(\mathbb{R})$, the family

$$\psi_{j,k}(x) \equiv 2^{j/2}\psi(2^j x - k); \quad j,k \in \mathbb{Z},$$

constitutes an (orthonormal) basis of $L^2(\mathbb{R})$ (Daubechies, 1992). Associated with each mother wavelet ψ is a scaling function ϕ that together yield a multiresolution analysis in $L^2(\mathbb{R})$. That is, after choosing an initial scale J_0, any $f \in L^2(\mathbb{R})$ can be expanded as:

$$f = \sum_{k \in \mathbb{Z}} c_{J_0,k}\phi_{J_0,k} + \sum_{j=J_0}^{\infty} \sum_{k \in \mathbb{Z}} d_{j,k}\psi_{j,k}.$$

Spatial wavelets in $\mathbb{R}^d$ are an easy generalization of one-dimensional wavelets (Mallat, 1989). For the purpose of illustration, we consider the case $d = 2$, where wavelet analysis of two-dimensional images is an important application. A two-dimensional scaling function can be defined in the following separable form:

$$\Phi(x,y) = \phi(x)\phi(y); \quad x,y \in \mathbb{R},$$

and there are three wavelet functions given by,

$$\Psi^{(1)}(x,y) = \phi(x)\psi(y), \ \ \Psi^{(2)}(x,y) = \psi(x)\phi(y), \ \ \Psi^{(3)}(x,y) = \psi(x)\psi(y).$$

For $j, k_1, k_2 \in \mathbb{Z}$, write

$$\begin{aligned}
\Phi_{j,k_1,k_2}(x,y) &\equiv 2^j \Phi(2^j x - k_1, 2^j y - k_2), \\
\Psi^{(m)}_{j,k_1,k_2}(x,y) &\equiv 2^j \Psi^{(m)}(2^j x - k_1, 2^j y - k_2); \quad m = 1,2,3.
\end{aligned}$$

Then any function $g \in L^2(\mathbb{R}^2)$ can be expanded as

$$
\begin{aligned}
&g(x,y) \\
&= \sum_{k_1,k_2} c_{k_1,k_2} \Phi_{J_0,k_1,k_2}(x,y) + \sum_{j=J_0}^{\infty} \sum_{k_1,k_2} \sum_{m=1}^{3} \left\{ d^{(m)}_{j,k_1,k_2} \Psi^{(m)}_{j,k_1,k_2}(x,y) \right\}.
\end{aligned}
$$

Because of this direct connection between one-dimensional wavelets and spatial wavelets, we shall present most of the methodological development in $\mathbb{R}$. However, in a subsequent section, we do give an application of our wavelet methodology to two-dimensional spatial prediction of an image.

Wavelets have proved to be a powerful way of analyzing complicated functional behavior because in wavelet space, most of the "energy" tends to be concentrated in only a few of the coefficients $\{c_{J_0,k}\}, \{d_{j,k}\}$. It is interesting to look at the statistical properties of wavelet expansions; that is, if $f(\cdot)$ is a random function in $L^2(\mathbb{R})$, what is the law of its wavelet coefficients? We shall formulate this question more specifically in terms of the discrete wavelet transform, which we shall now discuss.

Suppose that we observe $Y(\cdot)$ at a discrete number $n = 2^J$ points; that is, we have data $\boldsymbol{Y} = (Y_1, \ldots, Y_n)$, where $Y_i = Y(t_i)$ and $t_i = i/n; i = 1, \ldots, n$. The discrete wavelet transform matrix $\mathcal{W}_n$ of $\boldsymbol{Y}$ is an orthogonal matrix such that

$$w \equiv ((w^*_{J_0})', w'_{J_0}, \ldots, w'_{J-1})' = \mathcal{W}_n \boldsymbol{Y} \tag{1}$$

is a vector of scaling function coefficients at scale J_0 and wavelet coefficients at scales $J_0, \ldots, J-1$ (Mallat, 1989). Thus, if $\boldsymbol{Y}$ is random, so too is w. In all that is to follow, we shall construct probability models directly for w, although it should be noted that if $Y(\cdot)$ is a stationary process, then $w^*_{J_0}$ and $\{w_j : j = J_0, \ldots, J-1\}$ are also stationary processes, except for some points near the boundary (Cambanis and Houdré, 1995).

We assume the following Bayesian model:

$$
\begin{aligned}
w \,|\, \beta, \sigma^2 &\sim \ Gau(\beta, \sigma^2 I), \tag{2} \\
\beta \,|\, \mu, \theta &\sim \ Gau(\mu, \Sigma(\theta)), \tag{3}
\end{aligned}
$$

where $\Sigma(\boldsymbol{\theta})$ is an $n \times n$ covariance matrix with structure (depending on parameters $\boldsymbol{\theta}$) to be specified. In a like manner to the definition of $\boldsymbol{w}$ in (1), we write $\boldsymbol{\beta} = \left((\boldsymbol{\beta}_{J_0}^*)', \boldsymbol{\beta}'_{J_0}, \ldots, \boldsymbol{\beta}'_{J-1} \right)'$ and $\boldsymbol{\mu} = \left((\boldsymbol{\mu}_{J_0}^*)', \boldsymbol{\mu}'_{J_0}, \ldots, \boldsymbol{\mu}'_{J-1} \right)'$. Notice that there are hyperparameters $\boldsymbol{\mu}, \sigma^2, \boldsymbol{\theta}$ still to be dealt with in the Bayesian model.

A couple of comments are worth making. The first level of the Bayesian model is the so-called data model that incorporates measurement error; indeed, we can write (2) equivalently as,

$$w = \beta + \epsilon, \tag{4}$$

where $\epsilon \sim Gau(\mathbf{0}, \sigma^2 I)$. Hence β is the signal, which we do not observe because it is convolved with the noise ϵ. Our goal is prediction of β, which we assume has a prior distribution given by (3). This prior is different from other priors used in the literature, in that we assume it to have *nonzero* $(\boldsymbol{\mu}'_{J_0}, \ldots, \boldsymbol{\mu}'_{J-1})$. We regard $\boldsymbol{\mu}$ to be a prior parameter to be specified, which represents the large-scale variation in β. Thus, we may write

$$\beta = \mu + \eta, \tag{5}$$

where $\boldsymbol{\mu}$ is deterministic and $\eta \sim Gau(\mathbf{0}, \Sigma(\boldsymbol{\theta}))$ is the stochastic component representing the small-scale variation.

The optimal predictor of β is $E(\beta \,|\, w)$, which we would like to transform back to the original data space. The inverse transform of $\mathcal{W}_n$ is $\mathcal{W}'_n$, since $\mathcal{W}_n$ is an orthogonal matrix. Hence (4) becomes

$$\mathcal{W}'_n w = \mathcal{W}'_n \beta + \mathcal{W}'_n \epsilon,$$

and because $\mathcal{W}'_n \epsilon$ is white-noise measurement error in the data space, $S \equiv \mathcal{W}'_n \beta$ represents the signal that we would like to predict. Because of linearity, the optimal predictor of $S \equiv (S_1, \ldots, S_n)$ is,

$$E(S \,|\, Y) = E(\mathcal{W}'_n \beta \,|\, w) = \mathcal{W}'_n E(\beta \,|\, w).$$

Thus, it is a simple matter to predict optimally the signal S once $E(\beta \,|\, w)$ has been found.

The Gaussian assumptions (1) and (2) make the calculation of

$$\hat{\boldsymbol{\beta}}(\boldsymbol{\mu}, \sigma^2, \boldsymbol{\theta}) \equiv E(\beta \,|\, w)$$

a very simple exercise (Section 3). Most of this chapter is concerned with specification of the hyperparameters $\boldsymbol{\mu}, \sigma^2$, and $\boldsymbol{\theta}$. Our approach is empirical, but we show that it offers improvements over the *a priori* specification of $(\boldsymbol{\mu}'_{J_0}, \ldots, \boldsymbol{\mu}'_{J-1})' = \mathbf{0}$ and previous methods of estimating σ^2. In Section 3, we outline our methodology for estimating $\boldsymbol{\mu}$ based on Q-Q plots and for estimating σ^2 based on the variogram. Section 4 contains a brief discussion of different covariance models for $\Sigma(\theta)$, although in all the applications of this chapter we use a simple model that assumes independence and homoskedasticity within a scale but heteroskedasticity (and independence) across scales. Estimation of θ is also discussed in Section 4. Based on estimates $\hat{\mu}$ (Section 3), $\hat{\sigma}^2$ and $\hat{\theta}$ (Section 4), we use the empirical Bayes spatial predictor,

$$\hat{S} \equiv \mathcal{W}_n' \, \hat{\beta}(\hat{\mu}, \hat{\sigma}^2, \hat{\theta})$$

to make inference on the unobserved signal S. Section 5 contains a small simulation study showing the value of our empirical Bayesian spatial prediction approach applied to a few test functions and an application to a two-dimensional image. Discussion and conclusions are given in Section 6.

14.2   Wavelet Shrinkage

In a series of papers, Donoho and Johnstone (1994, 1995, 1998), and Donoho et al. (1995) developed the wavelet shrinkage method for reconstructing signals from noisy data, where the noise is assumed to be Gaussian white noise. The wavelet shrinkage method proceeds as follows. First, the data Y are transformed using a discrete wavelet transform, yielding the empirical wavelet coefficients w. Next, to suppress the noise, the empirical wavelet coefficients are "shrunk" toward zero based on a shrinkage rule. Usually, wavelet shrinkage is carried out by thresholding the wavelet coefficients; that is, the wavelet coefficients that have an absolute value below a pre-specified threshold are replaced by zero. Finally, the processed empirical wavelet coefficients are transformed back to the original domain using the inverse wavelet transform. In practice, the discrete wavelet transform and its inverse transform can be computed very quickly in only $O(n)$ operations using the pyramid algorithm (Mallat, 1989).

With a properly chosen shrinkage method, Donoho and Johnstone (1994, 1995, 1998), and Donoho et al. (1995) show that the resulting estimate of the unknown function is nearly minimax over a large class of function spaces and for a wide range of loss functions. More importantly, it is computationally fast, and it is automatically adaptive to the smoothness of the corresponding true function, without the need to adjust a "bandwidth" as in the kernel smoothing method.

Of course, the crucial step of this procedure is the choice of a thresholding (or shrinkage) method. A number of approaches have been proposed including minimax (Donoho and Johnstone, 1994, 1995, 1998), cross-validation (Nason, 1995, 1996), hypotheses testing (Abramovich and Benjamini, 1995, 1996; Ogden and Parzen, 1996a, 1996b), and Bayesian methods (Vidakovic, 1998; Clyde et al., 1996, 1998; Chipman et al., 1997; Crouse et al., 1998; Abramovich et al., 1998; Ruggeri and Vidakovic, 1999).

Donoho and Johnstone (1994) proposed the hard-thresholding and soft-thresholding strategies. For a wavelet coefficient $w_{j,k}$ and a threshold λ, the hard-thresholding value is given by

$$T_H(w_{j,k}) = \begin{cases} w_{j,k}; & \text{if } |w_{j,k}| > \lambda, \\ 0; & \text{if } |w_{j,k}| \leq \lambda, \end{cases}$$

and the soft-thresholding value is given by

$$T_S(w_{j,k}) = \begin{cases} w_{j,k} - \lambda; & \text{if } w_{j,k} > \lambda, \\ 0; & \text{if } |w_{j,k}| \leq \lambda, \\ w_{j,k} + \lambda; & \text{if } w_{j,k} < -\lambda. \end{cases}$$

For these thresholding rules, the choice of the threshold parameter λ is important. The VisuShrink method proposed by Donoho and Johnstone (1994) uses the universal threshold $\lambda = \sigma\sqrt{2\log n}$ for all levels. Donoho and Johnstone (1995) also proposed the SureShrink method by selecting the threshold parameter in a level-by-level fashion. For a given resolution level j, the threshold is chosen to minimize Stein's unbiased risk estimate (SURE), provided the wavelet representation at that level is not too sparse; the sparsity condition is given by

$$\frac{1}{2^j} \sum_{k=0}^{2^j-1} \frac{(w_{j,k})^2}{\sigma^2} \leq 1 + \frac{j^{3/2}}{2^{j/2}}.$$

Otherwise, the threshold $\lambda_j = \sigma\sqrt{2\log(2^j)}$ is chosen.

In practice, as suggested by Donoho and Johnstone (1994, 1995), the scaling-function coefficients $w_{J_0}^*$ are not shrunk. Usually the noise parameter σ is unknown, in which case Donoho et al. (1995) proposed a robust estimator, the (standardized) median of absolute deviations (MAD) of wavelet coefficients at the highest resolution:

$$\tilde{\sigma} = \text{MAD}\left(\{w_{J-1,k}\}\right) \equiv \frac{\text{median}\left(|w_{J-1,k} - \text{median}\,(w_{J-1,k})|\right)}{0.6745}. \tag{6}$$

A Bayesian wavelet shrinkage rule is obtained by specifying a certain prior for both $\boldsymbol{\beta}$ and σ^2 based on (4). Vidakovic (1998) assumes that $\{\beta_{j,k}\}$ are independent and identically t-distributed with n degrees of freedom and σ is independent of $\{\beta_{j,k}\}$ with an exponential distribution. However, their wavelet shrinkage rule, either based on the posterior mean or via a Bayesian hypotheses testing procedure, requires numerical integration.

Chipman et al. (1997) also assume an independent prior for $\{\beta_{j,k}\}$. Since a signal is likely to have a sparse wavelet distribution with a heavy tail, they consider a mixture of two zero-mean normal components for $\{\beta_{j,k}\}$; one has a very small variance and the other has a large variance. Treating σ^2 as a hyperparameter, their shrinkage rule based on the posterior mean has a closed-form representation. Both Clyde *et al.* (1998) and Abramovich *et al.* (1998) consider a mixture of a normal component and a point mass at zero for wavelet coefficients $\{\beta_{j,k}\}$. Clyde *et al.* (1998) assume that the prior distribution for σ^2 is inverse gamma, and $\{\beta_{j,k}\}$ are independent, conditioned on σ^2. They use the stochastic search variable selection (SSVS) algorithm (George and McCulloch, 1993, 1997) to search for nonzero wavelet coefficients of the signal, and use the Markov chain Monte Carlo technique to obtain the posterior mean by averaging over all selected models. Moreover, closed-form approximations to the posterior mean and the posterior variance are also provided. Abramovich *et al.* (1998) consider a sum of weighted absolute errors as their loss function, resulting in a thresholding rule (i.e., coefficients with an absolute value below a certain threshold level are replaced by zero) that is Bayes, rather than a shrinkage rule, which is obtained from a Bayesian approach using squared error loss. Their thresholding rule based on the posterior median also has a closed-form representation, under the assumption that σ^2 is known. Lu *et al.* (1997) apply a nonparametric mixed-effect model, where the scaling-function coefficients are assumed to be fixed and the wavelet coefficients are assumed to be random with zero mean. Their empirical Bayes estimator is shown to have a Gauss-Markov type optimality.

Though the discrete wavelet transform is an excellent decorrelator for a wide variety of stochastic processes, it does not yield completely uncorrelated wavelet coefficients. In practice, the wavelet coefficients of an observed process are still somewhat correlated, especially for coefficients that are closer together in the same scale, or at nearby scales around the same locations. To describe this structure, Crouse *et al.* (1998) consider m-state Gaussian mixture models for wavelet coefficients. The state variables are linked via a tree structure in a Markovian manner. An empirical Bayes approach is taken and the prior parameters are estimated using an EM algorithm.

Vidakovic and Müller (1995) consider a conjugate-normal-inverse-gamma prior on (β, σ^2). That is, they assume (2) holds,

$$\beta \,|\, \sigma^2 \sim Gau(0, \sigma^2 \Sigma), \tag{7}$$

and σ^2 has an inverse-gamma distribution. The resulting shrinkage rule is,

$$E(\beta \,|\, w) = (I + \Sigma^{-1})^{-1} w. \tag{8}$$

Note that different choices of Σ can lead to various shrinkage procedures. Comparing (3) with (7), the main difference of our method from their method is that we apply a prior with a *nonzero* mean for the signal β. We also provide several parametric model classes for Σ. Details of our method will be given in the next two sections.

14.3   The DecompShrink Method

Recall that our approach is empirical Bayesian, which requires estimation of hyperparameters of the Bayesian model. First, we estimate the noise parameter σ by using the variogram method proposed by Huang and Cressie (1997a). Specifically, we estimate σ by

$$\hat{\sigma} = \begin{cases} \{2\hat{\gamma}(1) - \hat{\gamma}(2)\}^{1/2}\,; & \text{if } 2\hat{\gamma}(1) \geq \hat{\gamma}(2) \geq \hat{\gamma}(1), \\ \{(\hat{\gamma}(1) + \hat{\gamma}(2))/2\}^{1/2}\,; & \text{if } \hat{\gamma}(2) < \hat{\gamma}(1), \\ 0; & \text{otherwise,} \end{cases} \tag{9}$$

where

$$\hat{\gamma}(k) \equiv \left(\mathrm{MAD}\left(\{Y_{i+k} - Y_i : i = 1, \dots, n - k\}\right)\right)^2/2; \quad k = 1, 2,$$

is a robust estimator of the semi-variogram at lag k, and $\mathrm{MAD}(\cdot)$ is defined in (6). Note that for a stationary process $Z(t)$; $t \in \mathbb{R}$, the semi-variogram of Z at lag h is defined as:

$$\gamma(h) \equiv var\left(Z(t+h) - Z(t)\right)/2; \quad h \in \mathbb{R}.$$

We then estimate the deterministic signal $\mu \equiv \left((\mu_{J_0}^*)', \mu_{J_0}', \dots, \mu_{J-1}'\right)'$. Since the scaling-function coefficients $w_{J_0}^*$ correspond to large-scale features of the signal, we estimate $\mu_{J_0}^*$ by

$$\hat{\mu}_{J_0}^* = w_{J_0}^*. \tag{10}$$

Further, we shall assume that $\eta_{J_0}^* \equiv 0$ for modeling the stochastic part of the scaling function. That is, we declare the scaling-function coefficients $w_{J_0}^*$ to be purely deterministic. For the wavelet coefficients at the j-th level, the deterministic trend μ_j could be considered coming from components that are potential outliers in the normal probability plot of w_j. For $k = 0, 1, \dots, 2^j - 1$, let $q_{j,k}$ be the corresponding normal quantile of $w_{j,k}$. We estimate the slope of the fitted line in the normal probability plot by

$$\hat{\tau}_j \equiv \max\left\{\mathrm{MAD}(w_j), \hat{\sigma}\right\},$$

and estimate $\mu_{j,k}$ for $k = 0, 1, \dots, 2^j - 1$ by a soft-thresholding function:

$$\hat{\mu}_{j,k} = \begin{cases} w_{j,k} - \lambda_j; & \text{if } w_{j,k} > \lambda_j, \\ 0; & \text{if } |w_{j,k}| \leq \lambda_j, \\ w_{j,k} + \lambda_j; & \text{if } w_{j,k} < -\lambda_j, \end{cases} \tag{11}$$

where the threshold parameter λ_j is determined by

$$\lambda_j \equiv \hat{\tau}_j \times \max\left\{|q_{j,k}| : |w_{j,k}| < \hat{\tau}_j|q_{j,k}|\right\}.$$

Consequently, points in the normal probability plot with wavelet coefficients larger than λ_j, all fall above the fitted line, and points with wavelet coefficients smaller than $-\lambda_j$ all fall below the fitted line. Note that if w_j are actually normally distributed, the threshold value λ_j will become large. Therefore, only few points will be estimated as deterministic trend components but, importantly, these values will be small.

The final set of parameters to estimate is $\boldsymbol{\theta}$, which is the vector of prior covariance parameters in (3). In the next section, we propose using a pseudo maximum likelihood estimator $\hat{\boldsymbol{\theta}}$.

The wavelet shrinkage rule, which we call DecompShrink, is given by

$$
\begin{aligned}
\hat{\beta} &= \hat{\mu} + \mathrm{E}(\boldsymbol{\eta} \,|\, w, \hat{\mu}, \hat{\sigma}^2, \hat{\boldsymbol{\theta}}) \\
&= \hat{\mu} + \Sigma(\hat{\boldsymbol{\theta}})\big(\Sigma(\hat{\boldsymbol{\theta}}) + \hat{\sigma}^2 I\big)^{-1}(w - \hat{\mu}),
\end{aligned} \tag{12}
$$

where $\hat{\mu}$ can be obtained from (10) and (11). Several parametric models for $\Sigma(\boldsymbol{\theta})$ are given in the next section. Hence the empirical Bayes spatial predictor of $S \equiv W_n' \beta$ can be written as:

$$
\hat{S} \equiv W_n'\Big(\hat{\mu} + \Sigma(\hat{\boldsymbol{\theta}})\big(\Sigma(\hat{\boldsymbol{\theta}}) + \hat{\sigma}^2 I\big)^{-1}(w - \hat{\mu})\Big).
$$

14.4   Prior Covariance $\Sigma(\boldsymbol{\theta})$

In this section, we consider multiscale models for the prior covariance $\Sigma(\boldsymbol{\theta})$ or, equivalently, the corresponding stochastic components,

$$
\beta - \mu = \boldsymbol{\eta} \equiv \big((\boldsymbol{\eta}_{J_0}^*)', \boldsymbol{\eta}_{J_0}', \ldots, \boldsymbol{\eta}_{J-1}'\big)'.
$$

But first we shall discuss briefly estimation of the prior covariance parameters $\boldsymbol{\theta}$.

14.4.1   Estimation of Prior Covariance Parameters

For a given prior covariance $\Sigma(\boldsymbol{\theta})$, the hyperparameter $\boldsymbol{\theta}$ can be estimated by a pseudo maximum likelihood estimator, based on the distribution $p(w|\hat{\mu}, \hat{\sigma}^2, \boldsymbol{\theta})$. That is,

$$
\begin{aligned}
\hat{\boldsymbol{\theta}} &= \underset{\boldsymbol{\theta}}{\arg\sup}\; p(w \,|\, \hat{\mu}, \hat{\sigma}^2, \boldsymbol{\theta}) \\
&= \underset{\boldsymbol{\theta}}{\arg\inf}\; \Big\{\log\big|\Sigma(\boldsymbol{\theta}) + \hat{\sigma}^2 I\big| + (w - \hat{\mu})'\big(\Sigma(\boldsymbol{\theta}) + \hat{\sigma}^2 I\big)^{-1}(w - \hat{\mu})\Big\}.
\end{aligned}
$$

This method of pseudo maximum likelihood estimation was made popular by Gong and Samaniego (1981).

14.4.2 Scale-Independent Models

First, we consider scale-independent models, which correspond to block diagonal matrices $\Sigma(\boldsymbol{\theta})$. Specifically, we assume that the wavelet coefficients $\boldsymbol{\eta}_j; j = J_0, \ldots, J-1$, are statistically independent (i.e., scale independence) with zero means. Therefore, we can model $\boldsymbol{\eta}_j$ at each scale or level j separately. Note that $\boldsymbol{\eta}_{J_0}^* \equiv \mathbf{0}$, which corresponds to the earlier assumption that all the scaling-function coefficients $\boldsymbol{w}_{J_0}^*$ are attributed to the deterministic trend component $\boldsymbol{\mu}_{J_0}^*$ (Section 3).

If $\boldsymbol{\eta}$ comes from a temporal process (i.e., $d = 1$), it is natural to assume a Gaussian autoregressive moving average (ARMA) model, independently for each $\boldsymbol{\eta}_j; j = J_0, \ldots, J - 1$. If $\boldsymbol{\eta}$ is a d-dimensional process, we could specify a Gaussian Markov random field model, independently for each $\boldsymbol{\eta}_j$; $j = J_0, \ldots, J - 1$.

If one further assumes that the wavelet coefficients are also independent within each scale with $\mathrm{var}(\boldsymbol{\eta}_j) = \sigma_j^2 \boldsymbol{I}; j = J_0, \ldots, J-1$, then $\Sigma(\boldsymbol{\theta})$ becomes a diagonal matrix and $\boldsymbol{\theta} = \left(\sigma_{J_0}^2, \ldots, \sigma_{J-1}^2\right)'$. Therefore, from (12), the DecompShrink rule based on this simple model can be written as:

$$\hat{\beta}_{jk} = \hat{\mu}_{jk} + \frac{\hat{\sigma}_j^2}{\hat{\sigma}_j^2 + \hat{\sigma}^2}(w_{jk} - \hat{\mu}_{jk}), \tag{13}$$

for $j = J_0, \ldots, J - 1$, $k = 0, \ldots, 2^j - 1$, where

$$\hat{\sigma}_j^2 = \max\left(\frac{1}{2^j}\left(\boldsymbol{w}_j - \hat{\boldsymbol{\mu}}_j\right)'\left(\boldsymbol{w}_j - \hat{\boldsymbol{\mu}}_j\right) - \hat{\sigma}^2, \, 0\right), \tag{14}$$

and $\hat{\sigma}^2$ is given by (9).

Note that the wavelet coefficients are usually sparse (i.e., most of the coefficients are essentially zero), and have a distribution which is highly non-Gaussian with heavy tails. We are able to consider a Gaussian distribution in (3) because the non-Gaussian components are accounted for by the mean $\boldsymbol{\mu}$.

14.4.3 Scale-Dependent Models

Though the discrete wavelet transform is an excellent decorrelator for a wide variety of stochastic processes, it does not yield completely uncorrelated wavelet coefficients. A natural way to describe this structure is to use scale-dependent multiscale models for covariances $\Sigma(\boldsymbol{\theta})$. These models take the dependencies, both within scales and across scales, into account. Moreover, if $\boldsymbol{\eta}$ is convolved with noise, the optimal predictor of $\boldsymbol{\eta}$ can be computed efficiently using the change-of-scale Kalman-filter algorithm (Huang and Cressie, 1997b).

A multiscale model consists of a series of processes, $\boldsymbol{\eta}_j; j = J_0, \ldots, J-1$, with the following Markovian structure:

$$\boldsymbol{\eta}_j = \boldsymbol{A}_j \boldsymbol{\eta}_{j-1} + \boldsymbol{\xi}_j; \quad j = J_0 + 1, \ldots, J - 1,$$

where

$$\eta_{J_0} \sim \mathrm{N}\left(\mathbf{0}, \sigma_{J_0}^2 \boldsymbol{I}\right),$$
$$\xi_j \sim \mathrm{N}\left(\mathbf{0}, \sigma_j^2 \boldsymbol{I}\right); \quad j = J_0 + 1, \ldots, J - 1,$$

all random vectors are independent from one another, and $\boldsymbol{A}_{J_0+1}, \ldots, \boldsymbol{A}_{J-1}$ are deterministic matrices describing the causal relations between scales. For more details on these models, the reader is referred to Huang and Cressie (1997b).

The performance of multiscale models as priors for wavelet coefficients will be considered elsewhere. In all that is to follow, we assume the simplest model of independence, both between and within scales. That is, $\boldsymbol{\Sigma}(\boldsymbol{\theta})$ is a diagonal matrix with $\boldsymbol{\theta} = \left(\sigma_{J_0}^2, \ldots, \sigma_{J-1}^2\right)'$, to be estimated in a manner described in Section 4.1.

14.5 Applications

14.5.1 Simulation Study

We consider three test signals S with sample size $n = 2^J = 2048$. First, consider the "blocks" signal, created by Donoho and Johnstone (1994), rescaled so that the sample standard deviation $SD(S) = 7$ (see Figure 1 (a1)); second, consider a Gaussian AR(1) stationary process with the autoregressive parameter equal to 0.95 and $SD(S) = 7$ (see Figure 2 (a1)); and third, consider a mixture of the two above with the sample variance of the "blocks" component equal to 35 and the variance of the AR(1) component equal to 14 (see Figure 3 (a1)). To each signal, standard Gaussian white noise is added, which yields Y. Since the noise variance $\sigma^2 = 1$, we have $SD(S)/\sigma = 7$. Our goal is to reconstruct the original signal $S \equiv (S_1, \ldots, S_n)'$ from data $Y \equiv (Y_1, \ldots, Y_n)'$, and we assume that the noise parameter σ is unknown.

We apply our DecompShrink method with the thresholding function given by (13), and compare it with two commonly used wavelet shrinkage methods, VisuShrink and SureShrink, described in Section 2. For these two methods, the noise parameter σ is estimated by (6) based on the finest-scale wavelet coefficients, as proposed by Donoho et al. (1995). We chose $J_0 = 5$ and a nearly symmetric wavelet with 4 vanishing moments from a family of wavelets called symmlets (Daubechies, 1992) for all cases.

The noisy signals and the reconstructions from the three methods for the "blocks", AR(1), and "blocks" + AR(1) signals are shown in Figure 1, Figure 2, and Figure 3, respectively.

From the figures, it can be seen that the DecompShrink method performs the best, with SureShrink second, and VisuShrink third in terms of smaller prediction errors. For example, in Figure 1, it is clear that the

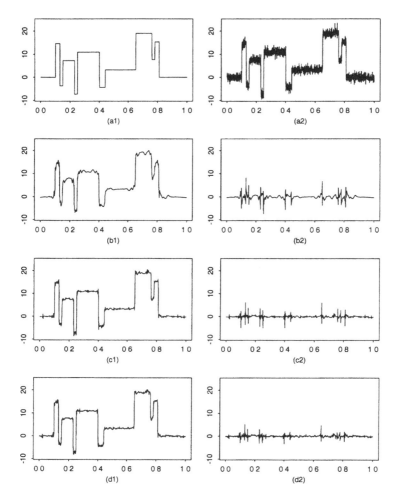

FIGURE 1. (a1) "blocks" signal ($n = 2048$, $SD(\boldsymbol{S}) = 7$); (a2) "blocks" signal + noise ($SD(\boldsymbol{S})/\sigma = 7$); (b1) VisuShrink reconstruction; (b2) VisuShrink prediction error; (c1) SureShrink reconstruction; (c2) SureShrink prediction error; (d1) DecompShrink reconstruction; (d2) DecompShrink prediction error.

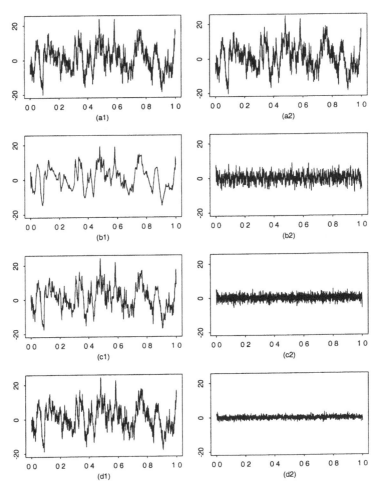

FIGURE 2. (a1) AR(1) signal ($n = 2048$, $SD(\boldsymbol{S}) = 7$); (a2) AR(1) signal + noise ($SD(\boldsymbol{S})/\sigma = 7$); (b1) VisuShrink reconstruction; (b2) VisuShrink prediction error; (c1) SureShrink reconstruction; (c2) SureShrink prediction error; (d1) DecompShrink reconstruction; (d2) DecompShrink prediction error.

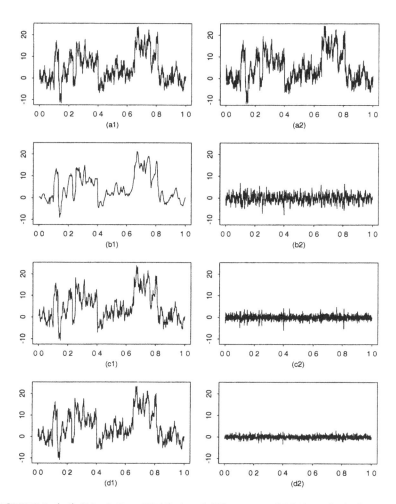

FIGURE 3. (a1) "blocks" + AR(1) signal ($N = 2048$, $SD(\boldsymbol{S}) = 7$); (a2) "blocks" + AR(1) signal + noise ($SD(\boldsymbol{S})/\sigma = 7$); (b1) VisuShrink reconstruction; (b2) VisuShrink prediction error; (c1) SureShrink reconstruction; (c2) SureShrink prediction error; (d1) DecompShrink reconstruction; (d2) DecompShrink prediction error.

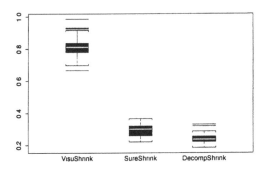

FIGURE 4. Boxplots of MSE performance of the "blocks" signal using various wavelet shrinkage techniques based on 100 replications ($n = 2048$, $SD(S) = 7$, $\sigma = 1$).

TABLE 1. Average MSE performance of simulated examples using various wavelet shrinkage techniques based on 100 replications ($n = 2048$, $SD(S) = 7$, $\sigma = 1$). The values in parentheses are the sample standard deviations of MSEs.

$n = 2048$, $SD(S) = 7$, $\sigma = 1$						
	Blocks		AR(1)		Blocks + AR(1)	
VisuShrink	0.809	(0.052)	6.946	(0.302)	3.131	(0.147)
SureShrink	0.292	(0.037)	2.006	(0.398)	0.963	(0.125)
DecompShrink	0.238	(0.025)	0.806	(0.054)	0.610	(0.039)

DecompShrink reconstruction is able to catch the jumps better than the reconstructions from the other two methods.

The performance of the various wavelet methods for recovering the signal S is compared using the mean-squared error (MSE) criterion:

$$\mathrm{MSE}(\hat{S}) \equiv \frac{1}{n} \sum_{i=1}^{n} \left(\hat{S}_i - S_i \right)^2.$$

Based on the random variation in the stochastic signal and noise, 100 replications of Y were obtained. Each replicate gives a $\mathrm{MSE}(\hat{S})$. Figure 4, Figure 5, and Figure 6 show the boxplots of these MSE values for the "blocks", AR(1), and "blocks + AR(1) signals, respectively, based on the 100 replications. The simulation results are also summarized in Table 1 by averaging over the 100 replications for each shrinkage method and for each test signal.

From Figure 4, Figure 5, Figure 6, and Table 1, we see that the MSE values obtained from the DecompShrink method have a distribution that is closer to zero than those from the other two methods. It is quite clear that the DecompShrink method is superior to the VisuShrink and the

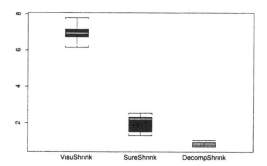

FIGURE 5. Boxplots of MSE performance of the AR(1) signal using various wavelet shrinkage techniques based on 100 replications ($n = 2048$, $SD(S) = 7$, $\sigma = 1$).

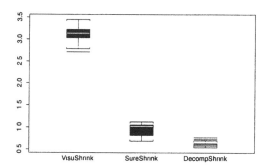

FIGURE 6. Boxplots of MSE performance of the "blocks" + AR(1) signal using various wavelet shrinkage techniques based on 100 replications ($n = 2048$, $SD(S) = 7$, $\sigma = 1$).

TABLE 2. Estimation of the noise parameter σ using MAD and the variogram methods based on 100 replications ($n = 2048$, $SD(S) = 7$, $\sigma = 1$).

	$n = 2048$, $SD(S) = 7$, $\sigma = 1$					
	MAD			Variogram		
	Mean	Bias	MSE	Mean	Bias	MSE
Blocks	1.008	0.008	0.0012	0.978	−0.022	0.0022
AR(1)	1.655	0.655	0.4324	1.054	0.054	0.0425
Blocks + AR(1)	1.244	0.244	0.0618	1.019	0.019	0.0094

SureShrink methods for recovery of these three signals.

Table 2 shows the estimation of the noise parameter for $n = 2048$, $SD(S) = 7$, and $\sigma = 1$, based on both the MAD method (Donoho *et al.*, 1995) and the variogram method (Huang and Cressie, 1997a). As we expect, $\hat{\sigma}$ based on the MAD method tends to overestimate the noise parameter σ, especially when the signal contains an AR(1) component, since an AR(1) signal is confounded with white noise in the finest-scale wavelet coefficients. The $\hat{\sigma}$ based on the variogram method is able to distinguish a continuous stochastic process and a discontinuous white noise process, hence is almost unbiased for the three test signals.

14.5.2   Image Denoising

In this section, we extend our DecompShrink method based on (12) to image denoising and compare it to the SureShrink method only. The prior covariance $\Sigma(\boldsymbol{\theta})$ is assumed to be based on a model of scale independence. We choose a two-dimensional symmlet wavelet, constructed from the one-dimensional symmlet wavelet with four vanishing moments, using the tensor-product method described briefly in Section 1.

We use the 256×256 pixel "Boat" image, shown in Figure 7 (a1), as our test image. The signal-to-noise ratio (SNR) is used as a quantitative measure of image quality, and is expressed as

$$SNR \equiv 10 \log_{10} \frac{\text{variance of signal}}{\text{variance of noise}},$$

in "db" unit. Figure 7 (b1) and Figure 7 (c1) show two boat images, degraded by adding Gaussian white noise (SNR = 5.5 db and 2 db, respectively).

The images reconstructed from Figure 7 (a1), (b1), (c1) based on the SureShrink method are shown in Figure 7 (a2), (b2), (c2), respectively, and the reconstructions based on our DecompShrink method are shown in Figure 7 (a3), (b3), (c3), respectively. Here, $J = 8$, and we chose $J_0 = 3$ for both methods.

The results show that for DecompShrink, the image reconstructed from the original image with no noise added is exactly the same as the original image (compare Figure 7 (a3) to Figure 7 (a1)), as we would hope for but would not always expect. Further, the SNR of the reconstructed images increases substantially over that of the noisy images. For example, for the boat images shown in Figure 7 (b1) and Figure 7 (b3), we see an increase in SNR from 5.5 db to 12.04 db and, for Figure 7 (c1) and Figure 7 (c3), we see an increase from 2 db to 10.51 db. Notice that the DecompShrink method performs somewhat better than the SureShrink method for recovering the noisy images, both in terms of SNR and the visual quality.

14.6   Discussion and Conclusions

In this chapter, we have investigated an empirical Bayesian spatial prediction approach for recovering signals proposed in Huang and Cressie (1997a). The DecompShrink method has an advantage over the current shrinkage methods in that it is adaptive to the smoothness of the signal regardless of whether the signal has a sparse wavelet representation, since our *nonzero-mean* Gaussian prior for the signal β not only catches the signal components with larger wavelet coefficients, but also catches the signal components with smaller wavelet coefficients based on a prior covariance $\Sigma(\theta)$. It is interesting to note that when the signal is piecewise smooth with sparse wavelet representation (e.g., "blocks"), we have $\hat{\sigma}_j^2 \approx \hat{\sigma}^2$; $j = J_0, \ldots, J - 1$ in (14). In this case, the DecompShrink function given by (13) is approximately equal to a soft-thresholding function.

Our simulation study shows that the method based on a simple independence model has superior MSE performance over the VisuShrink and the SureShrink methods for both deterministic and stochastic signals. It is not surprising that the VisuShrink and the SureShrink methods do not perform well for recovering stochastic signals, since shrinkage rules based on simple hard-thresholding or soft-thresholding functions rely heavily on a sparse wavelet representation, which typically is not the case for stochastic signals. It is encouraging that DecompShrink does well not only for stochastic signals but also for deterministic signals with sparse wavelet representation.

It may be advantageous to put a further prior on μ, or even apply a fully Bayesian analysis. It would also be interesting to see whether more general multiscale models that consider dependencies of wavelet coefficients both within scales and across scales (Huang and Cressie, 1997b) will yield better predictors. These are subjects to be explored in the future.

Acknowledgments: This research was supported by the Office of Naval Research under grant no. N00014-93-1-0001, and was partially carried out at Iowa State University.

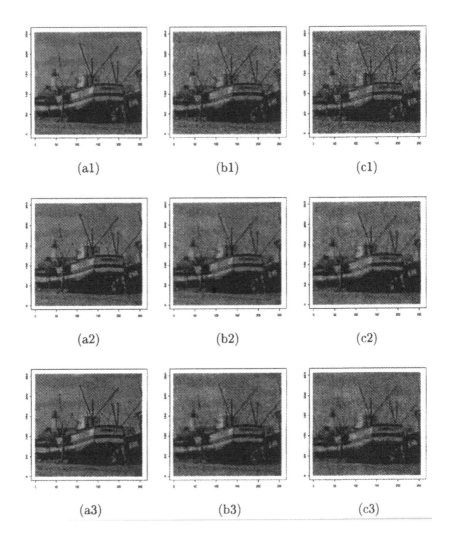

FIGURE 7. (a1), (b1), (c1): Boat images (256 × 256 pixels), where SNR = ∞ db, 5.5 db, and 2 db, respectively. (a2), (b2), (c2): SureShrink Reconstructions, where SNR = 31.96 db, 9.82 db, and 11.61 db, respectively. (a3), (b3), (c3): DecompShrink Reconstructions with SNR = ∞ db, 10.51 db, and 12.04 db, respectively.

References

Abramovich, F. and Benjamini, Y. (1995). Thresholding of wavelet coefficients as multiple hypotheses testing procedure. In Antoniadis, A. and Oppenheim, G., editors, *Wavelets and Statistics*, volume 103 of *Lecture Notes in Statistics*, pages 5–14. Springer-Verlag, New York.

Abramovich, F. and Benjamini, Y. (1996). Adaptive thresholding of wavelet coefficients. *Computational Statistics and Data Analysis*, 22:351–361.

Abramovich, F., Sapatinas, T., and Silverman, B. W. (1998). Wavelet thresholding via a Bayesian approach. *Journal of the Royal Statistical Society B*, 60:725–749.

Cambanis, S. and Houdré, C. (1995). On the continuous wavelet transform of second-order random processes. *IEEE Transactions on Information Theory*, 41:628–642.

Chipman, H. A., Kolaczyk, E. D., and McCulloch, R. E. (1997). Adaptive Bayesian wavelet shrinkage. *Journal of the American Statistical Association*, 92:1413–1421.

Clyde, M., Parmigiani, G., and Vidakovic, B. (1996). Bayesian strategies for wavelet analysis. *Statistical Computing and Statistical Graphics Newsletter*, 7:3–9.

Clyde, M., Parmigiani, G., and Vidakovic, B. (1998). Multiple shrinkage and subset selection in wavelets. *Biometrika*, 85:391–401.

Crouse, M., Nowak, R., and Baraniuk, R. (1998). Wavelet-based statistical signal processing using hidden Markov models. *IEEE Transactions on Signal Processing*, 46:886–902.

Daubechies, I. (1992). *Ten Lectures on Wavelets*. SIAM, Philadelphia.

Donoho, D. L. and Johnstone, I. M. (1994). Ideal spatial adaptation by wavelet shrinkage. *Biometrika*, 81:425–455.

Donoho, D. L. and Johnstone, I. M. (1995). Adapting to unknown smoothness via wavelet shrinkage. *Journal of the American Statistical Association*, 90:1200–1224.

Donoho, D. L. and Johnstone, I. M. (1998). Minimax estimation via wavelet shrinkage. *Annals of Statistics*, 26. Forthcoming.

Donoho, D. L., Johnstone, I. M., Kerkyacharian, G., and Picard, D. (1995). Wavelet shrinkage: Asymptopia? (with discussion). *Journal of the Royal Statistical Society B*, 57:301–369.

George, E. I. and McCulloch, R. (1993). Variable selection via Gibbs sampling. *Journal of the American Statistical Association*, 88:881–889.

George, E. I. and McCulloch, R. (1997). Approaches to Bayesian variable selection. *Statistica Sinica*, 7:339–373.

Gong, G. and Samaniego, F. J. (1981). Pseudo maximum likelihood estimation: Theory and applications. *Annals of Statistics*, 9:861–869.

Huang, H.-C. and Cressie, N. (1997a). Deterministic/stochastic wavelet decomposition for recovery of signal from noisy data. Technical Report 97-23, Department of Statistics, Iowa State University, Ames, IA.

Huang, H.-C. and Cressie, N. (1997b). Multiscale spatial modeling. In *ASA 1997 Proceedings of the Section on Statistics and the Environment*, pages 49–54, Alexandria, VA.

Lu, H. H.-C., Huang, S.-Y., and Tung, Y.-C. (1997). Wavelet shrinkage for nonparametric mixed-effects models. Technical report, Institute of Statistics, National Chiao-Tung University.

Mallat, S. (1989). A theory for multiresolution signal decomposition: The wavelet representation. *IEEE Transactions on Pattern Analysis and Machine Intelligence*, 11:674–693.

Nason, G. P. (1995). Choice of the threshold parameter in wavelet function estimation. In Antoniadis, A. and Oppenheim, G., editors, *Wavelets and Statistics*, volume 103 of *Lecture Notes in Statistics*, pages 261–280. Springer-Verlag, New York.

Nason, G. P. (1996). Wavelet shrinkage using cross-validation. *Journal of the Royal Statistical Society B*, 58:463–479.

Ogden, R. T. and Parzen, E. (1996a). Change-point approach to data analytic wavelet thresholding. *Statistics and Computing*, 63:93–99.

Ogden, R. T. and Parzen, E. (1996b). Data dependent wavelet thresholding in nonparametric regression with change-point applications. *Computational Statistics and Data Analysis*, 22:53–70.

Ruggeri, F. and Vidakovic, B. (1999). A Bayesian decision theoretic approach to wavelet thresholding. *Statistica Sinica*, 9. To appear.

Vidakovic, B. (1998). Nonlinear wavelet shrinkage with Bayes rules and Bayes factors. *Journal of the American Statistical Association*, 93:173–179.

Vidakovic, B. and Müller, P. (1995). Wavelet shrinkage with affine Bayes rules with applications. Discussion Paper 95-24, ISDS, Duke University, Durham, NC.

15

Geometrical Priors for Noisefree Wavelet Coefficients in Image Denoising

Maarten Jansen and Adhemar Bultheel

ABSTRACT   Wavelet threshold algorithms replace wavelet coefficients with small magnitude by zero and keep or shrink the other coefficients. This is basically a local procedure, since wavelet coefficients characterize the local regularity of a function. Although a wavelet transform has decorrelating properties, structures in images, like edges, are never decorrelated completely, and these structures appear in the wavelet coefficients. We therefore introduce a geometrical prior model for configurations of large wavelet coefficients and combine this with the local characterization of a classical threshold procedure into a Bayesian framework. The threshold procedure selects the large coefficients in the actual image. This observed configuration enters the prior model, which, by itself, only describes configurations, not coefficient values. In this way, we can compute for each coefficient the probability of being "sufficiently clean".

15.1   Introduction

Wavelet thresholding (Donoho and Johnstone, 1994; Donoho and Johnstone, 1995) is a popular method for the reduction of noise in images. It assumes that the original, non-corrupted image can be represented in a sparse way by a small number of large wavelet coefficients. In the case of an orthogonal transform, i.i.d. noise is spread out equally over all coefficients. Selecting the coefficients with the largest magnitude therefore removes most of the noise, while preserving the essential image information. In the case of correlated noise (Johnstone and Silverman, 1997) or a non-orthogonal transform, the multiresolution structure of a wavelet transform justifies the application of scale-dependent thresholds.

Whereas typical threshold algorithms are based on this heuristical approach that the largest coefficients capture the essential image features, *Bayesian* methods, including (Simoncelli and Adelson, 1996; Vidakovic, 1994; Ruggeri and Vidakovic, 1995) start from a full model for wavelet coefficients of the following type:

$$W = V + N$$

This is an additive model for wavelet coefficients where N is the noise vector, V is the uncorrupted wavelet coefficient vector, and W is the input (empirical) wavelet coefficient vector. Both the noise and the noise-free data are viewed as realizations of a probability distribution. A Bayesian method considers W as an observation from which it computes (in principle) the posterior distribution of V. This is used to estimate the uncorrupted data coefficients. For example, minimization of the L_2-loss leads to the posterior mean:

$$\hat{V} = \mathrm{E}(V|W)$$

For some models of V and N, Bayesian estimators can be interpreted as threshold procedures, although often more complicated than the simple hard- or soft-thresholding. Because of the complete modelling, this approach seems less heuristic. However, most models carry hyperparameters, that have to be chosen, often on a heuristic basis.

In this Chapter, we start from a wavelet threshold procedure and discuss why — especially for images — thresholding is a too local operation: important wavelet coefficients are concentrated around image structures, like edges. Section 15.3 therefore introduces a geometrical prior model for structures of important coefficients. This is typically a Gibbs Random Field. Using a geometrical prior that supports important coefficient *configurations*, we "correct" the result of threshold procedure. Our model introduces the notion of important and noisy coefficients. Both these classes have an according probability density. Other approaches with a similar *mixture model* for wavelet coefficients include (Malfait and Roose, 1997; Malfait, 1995; Chipman et al., 1997; Crouse et al., 1998; Clyde et al., 1998; Abramovich et al., 1997).

15.2   Wavelet Thresholding

15.2.1   Wavelet terminology

A discrete wavelet transform is a decomposition of an input vector into coefficients at different resolution levels. Each resolution represents image features of a specific scale. In two dimensions, we use a tensor-product like extension of a one-dimensional transform, the so called square wavelet-transform: at each level the coefficients belong to one of three *subbands* or *components*, corresponding to three orientations in the image: the coefficients carry vertical, horizontal, or diagonal information. Instead of the classical fast wavelet transform, we use the so called non-decimated, redundant, or stationary wavelet transform (Nason and Silverman, 1995; Pesquet et al., 1996; Malfait and Roose, 1997). While the fast transform has linear complexity, this alternative is of $\mathcal{O}(N \log N)$ complexity, both in computations and in memory requirements. On the other hand, the stationary transform is translation invariant and generates the same number of coefficients in each subband at each scale. This facilitates interscale coefficient comparisons. Moreover, the reconstruction procedure (Nason and Silverman, 1995) from this redundant data representation is of course not unique. For manipulated data, a linear combination of all possible reconstruction schemes causes an additional smoothing of the result (Nason and Silverman, 1995).

15.2.2 Threshold schemes

A wavelet threshold algorithm typically consists of three steps: first, the observational data are transformed into empirical wavelet coefficients. The next step is a manipulation of the coefficients and finally, an inverse transform of the modified coefficients yields the result.

Of course, the second step is the crucial one. Apart from the choice of the wavelet basis, all strategies and options are concentrated in this step. As mentioned before, heuristical arguments or at least assumptions about the noise-free image regularity, motivate threshold rules and threshold selection procedures. The manipulation is based on a *classification* of the coefficients: the most common threshold procedures use a binary classification: a coefficient is either dominated by noise *or* sufficiently important. The magnitude of the coefficient is the *criterion* of classification: we consider the absolute value of each coefficient as a measure of significance. This could of course be generalized: the measure of significance M is a function of the observation:

$$M = m(W),$$

Wavelet coefficients with a significance below a threshold δ, are classified as noisy. With each wavelet coefficient W_s, the algorithm associates a 'label' or 'mask' variable X_s such that:

$$X_s = \begin{cases} 0, & \text{if } W_s \text{ is noisy according} \\ & \text{to the criterion, i.e. if} M_s < \delta \\ 1, & \text{then } W_s \text{ is sufficiently clean, i.e. if} \\ & M_s \leq \delta \end{cases}$$

In these and following equations, s represents the 'multidimensional' index of a wavelet coefficient on a given resolution level j and for a given component (vertical, horizontal, or diagonal): $s = (k, l)$. To avoid overloaded notations we omit the resolution level j and the component m in our equations, and use the simple index s. So, if no confusion is possible, we write W_s instead of $W_{j;s}^m$ or $W_{j;k,l}^m$. This classification is followed by the modification step: If $W_{\delta s}$ denotes the modified coefficient, with subscript δ refering to the threshold value, we write:

$$W_{\delta s} = h(W_s, M_s, X_s),$$

for some *action* $h(W_s, M_s, X_s)$. The classic *hard-threshold* procedure corresponds to

$$h(W_s, M_s, X_s) = W_s X_s.$$

It keeps the 'uncorrupted' coefficients and replaces the noisy ones by zero.

15.2.3   Thresholding is local

Wavelet basis functions are localized in space and scale (frequency). As a consequence, manipulating a coefficient has a local effect, both in space and frequency. This is an important advantage of wavelet based methods.

On the other hand, usual *classification* rules are local too, and do not take into account all the correlations that exist among neighboring coefficients. Although a wavelet transform has decorrelating properties, this decorrelation is not complete (a wavelet transform is sometimes seen as an *approximation* of a Karuhnen-Loève-transform). We distinguish two types of correlations:

1. Important image features correspond to large coefficients at different scales: these coefficients are of course correlated. This correlation can be used as the significance criterion in the qualification of wavelet coefficients in a threshold procedure (Xu et al., 1994): coefficients with a small interlevel correlation are replaced by zero. Alternatively, this multiscale structure can be incorporated in a prior model (Crouse et al., 1998).

2. The second type of correlation is within one scale: especially in images, important coefficients tend to be clustered on the location of edges.

In this chapter, we neglect the first type of correlations, and we concentrate on the second type of clustering.

15.2.4   Threshold mask images

Figure 1 shows an image with artificial, additive, homoscedastic correlated noise. This noise was the result of a convolution of white noise with a FIR-highpass-filter (A FIR or finite impuls response filter has a finite number of taps). The signal-to-noise ratio is 4.97 dB, where we define signal-to-noise ratio as:

$$SNR = 10 \log_{10} \left(\frac{\frac{1}{N} \sum_{i=1}^{N} (f_i - \bar{f})^2}{\sigma^2} \right),$$

FIGURE 1. An image (Left) with artificial, additive correlated noise (Right), SNR = 4.97dB. The noise is the result of a convolution of white noise with a FIR-highpass-filter.

with $\overline{f} = \frac{1}{N} \sum_{i=1}^{N} f_i$. f_i is the gray intensity of pixel i. We apply a redundant wavelet transform. As wavelet filter, we use the variation on the CDF-(spline)-filters "with less dissimilar lengths" (Cohen et al., 1992; Antonini et al., 1992). We choose a basis with four primal and four dual vanishing moments. These wavelets are rather popular in image processing: they are smooth and have compact support.

Since we know the uncorrupted image, we can compute at each level the threshold that minimizes the mean square error (MSE) of the wavelet coefficients. Since we use a bi-orthogonal (i.e. not an orthogonal) transform, this is not exactly the same as a minimization of the MSE of the pixel values, although a *stable* basis (with Riesz constants close to each other (Daubechies, 1992)) guarantees a quasi-equivalence. Moreover, the eventual objective is visual quality, and there seems no reason why minimization in terms of original pixels is better than minimization in terms of wavelet coefficients. Of course, the user views images in the pixel domain, but we do not interpret an image pixel by pixel. Although a discussion about the human visual system is far beyond the scope of this text, a multiresolution analysis is said to be closer to the way we view an image. Note that MSE minimization is equivalent to SNR maximization. The black pixels in the binary image of Figure 2 indicate the empirical coefficients of the horizontal subband at the one but finest scale with magnitude larger than the minimum MSE threshold (The one but finest scale in the wavelet transform is two scales below the original image resolution). These coefficients are kept, while the white pixels correspond to coefficients that are replaced by zero. Although we recognize image structures, like edges, there also many isolated black pixels.

In practical situations, the minimum MSE threshold cannot be computed exactly. So, the next picture shows the same *mask* or *label* image for an *estimate* of the minimum MSE threshold, based on generalized cross validation (GCV) procedure (Weyrich and Warhola, 1995; Jansen et al., 1997). This method is as fast as the wavelet transform, requires no estimation for the noise deviation σ and it is asymptotically optimal, i.e. if the number of data is getting large, the estimated threshold approaches the minimum MSE threshold. These properties are preserved — on a level-dependent basis — if the noise is correlated in the wavelet representation (Jansen and Bultheel, 1998). Some other methods, like (Berkner and Wells, Jr., 1998) are based on a "probabilistic upper bound" for the noise in the wavelet domain. This upper bound and the corresponding threshold change if the noise on the wavelet coefficients is correlated. GCV-procedures, however, aim at a minimum MSE threshold and they can be used level by level without further modifications in the case of correlated noise and non-orthogonal transforms.

If we apply the same threshold to the noise-free coefficients, we get the left picture in Figure 3. We see that many of the isolated pixels have disappeared: they were due to noise.

However, the minimum MSE threshold is based on a global compromise between noise and data: this is not the best thing we can do: instead of applying one threshold for all coefficients at a given level, we would like to decide for each coefficient separately what is best: keeping or killing. If we know the noise deviation σ and the noise-free magnitude V_s, then the expected MSE is minimized by replacing W_s with zero if $|V_s| < \sigma$ and keeping it otherwise (Donoho and Johnstone, 1994). This is an ideal *diagonal projection*: instead of applying the minimum MSE threshold (or its estimate) to the noise-free coefficients, we choose a threshold on these uncorrupted coefficients equal to the noise deviation. The resulting label image is shown in Figure 3 (Right picture). Since neither V_s nor σ are known in practice, this is of course an ideal case.

To conclude this discussion, we compare the result from hard-thresholding with level- and subband-dependent GCV-threshold with the result from diagonal projection. Only the three finest scales were processed. Signal-to-noise ratio is respectively 18.64 dB and 21.05 dB.

15.2.5 Binary image enhancement methods

A comparison of the different label images clearly illustrates that thresholding each coefficient separately does not take into account the image structures. An obvious way to recover the optimal mask of Figure 3 (b), is trying classic enhancement methods. Figure 5 (a) shows the label image after applying a median filter to the approximate minimum MSE labels in Figure 2. Another possibility is the application of so called erosion-dilation methods: these methods proceed in two steps: in the first step, black pixels with less than, for instance, two black neighbors are removed. This erosion can be repeated several times, before going to the dilation. This second

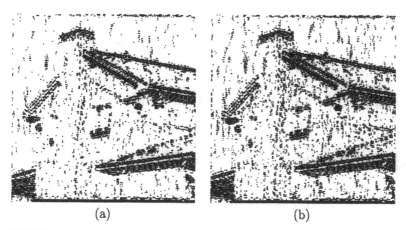

FIGURE 2. Mask or label images, corresponding to the horizontal component of the one but finest scale. Black pixels represent coefficients with magnitude above the threshold. Left: using the minimum MSE threshold. Right: using a GCV estimate of this threshold.

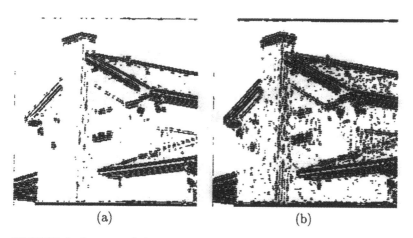

FIGURE 3. Same mask images as in Figure 2, here based on noise-free coefficients. Left: black pixels indicate noise-free coefficients with magnitude above the previous threshold. Right: black pixels indicate noise-free coefficients with magnitude larger than noise deviation. This corresponds to *diagonal projection*: if an "oracle" (Donoho and Johnstone, 1994) tells us whether or not a coefficient is dominated by noise, this is the best thing we can do.

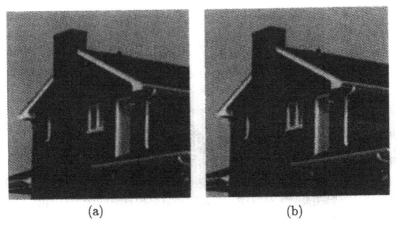

(a)   (b)

FIGURE 4. Left: the result from thresholding with level- and subband-dependent GCV-threshold. Only the three finest scales were processed. SNR = 18.64 dB. Right: result from the optimal (clairvoyant) diagonal projection. SNR = 21.05 dB.

step tries to restore the eroded objects by turning white background pixels into black object pixels, if there is already an object in the neighborhood. (This neighborhood is typically a 3×3 box containing the actual pixel in its center.) Figure 5 (b) contains the result of this operation. It is hard to

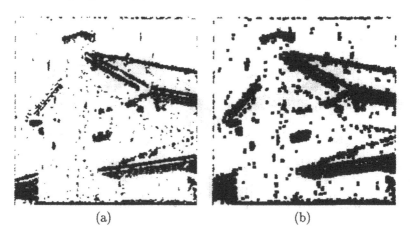

(a)   (b)

FIGURE 5. Result of elementary binary image enhancement methods on the approximate minimum MSE label image in Figure 2. Left: a median filter; Right: an erosion-dilation procedure.

preserve the fine edge structures, while removing the noisy pixels. These operations act on the label images only and forget about the background behind them: these pixels come from a wavelet coefficient classification. We would prefer a method that can deal with the geometry *and* the local

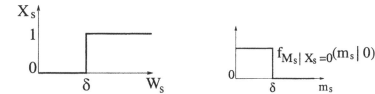

FIGURE 6. Left: The deterministic classification function for coefficient magnitude thresholding: if a coefficient magnitude M is below the threshold value δ, it is classified as noisy ($X = 0$), otherwise it is called sufficiently clean ($X = 1$). Right: this deterministic approach is a special case of the Bayesian model, where the conditional density is zero for coefficient magnitudes above the threshold if $X = 0$ and beneath the threshold if $X = 1$.

coefficient values at the *same* time. Bayes' rule tells us how we can do so.

15.3 A Bayesian procedure

15.3.1 A modified classification procedure

The classification in a threshold algorithm is a deterministic function of the empirical coefficients: thresholding on magnitudes corresponds to a simple step function, as illustrated in Figure 6. Recall that, in this text, the measure of significance is the coefficient magnitude: $M = |W|$. However, it would be interesting to examine measures based on interscale-correlations: e.g. $M_{j;s}^m = \prod_i W_{j+i;s}^m$, where j is the current resolution level and m the orientation ($m = \mathrm{HOR}, \mathrm{VER}, \mathrm{DIAG}$). Because we want to take the spatial configurations into account, we give up this tight relation between a coefficient value and its classification.

We introduce a *prior model* for coefficient classification configurations X. This prior should express the belief that clusters of noise-free coefficients have a higher probability than configurations with many isolated labels. In particular, edge-shaped clusters should be promoted.

Next, the *conditional model* (likelihood function) states that if the classification label for a coefficient equals one, this coefficient is *probably* above the threshold. A zero label means that the corresponding coefficient is probably small. The classical, deterministic approach can be seen as an extreme case of this probability model, where, for example, a label $X = 0$ tells that the coefficient is *certainly* in the range $[-\delta, \delta]$ (see right plot Figure 6).

If we have a prior distribution $P(X = x)$ and a conditional model $f_{M|X}(m|x)$, then Bayes' rule allows to compute the *posterior* probability:

$$P(X = x | M = m) = \frac{P(X = x) f_{M|X}(m|x)}{f_M(m)}$$

In a given experiment, where m is fixed, the denominator is a constant.

As we explain later, it is sufficient to know the *relative* probabilities of configurations, and therefore we write:

$$P(X = x | M = m) = C \cdot P(X = x) f_{M|X}(m|x)$$

15.3.2   The a priori model

As explained above, we are looking for a multivariate model for a binary image X. Expressing a probability function for all 2^{NM} possible values in a N by M label image may be cumbersome. We therefore construct the model starting from local descriptions of clustering. First we write the prior probability function as:

$$P(X = x) = \frac{1}{Z} \exp[-H(x)],$$

with the partition function Z:

$$Z = \sum_{x} \exp[-H(x)].$$

$H(x)$ is the *energy function* of a configuration x: the lower the energy, the higher the prior probability. To express that this energy comes from *local* interactions only, we first define for each pixel index s in the lattice S its set of neighbors, i.e. the set of indices $\partial s \subset S$ that interact with s. We assume that $s \notin \partial s$ and $s \in \partial t \Leftrightarrow t \in \partial s$. A set of indices that are all neighbors to each other is called a *clique*. The set of all cliques is

$$\mathcal{C} = \{C \in 2^{S} \mid \forall s, t \in C : t \in \partial s\}$$

The Gibbs Random Field (GRF) model (Besag, 1974; Geman and Geman, 1984; Winkler, 1995; Bello, 1994; Charbonnier et al., 1992; Luettgen et al., 1993) states that energy is the sum of clique potential functions:

$$H(x) = \alpha \sum_{C \in \mathcal{C}} U_C(x_C).$$

As the equation indicates, the clique potential U_C only depends on the values of x in the sites that belong to C.

A well-known example is the Ising model: the neighbors of a pixel with index s is a 3×3 submatrix, excluding s in its center. The model only considers pairs of neighbors. Other cliques in the system have no energy. The total energy is:

$$H(x) = \sum_{\{s,t\} \in \mathcal{C}} x_s x_t$$

For our experiments, we use a slightly different model, in a 5×5-neighborhood system. We only consider 3×3 cliques, i.e. the largest possible type in a 5 by 5 neighborhood system. The potentials of all other types of cliques are set to zero. For a 3 by 3 clique C we set:

$$U_C(\boldsymbol{x}_C) = \frac{\sum_{s \in C} x_s \sum_{t \in C \cap \partial s}(1 - 2x_s x_t)}{\sum_{s \in C} x_s},$$

i.e. for each $x_s = 1$, we subtract the number of neighbors within the clique with value one from the number of neighbors with zero value. The sum of these results is divided by the number of labels with value $x_s = 1$, to obtain a mean value. The idea behind this potential function is to compute a kind of "average degree of isolation" within the clique for the pixels with value one. The background pixels are considered to be neutral. Unlike the classical Ising model, this function is not symmetric for interchanges of ones and zeros.

15.3.3 The conditional model

We also need a conditional density $f_{M|X}(m|x)$. Whereas the prior describes the clustering of significant wavelet coefficients, this conditional model deals with the *local* significance measure. Therefore we write

$$f_{M|X}(\boldsymbol{m}|\boldsymbol{x}) = \prod_{s \in S} f_{M_s|X_s}(m_s|x_s).$$

This density expresses that if the label $X_s = 1$, i.e. if the corresponding wavelet coefficient is sufficiently noise-free, a large value of M_s is probable. Figure 7 shows how this density uses a threshold parameter δ. Another parameter is the transition width Δ between the high and the low density level. This parameter expresses the uncertainty of the classification. For instance, if $X_s = 1$, the coefficient is important, and its significance is normally above the threshold. Due to noise, it may happen to be smaller than the threshold. The more noise, the more probable this event. We therefore choose:

$$\Delta \sim \sigma$$

Since in practical cases, σ is unknown, we replace it with an estimate or another deviation-like quantity. For example, if we assume that most of the coefficients below the threshold are pure noise, we can compute the total amount of energy in these coefficients as a measure for the noise variance. We can write:

$$f_{M_s|X_s}(m_s|x_s) = \frac{1}{Y_s} \exp[-\beta V_s(m_s|x_s)],$$

where Y_s is a normalization factor and $V_s(m_s|x_s)$ is a conditional potential function, such as illustrated in figure 8. The conditional density now

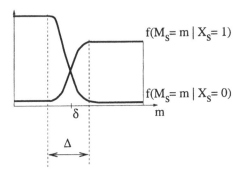

FIGURE 7. Conditional density $f(M_s|X_s)$. If $X_s = 1$, then large values of M_s are probable and vice versa. This function uses a threshold δ as a parameter.

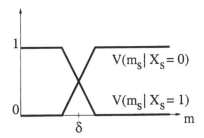

FIGURE 8. Conditional potential function.

becomes:

$$f_{M|X}(m|x) = \frac{1}{Y}\exp[-V(m|x)],$$

where

$$Y = \int_m \exp[-V(m|x)]\,dm = \prod_s Y_s,$$

and

$$V(m|x) = \beta \sum_{s\in S} V_s(m_s|x_s).$$

15.3.4 Posterior probabilities

From Bayes' rule, we can compute the posterior probabilities

$$P(X = x|M = m).$$

With these probabilities, a Bayesian decision rule leads to an estimation of the optimal label. The Maximum A Posteriori (MAP) procedure chooses the mask x with the highest posterior probability. The Maximal Marginal Posterior (MMP) rule is a more local approach: it computes in each site s

the marginal probabilities:

$$P(X_s = 1|M = m) = \sum_x x_s P(X = x|M = m),$$

and if this probability is more than 0.5, the pixel gets value 1. Both decision rules have a binary outcome: each coefficient is classified as noisy $(X = 0)$ or relatively uncorrupted $(X = 1)$. We would like to exploit the entire posterior probability: the posterior mean value

$$E(X_s|M = m) = P(X_s = 1|M = m)$$

preserves all information. It is a minimum least squares estimator. This classification leads to a posterior 'expected action':

$$E(W_{\delta s}|M) = h(W_s, m_s(W), 1)P(X_s = 1|M) \\ +h(W_s, m_s(W), 0)P(X_s = 0|M).$$

If $h(W_s, M_s, X_s) = X_s W_s$, this is:

$$E(W_{\delta s}|M) = W_s E(X_s|M = m) = W_s P(X_s = 1|M).$$

Unlike most thresholding methods, this is not a binary procedure: using the posterior probability leads to a more continuous approach.

15.3.5   Stochastic sampling

The computation of $P(X_s = 1|W)$ involves the probability of all possible configurations x. Because of the enormous number of configurations, this is an intractable task. The sum we have to compute is of the following form:

$$\mu_s = \sum_x f_s(x)P(X = x|M = m)$$

where in this case $f_s(x) = x_s$ and $\mu_s = P(X_s = 1|M = m)$.

To estimate this type of sum (or integral for random variables on a continuous line), one typically uses stochastic samplers. These methods generate subsequent samples $X^{(i)}$, not selected uniformly, but in proportion to their probability. This allows to approximate the matrix of required marginal probabilities by the mean value of the generated masks:

$$\hat{\mu}_s = \sum_i X_s^{(i)}.$$

Mostly, the samples are generated, not independently of each other, but in a chain, hence the name Markov Chain Monte Carlo (MCMC) estimation.

The next sample is generated, starting from the previous one. One advantage of this procedure is that knowledge of the *relative* probabilities of the candidates is sufficient. The probability ratio of two subsequent samples:

$$r^{(i)} = \frac{P(X^{(i+1)}|M)}{P(X^{(i)}|M)}$$

is the only quantity needed by the algorithm, as we explain below, and if

$$P(X = x|M) = \frac{1}{Z}\exp[-H(x)]\frac{1}{Y}\exp[-V(m|x)],$$

there is no need for the enormous computation of the partition function ZY.

We use the classic Metropolis MCMC sampler. The chain of states is started from an initial state $X^{(0)}$. The successive samples $X^{(i)}$ are then produced as follows: a candidate intermediate state is generated by a local random perturbation of the actual state. Then the probability ratio r of the actual state and its perturbation is computed. Since the GRF-model is based on local potential functions, only positions s whose mask labels are switched by the perturbation or which have a switched label in their neighborhood ∂s are involved in the computation. If the candidate has a higher probability than the actual state, i.e. if the probability ratio is larger than one, then the new state is accepted, otherwise it is accepted with probability equal to r. To generate a completely new sample, we repeat this local switching for all locations in the grid.

15.3.6 Algorithm overview

This is a schematic overview of the subsequent steps of the algorithm:

1. Compute the stationary wavelet transform W of the input.

2. At each level and for each component, select the appropriate threshold. This threshold generates an initial label image $X^{(0)}$.

3. Run a stochastic sampler to estimate for each coefficient at the given resolution level the probability $P(X_s|W)$. Use $X^{(0)}$ from the previous step as the starting sample. A Markov Chain Monte Carlo algorithm produces the sequence of samples.

4. $\hat{W}_{\delta s} \leftarrow W_s P(X_s = 1|W)$.

5. Inverse wavelet transform yields the result.

15.4 Results and discussion

The expression of the posterior energy:

$$H(x|m) = H(x) + V(m|x) + C^*,$$

where the constant C^* comes from the denominator in Bayes' rule and where:

$$H(x) = \alpha \sum_{C \in \mathcal{C}} U_C(x_C)$$

and

$$V(m|x) = \beta \sum_{s \in S} V_s(m_s|x_s),$$

leaves us with two parameters α and β. They determine the rigidity of prior and conditional model. The higher these numbers, the larger the energy difference between the two states. If the values are low, moving from one state to another is easy: a low β value indicates that the threshold in the conditional model is not a very strict boundary, while a small α means that clustering is not so important. The choice $\alpha = 0$ disregards spatial structures. It only considers the uncertainty in the conditional model, and leads to a kind of 'smoothed hard-thresholding'.

From our numerical experiments, we learned that the best results were obtained for $\alpha \approx 2$ and $\beta \approx 90$. Ten MCMC iterations were sufficient to obtain the labels in Figure 9, provided we use the labels of the threshold procedure as the initial state. More iterations (up to 100) did not improve the output quality. We used the minimum GCV-threshold. This image comes close to the labels obtained by applying the minimum MSE threshold to the noise-free coefficients in Figure 3(a). The right picture is the corresponding result of the algorithm, applied to all subbands in the three finest resolution levels. Signal-to-noise ratio is 18.47 dB.

Figure 10(a) illustrates the influence of the prior if β is set to 1 (instead of $\beta = 90$). The prior indeed causes edge-like structures, but without the aid of the conditional, these structures do not show a strong correspondence to the important image features.

For a good parameter choice, the resulting label image "comes close to" the mask obtained by applying the threshold to the noise-free coefficients. However, since this example shows a *decrease* in signal-to-noise ratio, one could wonder if starting a from threshold value, different from the minimum MSE threshold, could yield better results. Choosing the threshold $\delta = \sigma$ should lead to an approximation of the ideal diagonal projection method of Figure 3(b). The result appears in Figure 10(b). Of course, we do not know the exact value of σ in each subband, so we assume that the energy of the coefficients below the threshold is a good measure for the noise variance. SNR is now 18.78dB.

FIGURE 9. Left: label image after ten MCMC iterations. Consequently, this image has 11 grey values. A pixel value is an estimate of the marginal posterior probability $P(X_s = 1|M)$. We used the minimum GCV-threshold. Right: the algorithm output. Three resolution levels were processed. Signal-to-noise ratio is 18.47 dB.

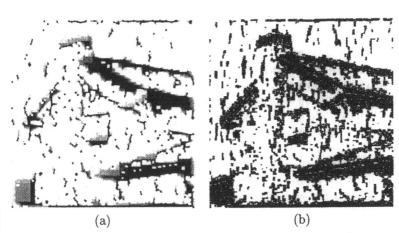

FIGURE 10. (a) The same label image as in the previous case, but now with conditional parameter $\beta = 1$ instead of $\beta = 90$. The influence of the prior causes edge-like structures, but without the aid of the conditional, these structures correspond too little with the really important features. (b) The labels of the horizontal component at the one but finest scale if we take $\delta = \sigma$.

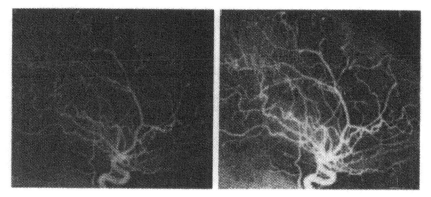

FIGURE 11. Digital subtraction angiogram of cerebral blood vessels with artificially added noise. SNR = 4.58dB. The picture on the right is the result of a contrast enhancement technique, called histogram equalization.

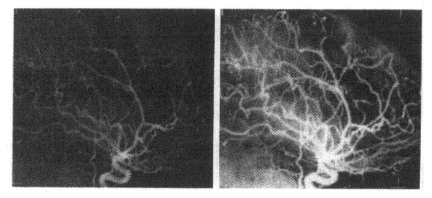

FIGURE 12. Result of classic threshold algorithm on DSA-image. SNR = 17.02dB.

Several possible improvements still need to be investigated. First, the choice of the hyperparameters α and β could maybe done on empirical basis. Also, these hyperparameters could be made subband-dependent: label images at different subbands show different characteristics. In general, there is an everlasting search for appropriate priors: priors based on local potential functions are attractive because of their easy computation, but at the same time, a prior should be selective on a more global region, to distinguish edges and other image features. A threshold selection based on interscale correlations could deal with interlevel dependencies.

We now apply the procedure to a DSA-image (Digital Subtraction Angiography) of cerebral blood vessels, with artificially added noise. The SNR value of the input image in Figure 11 is 4.58dB. Because of the fine image structure, we operate only on the highest resolution level. A classical threshold algorithm leads to Figure 12 with SNR = 17.02dB. A Bayesian

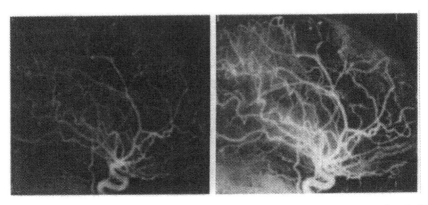

FIGURE 13. Result of Bayesian procedure. The initial labels were the threshold labels from the previous figure. SNR = 17.32dB.

correction on this image yields the result in Figure 13. SNR = 17.32dB. Some of the vessels are clearly better reproduced.

15.5   Conclusions

This chapter has introduced a geometrical prior model for configurations of important wavelet coefficients. This model can be used in combination with threshold algorithms. The threshold defines an initial binary "label image", where labels $X_s = 1$ indicate coefficients with large magnitude. Unlike more heuristic approaches, like erosion-dilation methods, Bayes' rule leads to a mathematically based procedure that cleans up the label images, taking into account prior beliefs and observed coefficients at the same time. The output are the posterior probabilities of the coefficients being sufficiently noise-free. These probabilities can be used to replace the binary threshold procedure with a more continuous approach. Although results are promising, further research for priors and for optimal hyperparameters remains necessary.

Acknowledgments: This paper presents research results of the Belgian Programme on Interuniversity Poles of Attraction, initiated by the Belgian State, Prime Minister's Office for Science, Technology and Culture. The scientific responsibility rests with its authors. The first author is financed by a grant from the Flemish Institute for the Promotion of Scientific and Technological Research in the Industry (IWT).

References

Abramovich, F., Sapatinas, F., and Silverman, B. W. (1997). Wavelet thresholding via a Bayesian approach. *Journal of the Royal Statistical Society, Series B*, 58.

Antonini, M., Barlaud, M., Mathieu, P., and Daubechies, I. (1992). Image coding using the wavelet transform. *IEEE Transactions on Image Processing*, 1(2):205–220.

Bello, M. G. (1994). A combined Markov Random Field and wave-packet transform-based approach for image segmentation. *IEEE Transactions on Image Processing*, 3(6):834–846.

Berkner, K. and Wells, Jr., R. O. (1998). Correlation-dependent model for denoising via nonorthogonal wavelet transforms. Technical report, C. M. L., Dept. Mathematics, Rice University.

Besag, J. E. (1974). Spatial interaction and the spatial analysis of lattice systems. *Journal of the Royal Statistical Society, Series B*, 36:192–236.

Charbonnier, P., Blanc-Féraud, L., and Barlaud, M. (1992). Noisy image restoration using multiresolution Markov Random Fields. *Journal of Visual Communication and Image Representation*, 3(4):338–346.

Chipman, H., Kolaczyk, E., and McCulloch, R. (1997). Adaptive Bayesian wavelet shrinkage. *J. Amer. Statist. Assoc.* To Appear.

Clyde, M., Parmigiani, G., and Vidakovic, B. (1998). Multiple shrinkage and subset selection in wavelets. *Biometrika*, 85:391–402.

Cohen, A., Daubechies, I., and Feauveau, J. (1992). Bi-orthogonal bases of compactly supported wavelets. *Comm. Pure Appl. Math.*, 45:485–560.

Crouse, M. S., Nowak, R., and Baraniuk, R. G. (1998). Wavelet-based signal processing using Hidden Markov Models. *IEEE Transactions on Signal Processing*, Special Issue on Wavelets and Filterbanks. To Appear.

Daubechies, I. (1992). *Ten Lectures on Wavelets*. CBMS-NSF Regional Conf. Series in Appl. Math., Vol. 61. Society for Industrial and Applied Mathematics, Philadelphia, PA.

Donoho, D. L. and Johnstone, I. M. (1994). Ideal spatial adaptation via wavelet shrinkage. *Biometrika*, 81:425–455.

Donoho, D. L. and Johnstone, I. M. (1995). Adapting to unknown smoothness via wavelet shrinkage. *J. Amer. Statist. Assoc.*, 90:1200–1224.

Geman, S. and Geman, D. (1984). Stochastic relaxation, Gibbs distributions, and the Bayesian restoration of images. *IEEE Transactions on Pattern Analysis and Machine Intelligence*, 6(6):721–741.

Jansen, M. and Bultheel, A. (1998). Multiple wavelet threshold estimation by generalized cross validation for images with correlated noise. *IEEE Transactions on Image Processing.* to appear.

Jansen, M., Malfait, M., and Bultheel, A. (1997). Generalized cross validation for wavelet thresholding. *Signal Processing,* 56(1):33–44.

Johnstone, I. M. and Silverman, B. W. (1997). Wavelet threshold estimators for data with correlated noise. *Journal of the Royal Statistical Society, Series B,* To Appear.

Luettgen, M. R., Karl, W. C., Willsky, A. S., and Tenney, R. R. (1993). Multiscale representations of Markov Random Fields. *IEEE Transactions on Signal Processing,* 41(12):3377–3395.

Malfait, M. (1995). Using wavelets to suppress noise in biomedical images. In Aldroubi, A. and Unser, M., editors, *Wavelets in Medicine and Biology,* Chapter 8, pages 191–208. CRC Press.

Malfait, M. and Roose, D. (1997). Wavelet based image denoising using a Markov Random Field a priori model. *IEEE Transactions on Image Processing,* 6(4):549–565.

Nason, G. P. and Silverman, B. W. (1995). The stationary wavelet transform and some statistical applications. In Antoniadis, A. and Oppenheim, G., editors, *Wavelets and Statistics,* Lecture Notes in Statistics, pages 281–299.

Pesquet, J.-C., Krim, H., and Carfantan, H. (1996). Time invariant orthonormal wavelet representations. *IEEE Transactions on Signal Processing,* 44(8):1964–1970.

Ruggeri, F. and Vidakovic, B. (1995). A Bayesian decision theoretic approach to wavelet thresholding. Preprint 95-35, Duke University, Durham, NC.

Simoncelli, E. P. and Adelson, E. (1996). Noise removal via Bayesian wavelet coring. In *proceedings 3rd International Conference on Image Processing.*

Vidakovic, B. (1994). Nonlinear wavelet shrinkage with Bayes rules and Bayes factors. Preprint 94-24, Duke University, Durham, NC.

Weyrich, N. and Warhola, G. T. (1995). De-noising using wavelets and cross validation. In Singh, S., editor, *Approximation Theory, Wavelets and Applications,* volume 454 of *NATO ASI Series C,* pages 523–532.

Winkler, G. (1995). *Image analysis, random fields and dynamic Monte Carlo methods.* Applications of Mathematics. Springer.

Xu, Y., Weaver, J. B., Healy, D. M., and Lu, J. (1994). Wavelet transform domain filters: a spatially selective noise filtration technique. *IEEE Transactions on Image Processing,* 3(6):747–758.

16

Multiscale Hidden Markov Models for Bayesian Image Analysis

Robert D. Nowak

ABSTRACT Bayesian multiscale image analysis weds the powerful modeling framework of probabilistic graphs with the intuitively appealing and computationally tractable multiresolution paradigm. In addition to providing a very natural and useful framework for modeling and processing images, Bayesian multiscale analysis is often much less computationally demanding compared to classical Markov random field models. This chapter focuses on a probabilistic graph model called the multiscale hidden Markov model (MHMM), which captures the key inter-scale dependencies present in natural signals and images. A common framework for the MHMM is presented that is capable of analyzing both Gaussian and Poisson processes, and applications to Bayesian image analysis are examined.

16.1   Introduction

The goal of image analysis is to extract some information of interest from image data. The information may simply be the underlying image intensities or the location and/or boundaries of objects, or it may be a high-level description of a scene. The common feature of the vast majority of these challenging problems is that they usually cannot be solved without including prior information or knowledge. Hence, many of the more successful image analysis tools are Bayesian. Arguably, the crucial element of Bayesian techniques is the choice of the prior probability model. The most common tools for Bayesian image analysis are Markov random field (MRF) models, which have been successfully applied in a host of problems including restoration, segmentation, and tomographic reconstruction. For examples, see Cross and Jain (1983), Geman and Geman (1984), and Chellappa and Jain (1993). Since edges and other inhomogeneities are key visual features (as is testified to by the huge amount of research devoted to the problem of edge detection), good image priors should be capable of representing them. Although this is possible within the MRF framework, the resulting inference criteria require, in general, computationally intensive Monte Carlo methods, like the version of the Metropolis algorithm proposed by Geman and Geman (1984).

Multiscale (or multiresolution) techniques have been another popular and successful approach to many image analysis problems. Beginning with the seminal work of Adelson and Burt (1981), multiscale image analysis has found application in a wide range of tasks, from low-level image processing to high-level machine vision, and today, it is the basis for most state-of-the-art image compression schemes. One of the advantages of the multiscale approach is its computational efficiency; multiscale analysis involves efficient schemes for passing intermediate results obtained at one analysis scale to the next scale of analysis. Remarkably, Field (1993) argues that similar processing mechanisms take place in the human visual system.

Recently, attempts have been made to develop multiscale image models, that combine the powerful modeling framework of MRFs with the intuitively appealing and computationally tractable multiscale analysis paradigm. Rather than specifying inter-pixel relationships directly in the spatial domain, multiscale models attempt to represent structural relationships more efficiently through *causal* relationships across scales of analysis. Along this line of thinking, various types of multiscale stochastic image models have recently been proposed by a number of researchers, *e.g.*, Charbonnier et al. (1992), Bouman and Shapiro (1994), Crouse et al. (1998), Luettgen et al. (1993), Malfait and Roose (1997), Simoncelli (1997), Timmermann and Nowak (1997), and Vidakovic (1998). These models have been shown to be useful and adequate for a wide range of problems, and lead to (signal and image) processing methods which are much less computationally demanding than those obtained from the classical MRF framework. The computational efficiency stems from the fact that the joint probability distribution associated with causal multiscale models can be specified in terms of conditional probability functions, instead of MRF clique potential functions which are generally much more difficult to work with. This chapter focuses on one multiscale model in particular, the multiscale hidden Markov model (MHMM), which is a generalization of the wavelet-domain HMM developed by Crouse et al. (1996, 1998) for Gaussian observation models. The MHMM is a probabilistic graph model constructed on a quadtree[1] (or binary tree in the case of one-dimensional signals) associated with multiscale image analysis. MHMMs capture the key inter-scale dependencies present in natural signals and images.

The relationship between classical MRFs and multiscale models has been extensively studied by Gidas (1989), Luettgen et al. (1993), and Pérez and Heitz (1996). It is well known that most multiscale models display *long-range* dependencies, and do not, in general, possess a local Markovian property at all scales as shown by Pérez and Heitz (1996). However,

---

[1] See Mallat (1998) for general information on the tree structures associated with wavelet and multiscale analysis.

since many natural signals do display long-range dependencies, the non-local behavior of multiscale models may be quite desirable. In fact, certain multiscale models generate $1/f$ random processes[2] which appear to be very well matched to the spectral characteristics of natural imagery as demonstrated in the comprehensive study of van der Schaaf and van Hateren (1996). See the work of Wornell (1996), Nowak (1998), and Timmermann and Nowak (1999) for more information on $1/f$ processes and multiscale analysis.

This chapter is organized as follows. Section 2 reviews the basic multiscale analysis of Gaussian and Poisson processes, which are two of the most commonly encountered data models in image processing. Section 3 considers simple "independent parameter" prior probability models for Bayesian multiscale analysis. Section 4 studies a more sophisticated prior model, the MHMM, that moves beyond the assumption of independence and that better reflects the characteristics of natural signals and images. To keep the presentation as simple as possible, Sections 2, 3, and 4 focus on the one-dimensional (signal) setting. Section 5 discusses some additional issues arising in the two-dimensional (image) setting. Section 6 examines two applications of this framework to image analysis. Conclusions are made in Section 7. In order to deal with both Gaussian and Poisson problems within the same general framework, multiscale analyses and models based on a Haar multiscale analysis are emphasized throughout the chapter. In the Gaussian case, an analogous MHMM framework was developed by Crouse et al. (1998) for multiscale analysis based on any orthogonal wavelet basis.

16.2   Multiscale Data Analysis

In signal and image analysis applications, two of the most common observation models are the Gaussian and Poisson models, see Castleman (1996) for further discussion. For simplicity, let us consider these two models in one dimension; extensions to two dimensions are discussed in Section 5. The Gaussian observation model is:

$$x_k = \mu_k + w_k, \quad k = 0, \ldots, 2^J - 1, \tag{1}$$

where $\mathbf{x} = \{x_k\}$ are observations, $\boldsymbol{\mu} = \{\mu_k\}$ are interpreted as signal samples, and $\{w_k\}$ are realizations of a Gaussian noise process. For convenience in the subsequent multiscale analysis, we adopt the usual convention that the length of the signal is a power of 2, however it is possible to deal with signals of arbitrary length in a multiscale framework. The $\{w_k\}$ are independent, identically distributed samples of a zero-mean Gaussian random

---

[2]A $1/f$ process is a random process whose power spectrum behaves like $1/|f|^\gamma$, for some power $\gamma > 0$.

variable with *known* variance σ^2, leading to the likelihood function

$$p(\mathbf{x}\,|\,\boldsymbol{\mu}) = \prod_{k=0}^{2^J-1} \mathcal{N}(x_k\,|\,\mu_k,\sigma^2), \tag{2}$$

where $\mathcal{N}(x\,|\,\mu,\sigma^2)$ denotes a Gaussian density with mean μ and variance σ^2 evaluated at the point x.

In the Poisson case, the data are (conditionally) independent

$$x_k \sim \mathcal{P}(x_k\,|\,\mu_k), \quad k = 0,\dots,2^J - 1, \tag{3}$$

where $\mathcal{P}(x\,|\,\mu)$ denotes the Poisson mass function with intensity μ evaluated at the point x. The likelihood function in this case is simply

$$p(\mathbf{x}\,|\,\boldsymbol{\mu}) = \prod_{k=0}^{2^J-1} \mathcal{P}(x_k\,|\,\mu_k). \tag{4}$$

Now let us consider a multiscale data analysis. In general, multiscale analysis refers to the study of behavior or structure in signals or data at various spatial and/or temporal resolutions. For further background information on multiscale signal analysis see Mallat (1998). Perhaps the simplest technique is the Haar multiscale analysis defined according to:

$$\begin{aligned}
x_{J,k} &\equiv x_k, \quad k = 0,\dots,2^J - 1 \\
x_{j,k} &= x_{j+1,2k} + x_{j+1,2k+1}, \quad k = 0,\dots,2^j - 1,\ 0 \le j \le J - 1.
\end{aligned}$$

The index j refers to the resolution of the analysis, 2^j; $j = J$ being the highest resolution (finest scale), and $j = 0$ being the lowest resolution (coarsest scale). This multiscale data analysis is organized into the binary data tree shown in Figure 1.

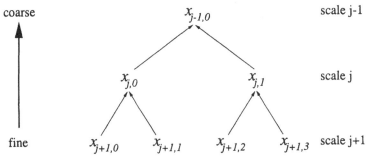

FIGURE 1. *Binary data tree associated with multiscale (fine-to-coarse) analysis. In this figure, analysis begins at fine scale $j + 1$ and produces coarser representations of data at scales j and $j - 1$.*

The data $\{x_{j,k}\}$ are the (unnormalized) Haar scaling coefficients of $\mathbf{x}$. The relationship between a "parent" (*e.g.*, $x_{j,k}$) and a "child" (*e.g.*, $x_{j+1,2k}$)

is of fundamental interest in multiscale data analysis. Specifically, given an observation model, e.g., Gaussian, we are interested in the conditional distribution of the child given the parent.

The parent-child relationship is expressed by the conditional likelihood $p(x_{j+1,2k} \mid x_{j,k}, \mu)$, which happens to have a very simple form for both the Gaussian and Poisson observation models. Note that it is unnecessary to consider the conditional likelihoods of both children due to the fact that $x_{j+1,2k+1}$ is uniquely determined by $x_{j+1,2k}$ and $x_{j,k}$. Before examining the conditional likelihoods, let us define a multiscale analysis of the parameter μ, analogous to that defined for the data $\mathbf{x}$:

$$\mu_{J,k} \equiv \mu_k, \quad k = 0, \dots, 2^J - 1$$
$$\mu_{j,k} = \mu_{j+1,2k} + \mu_{j+1,2k+1}, \quad k = 0, \dots, 2^j - 1, \ 0 \le j \le J - 1.$$

The parameters $\{\mu_{j,k}\}$ are the (unnormalized) Haar scaling coefficients of μ. With this definition in hand, and using standard conditional probability relationships between pairs of Gaussian and Poisson random variables, we have the following expressions for the parent-child conditional likelihoods.

- Gaussian model:

$$p(x_{j+1,2k} \mid x_{j,k}, \mu) = \mathcal{N}\left(x_{j+1,2k} \mid \frac{x_{j,k}}{2} + \frac{\mu_{j+1,2k} - \mu_{j+1,2k+1}}{2}, \frac{\sigma_{j+1}^2}{2}\right),$$

where $\sigma_j^2 = 2^{J-j}\sigma^2$; $\tag{5}$

- Poisson model:

$$p(x_{j+1,2k} \mid x_{j,k}, \mu) = \mathcal{B}\left(x_{j+1,2k} \mid x_{j,k}, \frac{\mu_{j+1,2k}}{\mu_{j,k}}\right), \tag{6}$$

where $\mathcal{B}(x \mid n, \theta) = \binom{n}{x}\theta^x(1-\theta)^{n-x}$, denotes the binomial distribution with parameters n and θ. From these expressions, we identify the *canonical* multiscale parameters associated with the two models. In the Gaussian case, the canonical parameter is $\theta_{j,k} = \mu_{j+1,2k} - \mu_{j+1,2k+1}$, which is simply the (unnormalized[3]) Haar wavelet coefficient of the mean (signal) μ at resolution 2^j and location k. In the Poisson case, the canonical parameter is $\theta_{j,k} = \frac{\mu_{j+1,2k}}{\mu_{j,k}}$, which can be viewed as a "splitting" factor[4] that governs the multiscale refinement of the intensity μ. This type of multiscale intensity analysis was introduced independently by Timmermann and Nowak (1997) and Kolaczyk (1998). Also see Timmermann and Nowak (1999) for further

---

[3]The normalized Haar wavelet coefficients are obtained by the mapping $\theta_{j,k} \mapsto 2^{(j-J)/2}\theta_{j,k}$.

[4]Note that the splitting factors are also closely related to the Haar wavelet coefficients since $\frac{\mu_{j+1,2k}}{\mu_{j,k}} = \frac{1}{2}\left(1 + \frac{\mu_{j+1,2k} - \mu_{j+1,2k+1}}{\mu_{j,k}}\right)$.

details. As discussed in Section 16.3, the canonical multiscale parameters suggest the use of special prior distributions that complement the observation model leading to tractable and highly efficient processing strategies.

The simplicity of these relationships (well-known parametric distributions) is quite exceptional; in general, under other observation models, the parent-child relationship can be much more complicated, and usually does not admit a standard parametric form. For this reason, one can argue that multiscale analysis is especially well suited to Gaussian and Poisson data. In particular, in either case, one can factorize the likelihood function as follows:

$$p(\mathbf{x} \,|\, \mu) \;\equiv\; p(x_{0,0} \,|\, \mu_{0,0}) \prod_{j=0}^{J-1} \prod_{k=0}^{2^j-1} p(x_{j+1,2k} \,|\, x_{j,k}, \theta_{j,k}), \qquad (7)$$

where $p(x_{j+1,2k} \,|\, x_{j,k}, \theta_{j,k})$ is given by (5) in the Gaussian case, and (6) in the Poisson case. This factorization is possible because the linear mapping of observations to scaling coefficients, $\mathbf{x} = \{x_k\} \mapsto \{x_{j,2k}\}$, has a unit Jacobian.

Note that in the Gaussian case the likelihood can also be expressed equivalently in terms of the wavelet coefficients of the observation $\mathbf{x}$. That is, the Gaussian likelihood of $\mathbf{x}$ given μ can be written as a product of univariate Gaussian likelihoods, each involving a single "data" wavelet coefficient $x_{j+1,2k} - x_{j+1,2k+1}$ given the corresponding signal wavelet coefficient $\theta_{j,k} = \mu_{j+1,2k} - \mu_{j+1,2k+1}$. This is a more standard likelihood factorization for the Gaussian case, and it is possible because of the orthogonality of the discrete wavelet transform and the fact that the Gaussian likelihood structure is preserved under orthogonal linear transformations. In fact, this alternative "wavelet-based" factorization can be employed in conjunction with any orthogonal wavelet system, the Haar being one special case.

A similar wavelet-based factorization is not possible in the Poisson case; the difficulty lies in the fact that the Poisson distribution reproduces under straight (unweighted) summation (the sum of Poisson random variables is still Poisson), but not under rescaling. Therefore we use the factorization (7) above in order to treat both Gaussian and Poisson models within a common framework. For more information on this likelihood factorization and a discussion of its fundamental role in multiscale statistical analysis in general, see Kolaczyk (1999) in this volume.

The likelihood factorization also greatly facilitates multiscale analysis and modeling. For example, estimates of the multiscale parameters $\theta = \{\theta_{j,k}\}$ can be used to reconstruct an estimate of the underlying signal or intensity. In Section 16.6, we will look at two Bayesian image analysis applications based on the multiscale parameters. In general, Bayesian inference based on the multiscale parameters θ requires; (1) specification of a suitable prior probability model for θ; (2) determination of the posterior probability distribution of θ resulting from the likelihood and prior. The next two sections examine two types of prior probability models and their respective posterior distributions. Section 16.3 considers a simple approach in which $\boldsymbol{\theta}$ are modeled as independent random variables with prior probability densities designed to reflect the characteristics of natural signals and images and that lead to simple expressions for the posterior density. Section 16.4 examines a more sophisticated prior model, the MHMM, that moves beyond the assumption of independence and captures the parent-child dependencies also encountered in practice. To keep the notation and derivations as simple and clear as possible, throughout the remainder of the chapter we assume that the scaling coefficient at the coarsest scale, $\mu_{0,0}$, is known. The Bayesian modeling and analysis methods described next can be easily extended to include prior models and inference schemes that include $\mu_{0,0}$ as well.

16.3   Independent Parameter Models

Let us now consider prior probability models for the (unknown) canonical multiscale parameters $\boldsymbol{\theta}$. *Conjugate priors* are advantageous for computational reasons since the posterior distribution is obtained by simply "updating" the parameters of the prior based on the observations; see (Robert, 1994, pp. 97-111) for further information. Moreover, we will see that conjugate priors can provide very plausible models for the multiscale parameters. For the Gaussian likelihood function, the *natural* conjugate prior is also Gaussian. Hence, a simple approach is to model each multiscale parameter as an independent Gaussian $\theta \sim \mathcal{N}(\theta \,|\, 0, \tau^2)$. In the Poisson case, the natural conjugate prior for the binomial distribution is a beta density. Therefore, we model each multiscale parameter as an independent beta distributed random variable, $\theta \sim \mathcal{B}e(\theta \,|\, \alpha, \beta) = \frac{\theta^{\alpha-1}(1-\theta)^{\beta-1}}{B(\alpha,\beta)}, \ 0 \leq \theta \leq 1$, where $B(\alpha, \beta)$ denotes the standard beta function. In this chapter, we will only use symmetric beta priors of mean $1/2$, characterized by $\alpha = \beta$. Note that the Gaussian is also a symmetric prior (about its mean, in this case, zero). In both cases, we choose a symmetric prior for θ since there is no *a priori* support for asymmetry. Also, in general, the variance τ^2 or parameter α may depend on the resolution 2^j. Here, as in most related approaches, the parameters do not depend on the location k, since location dependent signal characteristics are usually not known *a priori*. Thus, a simple model for the unknown parameters $\boldsymbol{\theta}$ is

$$p(\boldsymbol{\theta}) \;=\; \prod_{j=0}^{J-1} \prod_{k=0}^{2^j-1} p(\theta_{j,k}), \tag{8}$$

with $p(\theta_{j,k})$ equal to $\mathcal{N}(\theta_{j,k} \,|\, 0, \tau_j^2)$ or $\mathcal{B}e(\theta_{j,k} \,|\, \alpha_j, \alpha_j)$ for the Gaussian or Poisson case, respectively.

   Although this "independent parameter" prior may appear too simplistic, in certain cases it is quite reasonable because multiscale decompositions tend to decorrelate signals and images. For example, if the signal is a fractional Brownian motion or $1/f$ process, then the correlations between Haar wavelet coefficients decay very rapidly across scale and space as demonstrated by Flandrin (1992) and Wornell (1996). Moreover, it can be shown that with special choices of $\{\tau_j^2\}$ or $\{\alpha_j\}$ the prior above (8) displays $1/f$-like behavior, see Nowak (1998) and Timmermann and Nowak (1999) for more information.

In both the Gaussian and Poisson cases, combining the prior (8) with the likelihood (7) produces a posterior density

$$p(\boldsymbol{\theta}\,|\,\mathbf{x}) \;=\; \prod_{j=0}^{J-1} \prod_{k=0}^{2^j-1} p(\theta_{j,k}\,|\,x_{j,k},x_{j+1,2k}), \qquad (9)$$

where

- Gaussian model:

$$p(\theta_{j,k}\,|\,x_{j,k},x_{j+1,2k}) \;=\; \mathcal{N}\left(\theta_{j,k}\,\Big|\,\frac{\tau_j^2\,(2x_{j+1,2k}-x_{j,k})}{\tau_j^2+\sigma_j^2},\,\frac{\sigma_j^2\tau_j^2}{\tau_j^2+\sigma_j^2}\right);$$

- Poisson model:

$$p(\theta_{j,k}\,|\,x_{j,k},x_{j+1,2k}) \;=\; \mathcal{B}e\left(\theta_{j,k}\,|\,\alpha_j+x_{j+1,2k},\alpha_j+x_{j,k}-x_{j+1,2k}\right).$$

The marginal posteriors above can be easily derived; also see (Robert, 1994, p. 104) for general forms of the posterior distributions resulting from these conjugate priors). The factorization of the posterior shows that inferences can be made on each multiscale parameter individually, instead of requiring a complicated high dimensional analysis. For example, it is straightforward to obtain the posterior mean or maximum *a posteriori* (MAP) estimate of each individual parameter, based on the one-dimensional Gaussian or beta posterior densities above. Similarly, other meaningful quantities such as posterior variances and confidence regions can also be easily computed because the posterior factorizes into one-dimensional parametric densities. Some specific examples in image analysis are considered in Section 16.6. Finally, notice that the analysis presented for the Gaussian case can clearly be generalized to other orthogonal wavelet bases following the work of Crouse et al. (1998).

16.3.1 Mixture Density Priors

A richer class of priors, more suitable for modeling the multiscale parameters of natural signals and images, is provided by mixture densities. Specifically, one can build larger classes of priors using mixtures of the *elementary* conjugate densities mentioned above; these mixture priors are still conjugate, with all the associated computational convenience; see (Robert, 1994,

p. 108) for more details. For the Gaussian observation model, Gaussian mixtures are often used as in Abramovich et al. (1998), Chipman et al. (1997), and Crouse et al. (1998), while for the Poisson likelihood, beta mixtures are adopted as in Timmermann and Nowak (1999). Formally:

- Gaussian model:

$$p(\theta_{j,k}) = \sum_{m=0}^{M-1} \rho_j(m)\, \mathcal{N}\left(\theta_{j,k}\,|\,0, \tau_{j,m}^2\right); \qquad (10)$$

- Poisson model:

$$p(\theta_{j,k}) = \sum_{m=0}^{M-1} \rho_j(m)\, \mathcal{B}e\left(\theta_{j,k}\,|\,\alpha_{j,m}, \alpha_{j,m}\right), \qquad (11)$$

where $\{\rho_j(m)\}_{m=0}^{M-1}$ denote the *a priori* probabilities of each component at scale j. To keep the notation simple, we will use M component mixtures at all scales.

The motivation for using mixtures is based on the following reasoning. If we believe that the underlying signal μ is generally smooth, except for (possibly) a few large singularities, then, for example, a mixture consisting of a highly probable low-variance component (to model the smooth areas of the signal) and a relatively low probability high-variance component (to model the possible singularities) is intuitively reasonable. A state (also called *latent* or *indicator*) variable $s_{j,k}$ is usually associated with each parameter $\theta_{j,k}$. The state takes values that indicate which component of the mixture is in effect; for example, in the Gaussian model, $p(\theta_{j,k}\,|\,s_{j,k} = m) = \mathcal{N}\left(\theta_{j,k}\,|\,0, \tau_{j,m}^2\right)$. The prior probabilities for each state are the *a priori* mixture weights, *i.e.*, $p(s_{j,k} = m) = \rho_j(m)$. These probabilities, along with the density shape parameters associated with them (i.e., the values of $\tau_{j,m}^2$ and $\alpha_{j,m}$), can be chosen based on prior beliefs about the regularity of the class of signals/images in question following the work of Abramovich et al. (1998), or can be inferred from the observed data through a hierarchical Bayes setting (possibly via an empirical Bayes approach) as in the work of Chipman et al. (1997), Crouse et al. (1998), Timmermann and Nowak (1997, 1999), and Kolaczyk (1998).

The posterior distribution has the same factorized form as (9) with mixture densities in place of the corresponding single component densities. Let $\mathbf{s} = \{s_{j,k}\}$ denote the set of state variables and $\mathbf{m} = \{m_{j,k}\}$ denote a set of state values. We can then write

$$p(\boldsymbol{\theta}, \mathbf{s} = \mathbf{m}\,|\,\mathbf{x}) = p(\boldsymbol{\theta}\,|\,\mathbf{s} = \mathbf{m}, \mathbf{x})\, p(\mathbf{s} = \mathbf{m}\,|\,\mathbf{x}) \qquad (12)$$

and

$$p(\boldsymbol{\theta}\,|\,\mathbf{x}) = \sum_{\mathbf{m}} p(\boldsymbol{\theta}\,|\,\mathbf{s} = \mathbf{m}, \mathbf{x})\, p(\mathbf{s} = \mathbf{m}\,|\,\mathbf{x}), \qquad (13)$$

where the sum is over all possible sets of state values. The "state-conditional" density $p(\theta \mid s = m, x)$ factorizes just as in (9)

$$p(\theta \mid s = m, x) = \prod_{j=0}^{J-1} \prod_{k=0}^{2^j-1} p(\theta_{j,k} \mid x_{j,k}, x_{j+1,2k}, s_{j,k} = m_{j,k}), \qquad (14)$$

The posterior state probabilities can be very efficiently computed due to the likelihood factorization (7). Note that

$$p(s = m \mid x) = \int p(s = m, \theta \mid x)\, d\theta$$

$$\propto \int p(x \mid s = m, \theta) p(\theta \mid s = m)\, p(s = m)\, d\theta$$

$$\propto \prod_{j=0}^{J-1} \prod_{k=0}^{2^j-1} \int p(x_{j+1,2k} \mid x_{j,k}, \theta_{j,k})$$

$$\times\, p(\theta_{j,k} \mid s_{j,k} = m_{j,k})\, p(s_{j,k} = m_{j,k})\, d\theta_{j,k}.$$

This shows that

$$p(s_{j,k} = m \mid x) = \frac{\rho_j(m)\, L_{j,k}(m)}{\sum_{m=0}^{M-1} \rho_j(m)\, L_{j,k}(m)}, \qquad (15)$$

where

$$L_{j,k}(m) \equiv \int p(x_{j+1,2k} \mid x_{j,k}, \theta_{j,k})\, p(\theta_{j,k} \mid s_{j,k} = m)\, d\theta_{j,k}. \qquad (16)$$

$L_{j,k}(m)$ is a marginal likelihood, i.e., $L_{j,k}(m) = p(x_{j+1,2k} \mid x_{j,k}, s_{j,k} = m)$, and it has a simple closed-form expression in both the Gaussian and Poisson cases:

- Gaussian model:

$$L_{j,k}(m) \propto \frac{1}{\left(\sigma_j^2 + \tau_{j,m}^2\right)^{1/2}} \exp\left(-\frac{(2x_{j+1,2k} - x_{j,k})^2}{2\left(\sigma_j^2 + \tau_{j,m}^2\right)}\right);$$

- Poisson model:

$$L_{j,k}(m) \propto \frac{B\left(x_{j+1,2k} + \alpha_{j,m}, x_{j,k} - x_{j+1,2k} + \alpha_{j,m}\right)}{B\left(\alpha_{j,m}, \alpha_{j,m}\right)}.$$

Therefore, the posterior density $p(\theta \mid x)$ is given by

$$p(\theta \mid x) = \prod_{j=0}^{J-1} \prod_{k=0}^{2^j-1} \sum_{m=0}^{M-1} p(s_{j,k} = m \mid x)\, p(\theta_{j,k} \mid x_{j,k}, x_{j+1,2k}, s_{j,k} = m),$$

$$(17)$$

where $p(\theta_{j,k} \mid x_{j,k}, x_{j+1,2k}, s_{j,k} = m)$ denotes a Gaussian or Beta density of the forms given in (9), with shape parameter $\tau_{j,m}^2$ or $\alpha_{j,m}$, respectively. Again, the factorization of the posterior and the simple parametric form of each factor shows that inferences can be easily made on each multiscale parameter individually. For example, mixture prior densities typically lead to posterior mean or MAP estimators, as in Abramovich et al. (1998), Chipman et al. (1997), Crouse et al. (1998), Timmermann and Nowak (1997, 1999), that resemble the non-linear shrinkage/thresholding estimators encountered in standard non-Bayesian approaches to wavelet-based noise removal like the now classical denoising methods developed by Donoho and Johnstone (1994).

16.4   Multiscale Hidden Markov Models

The multiscale hidden Markov model (MHMM) is a graphical model based on the binary tree (or quadtree in the case of image data) associated with multiscale signal analysis. One instance of the MHMM is the wavelet domain HMM developed in Crouse et al. (1996, 1998) for Gaussian observation models. Here, we focus on the Haar multiscale analysis discussed above, and develop a more general framework encompassing both the Gaussian and Poisson models. The priors described in Section 16.3 modeled the multiscale parameters $\boldsymbol{\theta}$ as independent mixture random variables. The MHMM moves beyond this simple prior, by specifying probabilistic dependencies between the states underlying the mixtures of parent and child multiscale parameters.

The MHMM is a *directed acyclic graph* (or *Bayesian network*), as depicted in Figure 2; see Pearl (1988) for more information on graphical models in general. The MHMM has a causal coarse-to-fine[5] scale structure indicated by the direction of the arrows. Specifically, the MHMM is based on the assumption that the value of each state $s_{j,k}$ is *caused* by the value of its parent state $s_{j-1,\lfloor k/2 \rfloor}$, where $\lfloor k/2 \rfloor$ is the largest integer less than or equal to $k/2$. This means that, given the value of its parent's state, $s_{j,k}$ is independent of all other states at scales $i \leq j$ (at same level and above) in the tree. This enables the following factorization of the joint state probability function

$$p(\mathbf{s}) \;=\; \prod_{j=0}^{J-1} \prod_{k=0}^{2^j-1} p\left(s_{j,k} = m_{j,k} \mid s_{j-1,\lfloor k/2 \rfloor} = m_{j-1,\lfloor k/2 \rfloor}\right),$$

with the convention $p(s_{0,0} \mid s_{-1,0}) \equiv p(s_{0,0})$. This is a more structured alternative to the independent parameter model previously considered. Another important property of the MHMM is that, given their respective state val-

---

[5] Alternatively, we could construct a causal fine-to-coarse model.

ues, all parameters $\boldsymbol{\theta}$ are conditionally independent. That is,

$$p(\boldsymbol{\theta} \mid \mathbf{s} = \mathbf{m}) = \prod_{j=0}^{J-1} \prod_{k=0}^{2^j-1} p(\theta_{j,k} \mid s_{j,k} = m_{j,k}). \qquad (18)$$

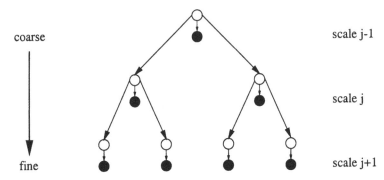

FIGURE 2. *Multiscale HMM graph. Each multiscale parameter of the signal (or intensity) μ is modeled as a mixture random variable (black node) with an associated hidden state variable (white node). To match the inter-scale dependencies, we link the hidden states across scale. The MHMM is a directed graph that synthesizes a signal in a coarse-to-fine fashion indicated by the direction of the arrows. These connections capture the key inter-scale dependencies present in many natural signals and images.*

This MHMM structure provides a mechanism for sharing (and exploiting) relevant inter-scale information and appears to be well justified by the empirical studies of Shapiro (1993) and Crouse et al. (1998); in fact, it is similar to the principles underlying some of the most successful wavelet based image compression algorithms known today such as the "zerotree" algorithm of Shapiro (1993). The MHMM captures the key inter-scale dependencies present in natural signals and images. These dependencies, termed *clustering* and *persistence-across-scale* by Crouse et al. (1998), refer to the fact that "high energy" multiscale parameters (*e.g.*, Haar wavelet coefficients) tend to cluster near edges in images, and that similar clusters are seen at multiple analysis scales indicating persistence-across-scale, as illustrated in Figure 3.

As a concrete example, consider the Gaussian case and suppose that there are two states associated with each multiscale parameter (Haar wavelet coefficient); state '0' corresponds to a low-variance Gaussian modeling the absence of a singularity, while state '1' is a high-variance Gaussian expressing the possible presence of a singularity (or edge). Because singularities, like edges in images, tend to persist across scales, if the parent state $s_{j-1,\lfloor k/2 \rfloor} = 1$, then it is probable that the child state $s_{j,k} = 1$, as well. Likewise, if $s_{j-1,\lfloor k/2 \rfloor} = 0$, indicating that signal is fairly smooth in the re-

(a)   (b)

FIGURE 3. *Persistence and clustering in an image. (a) Test image. (b) Magnitude of 2-d Haar DWT of test image. Each sub-image in (b) depicts the set of wavelet coefficients (white denotes largest magnitudes) at a specific scale and orientation (horizontal, vertical, or diagonal). For example, the three largest sub-images depict the wavelet coefficients at scale $J - 1$ and at horizontal (lower-left), vertical (upper-right), and diagonal (lower-right) orientations. The smaller sub-images depict the corresponding sets of wavelet coefficients at progressively coarser scales.*

gion corresponding to $\theta_{j-1,\lfloor k/2 \rfloor}$, then with high probability the sub-region corresponding to $\theta_{j,k}$ is also smooth and $s_{j,k} = 0$.

Now let us determine the posterior density associated with the MHMM. The posterior $p(\boldsymbol{\theta} \,|\, \mathbf{x})$ takes a form similar to (17), except that in this case the posterior state probabilities cannot be determined independently. However, $p(\boldsymbol{\theta} \,|\, \mathbf{x})$ can be efficiently computed as follows. As in (13), given $p(\mathbf{s} = \mathbf{m} \,|\, \mathbf{x})$, we compute

$$p(\boldsymbol{\theta} \,|\, \mathbf{x}) = \sum_{\mathbf{m}} p(\boldsymbol{\theta} \,|\, \mathbf{s} = \mathbf{m}, \mathbf{x})\, p(\mathbf{s} = \mathbf{m} \,|\, \mathbf{x}), \tag{19}$$

where, because of (18), $p(\boldsymbol{\theta} \,|\, \mathbf{s} = \mathbf{m}, \mathbf{x})$ is the state-conditional posterior density given by (14). So all that remains is to compute $p(\mathbf{s} = \mathbf{m} \,|\, \mathbf{x})$:

$$
\begin{aligned}
p(\mathbf{s} = \mathbf{m} \,|\, \mathbf{x}) \;&=\; \int p(\mathbf{s} = \mathbf{m}, \boldsymbol{\theta} \,|\, \mathbf{x})\, d\boldsymbol{\theta} \\[4pt]
&\propto\; \int p(\mathbf{x} \,|\, \mathbf{s} = \mathbf{m}, \boldsymbol{\theta}) p(\boldsymbol{\theta} \,|\, \mathbf{s} = \mathbf{m})\, p(\mathbf{s} = \mathbf{m})\, d\boldsymbol{\theta} \\[4pt]
&\propto\; \prod_{j=0}^{J-1} \prod_{k=0}^{2^j-1} \int p(x_{j+1,2k} \,|\, x_{j,k}, \theta_{j,k}, s_{j,k} = m_{j,k}) \\[2pt]
&\qquad\times\; p(\theta_{j,k} \,|\, s_{j,k} = m)\, p(s_{j,k} = m_{j,k} \,|\, s_{j-1,\lfloor \frac{k}{2} \rfloor} = m_{j-1,\lfloor \frac{k}{2} \rfloor})\, d\theta_{j,k} \\[4pt]
&=\; \prod_{j=0}^{J-1} \prod_{k=0}^{2^j-1} p(s_{j,k} = m_{j,k} \,|\, s_{j-1,\lfloor \frac{k}{2} \rfloor} = m_{j-1,\lfloor \frac{k}{2} \rfloor})\, L_{j,k}(m_{j,k}),
\end{aligned}
$$

where $L_{j,k}(m_{j,k})$ is defined in (16).

With this expression we can calculate the (marginal) posterior state probabilities using an *upward-downward probability propagation algorithm* (also called a forward-backward algorithm). Then, with posterior state probabilities in hand, we can compute the posterior density of the multiscale parameters according to (17). In the upward-downward algorithm, the **Up Step** recursively marginalizes (sub-tree by sub-tree) the joint posterior state probability beginning at the finest scale $j = J - 1$ (bottom of tree) up to the coarsest scale $j = 0$ (top "root" of tree). This provides us with the posterior state probabilities $\{p(s_{0,0} = m \,|\, \mathbf{x})\}_{m=0}^{M-1}$. The **Down Step** computes the marginal posterior state probabilities for each $s_{j,k}$ in a recursive fashion, making use of partial (sub-tree) marginalizations previously calculated in the **Up Step**. See Frey (1998) for a general overview of upward-downward (forward-backward) algorithms, their extensions, and connections to other inference methods in graphical models. For notational convenience, define $\rho_{j,k}(m|n) \equiv p(s_{j,k} = m \,|\, s_{j-1,\lfloor k/2 \rfloor} = n)$.

Upward-Downward Propagation Algorithm

Up Step

Beginning at $j = J - 1$ compute $q_{j,k}(n) = \sum_{m=0}^{M-1} \rho_{j,k}(m|n)L_{j,k}(m)$. Then for $j = J - 2, \ldots, 1$

$$q_{j,k}(n) = \sum_{m=0}^{M-1} q_{j+1,2k}(m)\, q_{j+1,2k+1}(m)\, \rho_{j,k}(m|n)L_{j,k}(m) \qquad (20)$$

and for $j = 0$, $q_{0,0}(n) = q_{1,0}(n)\, q_{1,1}(n)\, \rho_{0,0}(n)L_{0,0}(n)$, where $\rho_{0,0}(n) = p(s_{0,0} = n)$.

Note that $q_{j,k}(n)$ is the partial marginalization at node j, k over all states in the subtree beneath it. Hence, the final quantities $\{q_{0,0}(m)\}$ are the (unnormalized) posterior state probabilities $\{p(s_{0,0} = m \,|\, \mathbf{x})\}_{m=0}^{M-1}$ at the top (root) of the graph.

Down Step

Beginning the posterior states probabilities at scale $j = 0$, set $p_{0,0}(m) = q_{0,0}(m)$. Then for $j = 1, \ldots, J - 2$

$$p_{j,k}(m) = \sum_{n=0}^{M-1} \frac{p_{j-1,\lfloor \frac{k}{2} \rfloor}(n)\, \rho_{j,k}(m|n)\, q_{j+1,2k}(m)\, q_{j+1,2k+1}(m)\, L_{j,k}(m)}{q_{j,k}(n)}$$

$$(21)$$

and for $j = J - 1$,

$$p_{j,k}(m) = \sum_{n=0}^{M-1} \frac{p_{j-1,\lfloor \frac{k}{2} \rfloor}(n)\, \rho_{j,k}(m|n)\, L_{j,k}(m)}{q_{j,k}(n)}.$$

The final quantities $\{p_{j,k}(m)\}_{m=0}^{M-1}$ are the (unnormalized) posterior state probabilities $\{p(s_{j,k} = m \,|\, \mathbf{x})\}_{m=0}^{M-1}$.

Although the MHMM posterior density is more complicated than that of the independent multiscale parameter models considered in Section 16.3, the upward-downward algorithm provides us with a very efficient way of calculating the marginal posterior probabilities of the states. With these probabilities in hand, the factorization of the state-conditional posterior density (14) for the multiscale parameters shows that inference can be carried out on each multiscale parameter individually, just as in the independent multiscale parameter cases. The total computational complexity is $O(2^J)$, where 2^J is the number of data.

Finally, note that other graph structures (such as those including intra-scale dependencies) may be considered. However, computationally efficient algorithms may not be available; it is known that exact inference on general graphs is an NP-hard problem as discussed by Cooper (1990).

16.5   Extensions to Two Dimensions

The multiscale analyses and MHMMs can be easily extended to two dimensions. In two dimensions (2-d) the unnormalized Haar multiscale data analysis is as follows. We begin with data $\{x_{k,l}\}$, $k, l = 0, \ldots, 2^J - 1$, and define

$$
\begin{aligned}
x_{J,k,l} &\equiv x_{k,l}, \quad k, l = 0, \ldots, 2^J - 1 \\
x_{j,k,l} &= x_{j+1,2k,2l} + x_{j+1,2k+1,2l} + x_{j+1,2k,2l+1} + x_{j+1,2k+1,2l+1}, \\
&\qquad k, l = 0, \ldots, 2^j - 1, \; 0 \leq j \leq J - 1.
\end{aligned}
$$

Again, the index j refers to the resolution of the analysis, 2^j; $j = J$ and $j = 0$ are the highest (finest) and lowest (coarsest) resolutions (scales), respectively. This multiscale data analysis is organized into a so-called "quadtree" (the obvious generalization of the binary data tree in Figure 1); see Crouse et al. (1998) for information on quadtree representations.

We can also construct an analogous 2-d multiscale analysis of an image μ. However, there are some additional issues faced in 2-d that distinguish the Gaussian and Poisson case. The standard (unnormalized) 2-d Haar wavelet coefficients are computed as follows. Let $\{\mu_{j,2k+m,2l+n}\}_{m,n=0}^{1}$ denote four neighboring scaling coefficients at scale j in a 2-d Haar analysis of an image μ. The (unnormalized) Haar scaling coefficient and wavelet coefficients at scale $j - 1$ are

$$
\begin{aligned}
\mu_{j-1,k,l} &= \mu_{j,2k,2l} + \mu_{j,2k,2l+1} + \mu_{j,2k+1,2l} + \mu_{j,2k+1,2l+1}, \\
\theta^1_{j-1,k,l} &= \mu_{j,2k,2l} + \mu_{j,2k,2l+1} - \mu_{j,2k+1,2l} - \mu_{j,2k+1,2l+1},
\end{aligned}
$$

$$\theta^2_{j-1,k,l} = \mu_{j,2k,2l} - \mu_{j,2k,2l+1} + \mu_{j,2k+1,2l} - \mu_{j,2k+1,2l+1},$$

$$\theta^3_{j-1,k,l} = \mu_{j,2k,2l} - \mu_{j,2k,2l+1} - \mu_{j,2k+1,2l} + \mu_{j,2k+1,2l+1}, \qquad (22)$$

where the superscripts 1, 2, and 3 refer to the horizontal, vertical, and diagonal differences, respectively. The Haar wavelet coefficients $\{\theta^i_{j-1,k,l}\}^3_{i=1}$ are the *additive* refinements required to split the coarse scaling coefficient $\mu_{j-1,k,l}$ into the four finer scaling coefficients $\{\mu_{j,2k+m,2l+n}\}^1_{m,n=0}$. Due to the orthogonality of the mapping

$$\{\mu_{j,2k+m,2l+n}\}^1_{m,n=0} \mapsto \mu_{j-1,k,l}, \{\theta^i_{j-1,k,l}\}^3_{i=1},$$

in the Gaussian case, taking these (standard) Haar wavelet coefficients as multiscale parameters leads to a factorized likelihood.

The Poisson case is more complicated because orthogonality does not imply independence. However, an alternative set of multiscale parameters can be used in the Poisson case, that does lead to a factorized likelihood. Specifically, we take the 2-d multiscale parameters to be the factors corresponding to the *multiplicative* refinement of a coarse scaling coefficient (intensity) into four finer scaling coefficients by first splitting it horizontally (vertically) into two halves, then next vertically (horizontally) splitting each half into two quarters as described by Timmermann and Nowak (1999). That is, take

$$\mu_{j-1,k,l} = \mu_{j,2k,2l} + \mu_{j,2k,2l+1} + \mu_{j,2k+1,2l} + \mu_{j,2k+1,2l+1},$$

$$\theta^1_{j-1,k,l} = \frac{\mu_{j,2k,2l} + \mu_{j,2k,2l+1}}{\mu_{j,2k,2l} + \mu_{j,2k,2l+1} + \mu_{j,2k+1,2l} + \mu_{j,2k+1,2l+1}},$$

$$\theta^2_{j-1,k,l} = \frac{\mu_{j,2k,2l}}{\mu_{j,2k,2l} + \mu_{j,2k,2l+1}},$$

$$\theta^3_{j-1,k,l} = \frac{\mu_{j,2k+1,2l}}{\mu_{j,2k+1,2l} + \mu_{j,2k+1,2l+1}}. \qquad (23)$$

Alternatively, it is possible to consider a fully 2-d refinement process in which we simultaneously split a coarse scaling coefficient into four finer coefficients. In this case the conditional parent-child likelihoods would be multinomially instead of binomial, and the natural conjugate prior would be the Dirichlet rather than the beta density, but otherwise the multiscale framework would be essentially the same.

The 2-d multiscale parameters defined in (22) and (23) can be modeled with the same conjugate prior probability density functions proposed for the 1-d case, (10) and (11), respectively. Furthermore, if the states are modeled independently, then the posterior density of the 2-d parameters also factorizes in a fashion similar to the 1-d case.

The 2-d MHMM is also similar to the 1-d development in Section 16.4, with a quadtree replacing the binary tree structure. To be precise, in the Gaussian case, the standard 2-d Haar wavelet analysis consists of three sets of wavelet coefficients, superscript 1, 2, or 3 in (22), at each scale, associated with horizontal, vertical, and diagonal differences. In the Poisson case, we have three sets of multiplicative splits, superscript 1, 2, or 3 in (23). A single quadtree MHMM (analogous to the binary tree depicted in Figure 2) is associated with each set of multiscale parameters. For example, in the Gaussian case, one quadtree structure is used to specify the parent-child relationships between vertical Haar wavelet coefficients. In the numerical example considered next in Section 16.6, the three quadtrees are modeled as mutually independent, although it may be possible (and desirable) to introduce dependencies among them. The quadtree upward-downward algorithm is essentially the same as the binary tree upward-downward algorithm, except that each parent has four children instead of two.

16.6   Applications to Image Analysis

16.6.1   Image Denoising

Suppose that we observe an image μ with additive Gaussian white noise:

$$\mathbf{x} = \mu + \mathbf{w},   \tag{24}$$

where μ is an array of image intensities and $\mathbf{w}$ is an array of independent realizations of a zero-mean Gaussian random variable. The goal of the denoising problem is to estimate μ given the data $\mathbf{x}$. If we specify a prior for multiscale parameters (Haar wavelet coefficients) of μ and formulate the image estimation problem under squared error or 0/1 loss, it can be shown that an optimal image estimate $\hat{\mu}$ is obtained from the posterior mean or MAP estimates of the multiscale parameters, respectively, as shown by Figueiredo and Nowak (1998). Later in this section we will consider a numerical example of this problem and compare the posterior mean estimates obtained from an independent parameter prior to that obtained with an MHMM. Examples of 2-d Poisson intensity estimation can be found in the work of Timmermann and Nowak (1999).

16.6.2   Image Edge Detection

Multiscale methods of edge detection are usually based on finding the local wavelet coefficient maxima, as developed by Mallat (1998). The multiscale models considered in this chapter offer an alternative Bayesian approach to edge detection. Again, let us consider the Gaussian observation model above (24), and recall the simple two-state mixture model. State '0' is associated with a low-variance Gaussian component, indicative of a region of smooth behavior, while state '1', corresponding to a high-variance Gaussian, is a cue for the existence of an edge. This interpretation of the state variables suggests testing for the presence of an edge using the Bayes factors of the states. That is, decide an edge is present at scale j, orientation

$i = 1, 2,$ or 3 (corresponding to horizontal, vertical, and diagonal, respectively), and position k, l if

$$BF(\mathbf{x}) = \frac{p(s^i_{j,k,l} = 1 \,|\, \mathbf{x}) \, p(s^i_{j,k,l} = 0)}{p(s^i_{j,k,l} = 0 \,|\, \mathbf{x}) \, p(s^i_{j,k,l} = 1)} > 1, \qquad (25)$$

where $p(s^i_{j,k,l} = 1)$ and $p(s^i_{j,k,l} = 0)$ are the prior probabilities of the state. As we will see in the numerical example considered next, the ability of the MHMM to propagate state information from coarse-to-fine scales results in significantly better edge detection performance compared to that resulting from the independent state model.

16.6.3 Numerical Example

Here we consider a simple numerical illustration of the ideas presented in this paper. Figure 4 (a) depicts a close-up of the test image shown in Figure 3. Figure 4 (b) shows the same image with additive Gaussian white noise of standard deviation $\sigma = 25$. A two-state MHMM was specified for this problem with the following parameter settings:

$$
\begin{aligned}
\tau_0 &= 0, \\
\tau_1 &= 250, \\
\rho_{0,0}(0) &= 0.9, \\
\rho_{j,k}(0|0) &= 0.9, \quad k = 0, \ldots, 2^j - 1, \ j = 1, \ldots, J - 1, \\
\rho_{j,k}(0|1) &= 0.25, \quad k = 0, \ldots, 2^j - 1, \ j = 1, \ldots, J - 1.
\end{aligned}
$$

These parameters were selected with the noise variance and basic parent-child dependencies in mind. One can, however, plug-in maximum likelihood or moment-based estimates of these (hyper) parameters, for a fully automatic procedure. For example, an expectation-maximization algorithm based on the upward-downward algorithm is derived by Crouse et al. (1998) to obtain maximum likelihood estimates of the mixture variances and the transition probabilities. Here, for comparative purposes, we also consider an analogous independent multiscale parameter model (all states, and hence parameters, mutually independent), whose prior state probabilities are the same as the marginal state probabilities of the MHMM specified above.

Posterior mean estimates of the image are obtained by computing inverse Haar wavelet transform of the the posterior means of the Haar wavelet coefficients obtained from the two models.[6] Figure 4 (c) shows the estimate based on the independent coefficient model (average squared pixel error = 171.68) and Figure 4 (d) shows the estimate based on the MHMM (average squared pixel error = 163.79). In comparison, the average square pixel

---

[6] As is usual in wavelet denoising, the "raw" scaling coefficients obtained directly from the noisy image were used in the inverse transform.

error is 625 in the noisy image shown in Figure 4 (b). The MHMM based estimate also appears to be subjectively better than that obtained from the independent parameter prior; there are less residual noise spikes in the smooth background region of the image, and the edges appear slightly sharper.

Image edges were detected from the noisy data by computing the Bayes factors of the states at finest scale. The "edge maps" obtained from the independent coefficient model are shown in Figure 4 (e) and those from the MHMM are shown in Figure 4 (f). To visualize the complete set of edges, each pixel in these two edge maps was set to "black" if one or more of the three (corresponding to the three possible orientations) Bayes factors at the finest scale ($j = J - 1$) and at the corresponding spatial position tested positive according to (25), and was set to "white" otherwise. The edge map resulting from the MHMM is vastly superior than that resulting from the independent parameter model. There are far fewer false edge detections in the background of the MHMM edge map and the continuity of the true edges is captured to a higher degree.

16.7 Conclusions

The MHMM framework described in this chapter appears to be a promising new approach to Bayesian image analysis. The MHMM captures the key inter-scale dependencies present in natural imagery, and, unlike classical MRF based methods that typically require computationally intensive stochastic optimization, the MHMM allows for simple inference algorithms based on probability propagation. Hence, the computational complexity of the MHMM framework is $O(N)$, where N is the number of data. A common framework for MHMMs, capable of analyzing Gaussian and Poisson processes, was presented; applications to Bayesian image denoising and edge detection were examined.

There are three important features being exploited in the Gaussian and Poisson observations models:

1. parametric parent-child conditional probabilities, (5) and (6);

2. likelihood factorization, (7);

3. conjugate priors for multiscale parameters, (10) and (11).

Without these features, multiscale analysis and modeling would be significantly more complicated. In particular, the likelihood factorization allows us to postulate an alternative multiscale observation (or data generation) model; a single (coarse-scale) Poisson count refined by independent binomial splits in the Poisson case, and in the Gaussian case we have independent Gaussian distributed Haar wavelet coefficients. Again, in the Gaussian case, a similar factorization (multiscale observation model) exists for orthogonal wavelet transforms in general, due to the orthogonality of the transformation; see Crouse et al. (1998) for details. In essence, it is the multiscale observation model that enables the graphical interpretation of the problem, and it is doubtful that a simple inference algorithm exists without such a factorization. Hence, it is natural to seek out other observation models that have a similar factorization property. Some other cases are investigated by Kolaczyk (1999), but it appears that the Gaussian and Poisson cases are quite exceptional and that other common models may not be amenable to the MHMM framework.

The connection between MHMMs and $1/f$ processes also deserves mention. It has been shown in the work of Nowak (1998) and Timmermann and Nowak (1999) that, in certain cases, the independent parameter priors discussed in Section 16.3 for Gaussian and Poisson models both have $1/f$ spectral characteristics. This is very relevant to image analysis since there is convincing empirical evidence that natural images have similar spectral characteristics; see the comprehensive study by van der Schaaf and van Hateren (1996). MHMM priors can also display this behavior. For example, in the Gaussian case, because the Markov structure of the MHMM is imposed on the variances underlying the zero-mean Gaussian mixtures instead of directly on the Haar wavelet coefficients, the coefficients are uncorrelated (but not independent), and hence the MHMM has the same second order correlation structure as the independent coefficient model. Of course, the higher order correlation behavior is (desirably) different for MHMMs. One can argue that the higher order structure is especially relevant in image analysis. For instance, perhaps more important than the number of edges in an image (roughly speaking, measured by the decay of the second order spectrum) is the arrangement and structure of edges (reflected in higher order correlations). These observations suggest avenues for future investigations of the properties and applications of the MHMM framework.

Acknowledgments: The author thanks Richard Baraniuk and Mário Figueiredo for carefully reading an early version of this manuscript and making several suggestions that helped to improve the presentation. Special thanks go to Eric Kolaczyk for many helpful discussions and suggestions. The author also thanks Matthew Crouse and Klaus Timmermann for their valuable contributions to many aspects of this work.

References

Abramovich, F., Sapatinas, T., and Silverman, B. W. (1998). Wavelet thresholding via a Bayesian approach. *J. Roy. Statist. Soc. Ser. B., 60, 725-749,* 60:725–749.

Adelson, E. and Burt, P. (1981). Image data compression with the Laplacian pyramid. In *Proc. Patt. Recog. Info. Proc. Conf.*, pages 218–223, Dallas, TX.

Bouman, C. and Shapiro, M. (1994). A multiscale random field model for Bayesian image segmentation. *IEEE Trans. Image Proc.*, 3(2):162–177.

Castleman, K. (1996). *Digital Image Processing.* Prentice-Hall, Englewood Cliffs, New Jersey.

Charbonnier, P., Blanc-Fèraud, L., and Barlaud, M. (1992). Noisy image restoration using multiresolution Markov random fields. *Journal of Visual Communication and Image Representation*, 3(4):338–346.

Chellappa, R. and Jain, A. (1993). *Markov Random Fields: Theory and Applications.* Academic Press, San Diego, CA.

Chipman, H., Kolaczyk, E., and McCulloch, R. (1997). Adaptive Bayesian wavelet shrinkage. *J. Amer. Statist. Assoc.*, 92:1413–1421.

Cooper, G. F. (1990). The computational complexity of probabilistic inference using Bayesian belief networks. *Artificial Intelligence*, 42:393–405.

Cross, G. and Jain, A. (1983). Markov random field texture models. *IEEE Trans. Patt. Anal. Mach. Intell.*, 5:25–39.

Crouse, M., Nowak, R., and Baraniuk, R. (1996). Hidden Markov models for wavelet-based signal processing. In *Proc. Thirtieth Asilomar Conf. Signals, Systems, and Comp.*, Pacific Grove, CA, pages 1029–1034. IEEE Computer Society Press.

Crouse, M., Nowak, R., and Baraniuk, R. (1998). Wavelet-based statistical signal processing using hidden Markov models. *IEEE Trans. Signal Processing*, 46:886–902.

Donoho, D. and Johnstone, I. (1994). Ideal adaptation via wavelet shrinkage. *Biometrika*, 81:425–455.

Field, D. (1993). Scale-invariance and self-similar 'wavelet' transforms: an analysis of natural scenes and mammalian visual systems. in *Wavelets, Fractals, and Fourier Transforms*, Claredon Press, Oxford:151–193.

Figueiredo, M. and Nowak, R. (1998). Bayesian wavelet-based signal estimation using non-informative priors. In *Proc. Thirty-Second Asilomar Conf. Signals, Systems, and Comp.*, Pacific Grove, CA. IEEE Computer Society Press.

Flandrin, P. (1992). Wavelet analysis and synthesis of fractional Brownian motion. *IEEE Trans. Inform. Theory*, 38(2):910–916.

Frey, B. (1998). *Graphical Models for Machine Learning and Digital Communication.* MIT Press, Cambridge, Massachusetts.

Geman, S. and Geman, D. (1984). Stochastic relaxation, Gibbs distribution and the Bayesian restoration of images. *IEEE Trans. Patt. Anal. Mach. Intell.*, 6(6):712–741.

Gidas, B. (1989). A renormalization group approach to image processing problems. *IEEE Trans. Patt. Anal. Mach. Intell.*, 11(2):164–180.

Kolaczyk, E. (1998). Bayesian multi-scale models for Poisson processes. *Technical Report 468, Dept. of Statistics, University of Chicago.*

Kolaczyk, E. (1999). Some observations on the tractability of certain multi-scale models. In *Bayesian Inference in Wavelet Based Models*. Springer-Verlag. Editors B. Vidakovic and P. Müller.

Luettgen, M., Karl, W., Willsky, A., and Tenney, R. (1993). Multiscale representations of Markov random fields. *IEEE Trans. Signal Proc.*, 41(12):3377–3395.

Malfait, M. and Roose, D. (1997). Wavelet based image denoising using Markov random field *a priori* model. *IEEE Transactions on Image Processing*, 6(4):549–565.

Mallat, S. (1998). *A Wavelet Tour of Signal Processing*. Academic Press, San Diego, CA.

Nowak, R. (no. 83, Bryce Canyon, UT, 1998). Shift invariant wavelet-based statistical models and $1/f$ processes. *Proc. IEEE Digital Signal Processing Workshop*.

Pearl, J. (1988). *Probabilistic Reasoning in Intelligent Systems: Networks of Plausible Inference*. Morgan Kaufmann, San Francisco, CA.

Pérez, P. and Heitz, F. (1996). Restriction of a Markov random field on a graph and multiresolution statistical image modeling. *IEEE Trans. Info. Theory*, 42(1):180–190.

Robert, C. (1994). *The Bayesian Choice: A Decision Theoretic Motivation*. Springer-Verlag, New York.

Shapiro, J. (1993). Embedded image coding using zerotrees of wavelet coefficients. *IEEE Trans. Signal Proc.*, 41(12):3445–3462.

Simoncelli, E. (1997). Statistical models for images: Compression, restoration and synthesis. In *Proc. Thirty-First Asilomar Conf. Signals, Systems, and Comp.*, Pacific Grove, CA, pages 673–678. IEEE Computer Society Press.

Timmermann, K. and Nowak, R. (1997). Multiscale Bayesian estimation of Poisson intensities. In *Proc. Thirty-First Asilomar Conf. Signals, Systems, and Comp.,*, pages 85–90. IEEE Computer Society Press.

Timmermann, K. and Nowak, R. (April, 1999). Multiscale modeling and estimation of Poisson processes with application to photon-limited imaging. *IEEE Transactions on Information Theory*, 45(3).

van der Schaaf, A. and van Hateren, J. (1996). Modelling the power spectra of natural images. *Vision Research*, 36(17):2759–2770.

Vidakovic, B. (ISDS, Duke University, 1998). Honest modeling in the wavelet domain. *Discussion Paper XX-98*.

Wornell, G. (1996). *Signal Processing with Fractals. A Wavelet-Based Approach*. Prentice Hall, Englewood Cliffs, New Jersey.

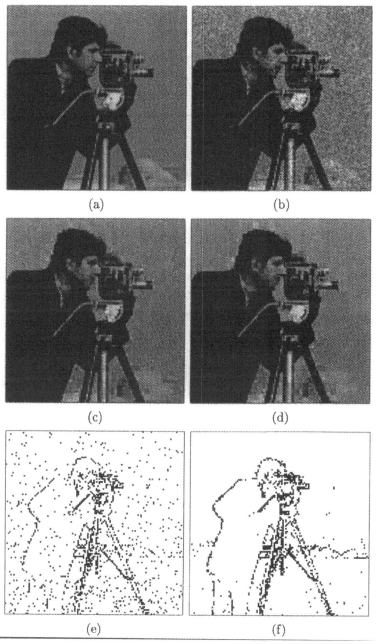

FIGURE 4. Bayesian image denoising and edge detection. (a) Close-up of original test image (full image shown in Figure 3). (b) Close-up of noisy image. (c) Close-up of posterior mean estimate based on independent multiscale parameter prior model. (d) Close-up of posterior mean estimate based on MHMM. (e) Edges detected from Bayes factors resulting from independent wavelet coefficient prior. (f) Edges detected from Bayes factors resulting from MHMM.

17

Wavelets for Object Representation and Recognition in Computer Vision

Luis Pastor, Angel Rodríguez and David Ríos Insua

ABSTRACT Finding efficient and powerful techniques for representing 2D and 3D entities is central in many areas, such as computer vision, graphics, CAD or data visualization. The large amount of data often involved and the small processing time required place strong demands on modeling techniques. This chapter gives an overview of the possibilities wavelets offer for object modeling, specially from the point of view of object recognition in computer vision environments. For the 2D case, techniques based on segmented contours or raw grey level information are described. For 3D data, both the cases of dense, regularly sampled volumetric information and sparse, irregularly sampled geometric surface information are considered. Ideas on how Bayesian methods may enhance wavelet based techniques for object representation and recognition are outlined.

17.1 Introduction

Representing and recognizing objects are two of the main goals of computer vision systems, resulting essential stages for understanding the environment's structure. Object representation or modeling aims at creating thorough descriptions of the entities that integrate the perceived scene. Sometimes modeling is an objective in itself, as in reverse engineering, where data acquired from real objects is used as an input to a CAD system. In other cases, the representation of an item is an intermediate stage of the vision system, yielding results used by other processes that perform more abstract operations on the data acquired from the scene objects. An important example is object recognition, which is one of the main applications of vision systems and may be also a key stage in many areas such as robotics, quality control, tracking, etc. In general, any computer vision system involved in high-level activities requires some sort of object recognition.

Traditionally, image processing and computer vision applications combine large data sets and small processing times. It is well known that starting by processing reduced resolution versions of the input image may result in large computational savings. There are also specific tasks such as edge detection or texture classification that benefit from analyzing input data at different scales. It is therefore not surprising that hierarchical and multiresolution techniques have always raised interest in the computer vision community [1], [2]. The development of the wavelet transforms theory has spurred new interest, providing also a more rigorous mathematical framework.

This chapter gives an overview of the current use of wavelet transforms for representing and recognizing 2D and 3D objects in computer vision, emphasizing the ability of wavelets to describe and represent shapes. The following section includes descriptions of modeling techniques using segmented object boundaries or direct image intensities for the 2D case, and using dense or sparse, regularly or irregularly sampled, volumetric density values or clouds of surface points for the 3D case. Section 17.3 gives a summary of how wavelet-based representations have been used for recognition, mostly in the 2D case. Section 4 outlines how Bayesian methods may enhance wavelet-based approaches in object modeling and recognition. In Section 17.5 we provide conclusions.

17.2   Wavelet methods for object representation

Depending on the system's purpose, the objective of the modeling stage can either be to collect a set of descriptors characterizing the acquired object with respect to the system's universe of discourse, or generate an exhaustive description that allows the object representation to be manipulated afterwards. We deal mainly with the former issue, although some examples of the latter are also included.

It is generally accepted that a good modeling system should have the following properties ([3], [4]):

- Expressive richness.

- Stability in presence of errors or noise in the input data.

- Ability to deal with occlusions, which implies some sort of representation for locality.

- Capability to reflect physical measures of the object's shape.

- Efficiency.

Many approaches have been followed to model 2D or 3D objects. For planar shapes, moments and Fourier descriptors have been used frequently [5]. For 3D systems, the use of invariants, aspect graphs and surface approximations are classical. Other techniques adopt a qualitative approach [4].

Wavelets emerge as a powerful alternative for object modeling yielding scale information similar to the frequency data supplied by Fourier techniques, with the additional advantage of locality. As an example, Figures 17.1–17.3 show how wavelets may aid in extracting different shape features from an object boundary at various resolution levels. They present a planar shape which is the superposition of a circular and a star-shaped object, with jagged contour, Figure 17.1(a). The contour is first parameterized, Figure 17.1(b), as $r(s)$, r being the distance between contour points and the object's centroid, and s the perimeter length. A wavelet transform is then applied to the resulting function. Just by selecting which detail levels will be used to reconstruct the contour, we may obtain:

1. The extremely coarse, almost circular global shape, see Figure 17.1(d), using the contour's approximation coefficients at decomposition level seven. Those coefficients give small variations around the boundary's average radius.

2. Intermediate form features, by including the extremely coarse, level $n - 7$ approximation coefficients and the most relevant detail coefficients, which are those at resolution level $n - 5$ (Fig. 17.2(b); n defines the resolution of the original signal, which is 2^n samples).

3. The smoothed object boundary, reconstructed with the approximation coefficients at level $n - 5$ (Fig. 17.2(d)).

4. Contour details, including high curvature points such as the star's tips, by using approximation coefficients from level $n - 7$ and detail coefficients from levels $n - 1$ to $n - 3$. (Fig. 17.3(b)). Note how well small, local variations have been captured, being in general possible to trace them in the processed contours.

Wavelets can also be used with the original object's grey levels, rather than with its segmented contour. Section 17.3.1 presents some examples of how Gabor transforms have been used for gathering exhaustive grey level information at neighborhoods of specific points for recognition purposes within 2D images.

From the point of view of object modeling and recognition, wavelet techniques were initially seen as a reformalization of multiresolution methods from the 80's. The developments that have taken place during the last decade have opened up new possibilities, resulting in new techniques and applications. The following sections describe some of them, grouped according to whether they take a 2D or a 3D perspective.

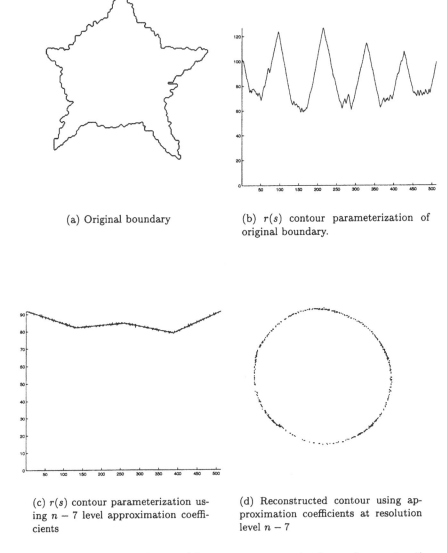

(a) Original boundary

(b) $r(s)$ contour parameterization of original boundary.

(c) $r(s)$ contour parameterization using $n - 7$ level approximation coefficients

(d) Reconstructed contour using approximation coefficients at resolution level $n - 7$

FIGURE 17.1. Multiresolution $r(s)$ contour parameterization and reconstruction using a biorthogonal 2.2 base up to decomposition level $n - 7$

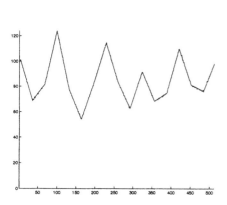

(a) $r(s)$ contour parameterization using the $n-7$ approximation coefficients and $n-5$ detail coefficients

(b) Reconstructed contour using the $n-7$ level approximation coefficients and the most relevant detail coefficients (resolution level $n-5$)

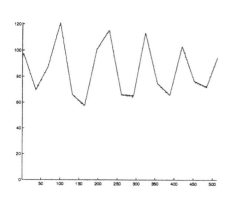

(c) $r(s)$ contour parameterization using only the $n-5$ level approximation coefficients

(d) Reconstructed contour using only the $n-5$ level approximation coefficients

FIGURE 17.2. Multiresolution contour $r(s)$ parameterization and reconstruction using a biorthogonal base 2.2.

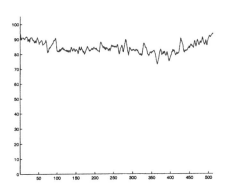

(a) Contour parameterization using the $n-7$ approximation and finest resolution detail coefficients (levels $n-1$ to $n-3$)

(b) Reconstructed contour using the $n-7$ approximation and finest resolution detail coefficients (levels $n-1$ to $n-3$)

FIGURE 17.3. Multiresolution contour $r(s)$ parameterization and reconstruction using a biorthogonal base 2.2.

17.2.1   Planar shapes/2D objects

Wavelets offer alternative domains for representing planar shapes. The transform coefficients -including the coarse approximation coefficients- reflect shape changes at a variety of detail or scale levels, as illustrated in Figures 17.1 to 17.3. In general, information at finer resolution levels will be more affected by noise, whereas at smaller resolution levels, approximation coefficients provide an almost constant function, hence providing little information to discriminate within a class.

Within modeling/recognition environments, substituting the original contour with low resolution approximations or with representations using different levels of detail may facilitate certain tasks. For example, in applications involving recognition, we wish to extract enough information from the perceived objects for the system to be able to discern the class to which the objetcs belong. Should intraclass variability be high, we might find it more appropriate to discard detail information, since details may serve better to distinguish among particular instances than to determine the common class to which all instances belong [6]–[8]. In other cases, rather than being interested in discriminating a general class, we look for the specific object which generated the pattern. Such situation arises, for example, in hand recognition: the system's purpose is to distinguish the hand owner, rather than whether a particular shape corresponds to the *human hand* class. In this case, enough detail information has to be included to allow the system to differentiate among slight shape variations [9].

Perhaps the biggest shortcoming which has prevented the application of wavelets to modeling and recognition in computer vision stems from their lack of shift invariance. In general, two approaches have been taken:

- Compute transforms that achieve some sort of position invariance

- Normalize the image data to mitigate position changes in the image data before the transform is computed.

Search for invariance

Several approaches have been followed to achieve translation, scale and rotation invariance. Mallat [10], following ideas that can be traced back to Marr [1], suggested using the zero-crossings or local extrema of certain wavelet transforms [1]. Choosing as wavelets the second derivative of a smoothing function [10], the transform's zero-crossings correspond to the signal's sharper variation points (its inflection points). In order to stabilize the representation provided by zero-crossings, [2] Mallat uses the piecewise constant function

$$Z_{2^j} f(x) = \frac{e_i}{z_i - z_{i-1}}, \text{ for } x \in [z_{i-1}, z_i], j \in \mathcal{Z} \tag{1}$$

$f(x)$ being the original function, z_i and z_{i-1} the location of two successive zero-crossings, and

$$e_i = \int_{z_{i-1}}^{z_i} W_{2^j} f(x) dx, \tag{2}$$

with $W_{2^j} f$ the wavelet transform of f.

Although it has been shown that this representation is not unique [11], Mallat's approach has often been used. For example, in [13], a wavelet transform zero-crossing representation for the human iris is used in a security system. A normalized iris signature is formed by projecting virtual concentric circles centered at the pupil centroid to extract the iris pattern at specific sample points. Then, a multilevel zero-crossing representation is generated from the normalized iris signature. It should be pointed out that even if zero-crossing representations are not unique, they can provide

---

[1]Detecting a signal's sharpest variation points may be done by detecting zero-crossings or local extrema of the signal's wavelet transforms using certain basis functions [10], [11]

[2]Following Mallat, in an unstable representation, large perturbations of the original signal can yield small perturbations in the signal's representation. It is possible to reconstruct quite different signals from similar zero-crossing representations [10], [12].

appropriate information for object recognition if the probability of having two similar representations for objects belonging to different classes is negligible.

An alternative approach to look for invariance has been studied by Simoncelli *et al.* [14], where the authors introduce properties that a base function must fulfill in order to reach *shiftable* multiscale transforms. Other invariant approaches that can be mentioned are Minh and Boles' [15] and Pesquet *et al.* [16].

Finally, Yu-Ping *et al.*, [17], [18] suggest representing object boundaries using the contour curvature function

$$K(u) = \dot{x}(u)\ddot{y}(u) - \dot{y}(u)\ddot{x}(u), \tag{3}$$

where u is the normalized arc length parameter, since $K(u)$ presents affine invariance. Analyzing curvature changes at different scales provides evidence for curve identification, since peaks in the transform coefficients correspond to contour corners. A compressed contour representation is achieved by selecting the dominant points at several scales and considering their relative positions.

Data normalization

Sometimes normalizing the object contour is a straightforward procedure, because there are points which can be easily identified within the boundary, and may act as a reference to compensate for data translation and rotation (scale compensation can be done by resampling). For example, Wang *et al.* [9] extract fingers in hand silhouettes by detecting points with maximal curvature. Once the finger contours are parameterized and normalized in scale, a wavelet transform is computed and coarse approximation and detail coefficients near the finger tips are used for classification purposes. Finger tips -the boundary points located farthest from the silouhette centroid- and finger extreme points -those with maximal curvature- are used for position normalization.

Note that the effects of some transformations on wavelet coefficients may be determined, therefore facilitating normalization. As an example, in [8], within a character recognition environment, character contours are initially parameterized according to $X(t)$ and $Y(t)$, coordinates and coefficients normalized to achieve invariance against translation, rotation and scaling, t being the normalized arc length. Coefficients are scaled by the same scaling factor; detail coefficients are invariant under translation, whereas approximation coefficients are translated by a weighted factor of the translation; finally, under rotation, the magnitude of coefficients remain invariant, whereas the phase is modified by the same angle, for polar coordinate versions of both approximation and detail coefficients. As a consequence, to achieve translation normalization we may subtract from approximation coefficients their centroid; to achieve scale normalization, we may divide the magnitude of the coefficients by their averaged value; and achieve rotation normalization by performing an inverse rotation with the average phase of wavelet vectors.

In many imaging applications, the input data set is usually large, and therefore, normalization operations are computationally expensive. A possibility suggested by Reissell [19] is to use the approximation coefficients, which constitute a reduced data set. These coefficients reflect the object's global shape, allowing to efficiently perform approximated input data realignments. The biorthogonal wavelet transform proposed is based on trigonometric polynomials denominated *pseudocoiflets*.

To conclude this section on wavelet techniques for 2D object modeling, we should point out that it is also possible to represent object features by working directly with image grey levels rather than with extracted contours. For example, some authors use the multiresolution information provided by Gabor transforms computed at a set of fixed image locations [20]–[23]. This group of techniques, specifically oriented to recognition, will be described in Section 17.3.1.

17.2.2   3D Models

Moving from 2D to 3D modeling entails a large complexity increase. Achieving accurate and compact 3D representations is difficult, both because of the growth in raw data size and because of the presence of new difficulties, inherent to the 3D case, including:

- Data incompleteness: 3D data are often incomplete, at least because of self-occlusion.

- Data unreliability: the *signal to noise* ratio of 3D data is often much lower than in calibrated 2D systems. This may be originated by various factors, such as sensor errors, increased difficulties for system calibration, complex mechanical setups, ambiguities in matching processes,...

- Data scarcity and sampling irregularity: many 3D acquisition techniques provide sparse measurements, irregularly spread over the object's external surface.

Data scarcity is typical of stereo vision systems. Only image reconstruction techniques such as CAT and NMR provide dense, regularly sampled 3D data. In these cases, objects are represented by spatial decompositions in 3D arrays of occupation cells, usually referred to as volume elements or voxels. Sampling regularity is important since it facilitates the extension of 1D or 2D techniques to the 3D case. For example, figures 17.4 and 17.5 present two volumetric data sets, a 3D tomographic reconstruction from a

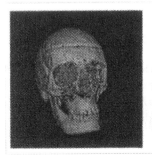

(a) Original volumetric data set. Resolution: 128^3

(b) Approximation coefficients. Resolution level $n-1$ (64^3).

(c) Approximation coefficients. Resolution level $n-2$ (32^3).

FIGURE 17.4. 3D Haar transform over a human skull of size 128^3. From left to right, the images represent the original data set and its $n-1$ (64^3) and $n-2$ (32^3) reduced resolution versions. Transform images are zoomed in for better visualization of details.

human skull and a ceramic piece of a car engine, together with the first and second step approximation coefficients of a 3D Haar transform. Wavelet decompositions can be used in this environment for data compression or for the efficient manipulation of volumetric data. On the other hand, range data provided by laser sensors, 3D digitizers or stereo systems comes usually as irregularly sampled clouds of points. To overcome this problem, it is possible to approximate the cloud with a surface mesh that can be later regularized to admit a subdivision scheme ([24], [25]). Another possibility is to start from a faceted surface meeting certain connectivity requirements, perform subdivisions up to a resolution level similar to the input cloud and, finally, adapt the surface to fit the shape of the input object ([26], [27]). Both approaches are described in more detail in the following sections.

Sometimes 3D input data can be projected onto a plane, which simplifies the regularization process. This is the case in [28], who consider 3D shapes such as human faces or terrains represented as $z = f(x, y)$, where z is the depth. Wavelet transforms are used there to reduce the complexity of the triangular meshes that approximate the objects' surfaces. The analysis of coefficient magnitudes allows the detection of regions with low spatial activity, and the removal of some of the triangulation vertices.

Mesh regularization

Eck *et al.* [24] were the first to suggest subdivision schemes for 3D irregular triangular meshes, an irregular mesh being one that does not meet

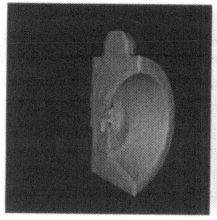 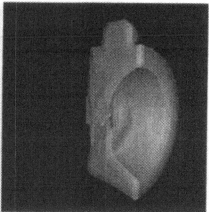

(a) Slice from original volumetric data set.

(b) Slice at resolution level $n - 1(64^3)$.

FIGURE 17.5. Visualization of a ceramic piece slice reconstructed by 3D tomographic techniques. Resolution of original data set is 128^3. The transform image is zoomed in.

the *subdivision connectivity* condition [3]. The process consists of two steps: a first stage that approximates the original irregular mesh M with another mesh M^J, meeting *subdivision connectivity*, and a second stage that computes a multiresolution representation for M^J [29]. M^J is constructed via a retiling process using Voronoi-Delaunay based spatial decomposition techniques. Once a low resolution partition of the original mesh M is obtained, the authors make a parameterization of the original facets fitting inside each of the new tiles in M^J using harmonic maps and topological constraints, so as to achieve a finer partition of the original mesh that supports a multiresolution representation with *subdivision connectivity*. A related approach using spline-based functions can be found in [25].

Regular mesh approximation.

Ikeuchi *et al.* [30] suggested first a process of deformation of an icosahedron (20-faced solid figure) to approximate clouds of 3D points. In [26], we introduce an alternative approach for adjusting a regular mesh to the input data by using a deformed icosahedron as an intermediate representation. Since the scattered point's surface approximation is carried out by deforming a surface with *subdivision connectivity*, there is no need to regularize

---

[3] A mesh meets the *subdivision connectivity* condition if it can be generated from a base mesh through a 1 to 4 subdivision process.

the mesh. Given a set of 3D points obtained from range data, our algorithm adapts the regular mesh to the input cloud as follows:

1. Construct an initial regular tessellated sphere which wraps around the set of 3D points defining the object. This involves three substeps:

   - Approximate the sphere by a 20-face icosahedron.

   - Recursively tessellate each of its faces into N small triangular faces, with N ($N = 4^k, k = 1, 2, ...$) the tessellation degree.

   - Define the final tessellation by taking the dual of the previous triangulation, yielding a geodesic dome with the same number of nodes.

2. Determine for each node (face) of the geodesic dome, the closest point in the data file corresponding to the object being modeled.

3. Perform an iterative deformation of the spherical mesh while the average sum of local errors exceeds a fixed threshold. Every local error is defined by the distance between a node of the sphere and its corresponding closest point, as determined in step 2. Each node is deformed to match the object, according to an approximation force F_o and a curvature force F_g, both defined for the actual position P_t of each node at time t. The new position of the node at time $t + 1$ is given by: $P_{t+1} = P_t + F_o + F_g + d(P_t - P_{t-1})$, where d is a damping coefficient affecting the convergence rate.

Mapping the achieved values over the geodesic dome, on the dual icosahedron, we construct a mesh that reflects the shape of an irregularly sampled cloud of points and meets *subdivision connectivity*. The analysis process followed on the skull of Figure 17.7 is based on the facets based subdivision scheme presented in [31]. Basically, it is the extension of the Haar transform to a topology domain formed by triangular facets. Figure 17.7 shows the results of applying this subdivision scheme to the human skull from 17.6 at several resolution levels.

Finally, we should point out that there are other approaches for 3D modeling developed within image synthesis environments. In these methods, wavelets offer good chances for data compression. The introduction given by Schröeder [32], the survey presented in [33] and the subdivision scheme proposed by Zorin *et al.* [34] give a representative idea of the state of the art.

17.3   Wavelet methods for object recognition

Object recognition within computer vision environments can be defined as the task of establishing a mapping $f(x) = y_i, x \in Input\_objects, y_i \in$

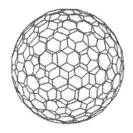

(a) Original point set of a human skull

(b) Geodesic dome

(c) Deformed mesh adjusted to original
the data set. Resolution: $20 \cdot 4^6$ facets

(d) Tesselated sphere

FIGURE 17.6. Deformation process of a mesh with *subdivision connectivity*

(a) First superficial approximation ($20 \cdot 4^5$ facets)

(b) Second superficial approximation ($20 \cdot 4^4$ facets)

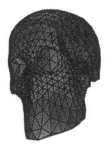

(c) Third superficial approximation ($20 \cdot 4^3$ facets)

(d) Fourth superficial approximation ($20 \cdot 4^2$ facets)

FIGURE 17.7. Approximation coefficients for the data set of fig. 17.6(c)

Object_database. The recognition system's performance is restricted by which classes are included in the object database and by how objects are represented within the system.

A recognition system should meet several requirements ([35], [4]):

- Objects should be quickly recognized, even when they are seen from unknown viewpoints, under the presence of light noise levels, or when they are partially occluded [4]. It is assumed that objects presented to the system are new instances of predefined object categories.

- The system should provide an acquired object-candidate class similarity measure.

- The process should be efficient, since it is common to find systems handling extensive object databases.

Recognizing the acquired objects, or selecting which is the class that best matches the input data, has been one of the most common applications of computer vision since its birth. Therefore it is not surprising that many different approaches have been used for this purpose [36], [37], such as those based on the use of classifiers. Other techniques are based on matching the input object's description with those from all of the classes included in the system's database, selecting the class that achieves the best matching. The strategy followed for the matching process depends strongly on the kind of information extracted during the modeling stage. Consequently, it is possible to find solutions based on production rules following syntactic approaches, graph matching techniques using either structural information [4] or different object views [38], or heuristic knowledge-based rules [36]. Indexing techniques can be used whenever necessary to speed up the class selection process. In general, modeling systems based on hierarchical and multiresolution representations are the most appropriate for supporting indexing operations [39].

The following sections describe a number of wavelet based recognition methods, again grouped in sections devoted to 2D and 3D environments.

17.3.1   2D Object Recognition

A number of 2D object recognition techniques using wavelets have been recently described in the literature. In general, they can be classified according to the kind of input data used: raw, direct grey level information or contours extracted after a segmentation stage. They will be studied here as methods using point or contour based representations.

---

[4]For recognizing partially occluded objects, the system should be able to perform some kind of local or partial matching between the unknown object and the predefined object classes.

Methods using point-based representations

These techniques are characterized by extracting exhaustive image information at the neighborhoods of specific points, usually by convolving the input image with Gabor kernels of different scale and orientation, all of them centered at the same image point. Recognition is often performed afterwards by graph-matching procedures.

These techniques have been frequently used within automated person identification/validation environments. For example, Daugman [23] has worked on iris recognition, basing his approach on carrying out a coarse phase quantization of Gabor filters' responses, computed at points (ρ, θ)[5] located within the iris. The local phase is quantized according to the signs of the real and imaginary filter outputs, achieving a very compact code (two bits per analyzed point; 256 bytes per eye).

Von der Malsburg *et al.* [20], [21], recognize human faces by performing a graph matching process, where graph nodes are labeled with the values produced by Gabor filters at different scales and orientations, computed at specific facial image points. Würtz [22] follows a similar approach, computing a hierarchical representation that allows a coarse to fine matching procedures. It has to be pointed out that even though faces are 3D structures, the methods described above use essentially 2D approaches.

Somehow related ideas have been used in texture analysis [40]–[42], and in *query by content* or *similarity retrieval* in image databases. For example, [42] proposes a scheme for classifying colored textures: wavelet correlation signatures are defined which contain the energies of each color plane, together with the cross-correlation between different planes. A k-nearest neighbor classifier is used to select the target.

In the case of *query by content*, the query and target images may be quite different, implying that retrieval techniques must allow for the existence of some distortions. Jacobs *et al.* [43] suggest using an *image query metric* that, for efficiency reasons, makes use of truncated, quantized versions of a Haar decomposition in a large database, referred to as signatures. Values are truncated by considering only between 40 and 60 of the most significant magnitud coefficients. Values are then quantizied by coding large positive values as +1, large negative values as -1, and the rest as 0. The signatures contain only the most significant information about each image. A search is implemented using a metric based on the number of matches between the signatures of the objects in the database and the input object.

Some improvements can be achieved by using energy signatures computed over approximation and detail coefficients at different resolution levels [44]. The main disadvantage of these methods comes from their lack of invariance against affine transforms.

---

[5] Here (ρ, θ) are polar coordinates of points in the iris relative to the center of the pupil, taking into consideration deformations of the iris.

Methods using boundary-based representations

Geometric shapes have been used exhaustively for 2D object recognition. In some cases, objects to be analyzed are either predominantly planar or have a limited number of stable positions. This allows the recognition process to be considered essentially a 2D task. In other applications such as character recognition the problem is truly 2D. In general, 2D recognition usually involves performing first an image segmentation stage that separates *object* and *background*. Later boundary extraction and parameterization processes provide the raw data from which a set of descriptors can be computed and compared to those extracted from the models included in the object database. In this environment, wavelets can be used both for preprocessing the input object's boundary and for extracting information useful for the recognition stage. The wavelets' ability to reflect signal changes at different scales into different transform levels has been used sometimes for extracting specific information from the input patterns. We mentioned in Section 17.2.1 that this idea has been used to extract the object's global shape, which is particularly interesting in cases when the input data presents high intraclass variability. Handprinted character recognition is a clear paradigm, since characters from different writers may show large variations depending on each person's writing style. On the other hand, characters' coarse approximations are more similar, leading to higher recognition rates. For example Laine *et al.* [6], [7] suggest using neural networks fed with characters' contour information compressed using wavelets. To improve the classifier's false acceptance ratio, they propose combining the outputs of different classifiers, fed with approximation coefficients at different resolution levels.

Low-frequency approximations do not give enough information whenever interclass variability is small. This circumstance arises frequently when the system has to discriminate between different varieties of one object class, a problem sometimes termed *object identification* [37], [38]. Typical examples are the systems described by Wang *et al.* [9] for human hand recognition, by Boles [13] for discriminating human iris, and by Tieng and Boles, [45] and [15], for generic shape recognition.

17.3.2 3D Object Recognition

Unlike the 2D case, where a number of techniques and applications have been described, the literature on wavelet-based 3D object recognition methods is much more limited. This might be due to the fact that 2D problems are more classical, and that many 3D acquisition systems provide sparse, irregularly sampled range data which prevent straightforward generalizations of 1D and 2D techniques, as discussed in Section 17.2.2.

A typical approach to the 3D recognition problem is to reduce its dimensionality by considering several 2D subproblems in place of the original

3D one. Each of these 2D cases could represent the 3D object's aspect when seen from a particular viewpoint. Following this idea, Wu and Bhanu describe in [46] a method using the point-based representations described in Section 17.3.1, representing object models and input images by a grid with their responses to multiscale Gabor filters (8 different directions and 7 different scales), computed at specific points. Grid nodes store node image coordinates together with filter responses, while grid edges represent neighborhood relationships, used as constraints during the matching process: grid edges are interpreted as elastic links, so that an edge can be deformed like a spring to make a model match a distorted object.

An alternative possibility suggested in [26] is to use the 3D objects' wavelet representation as an intermediate domain where certain shape features can be more easily extracted. For example, the object's coarse approximation can be fed into a qualitative shape analysis process [47]. Also, intermediate shape features and surface texture reflected in the transform's detail coefficients can provide additional data to be used in the recognition process.

17.4   Bayesian Methods

We have outlined various proposals for object modeling and recognition which use wavelets in various ways. Our own experience and the blossoming of research in this area show the potential of wavelets in computer vision. Any computer vision task with low quality data, for example because of low contrast images or high noise levels, is very challenging, especially in the presence of nontrivial shapes or scenes. Any approach that does not make use of prior knowledge of the objects of interest is likely to fail. The field calls therefore for Bayesian methods. While there is abundant literature in Bayesian computer vision, see [48], there is not much work with wavelets. We describe various attempts to couple Bayesian and wavelet methods, within the object modeling and recognition areas and outline other possible ideas.

As described in various chapters in this volume, a first possibility stems from viewing wavelets as filtering and/or data compression tools. In such case, we could, for example, parameterize the contour, perhaps after normalization, compute the wavelet transform, and discard small coefficients. We could model the object by storing the remaining coefficients, or use these coefficients for object recognition as input data to a classifying system. Discarding small coefficients may be given a Bayesian hypothesis testing or variable selection interpretation, see e.g. [49]. Note that these ideas may be extended to formalize some of the other wavelet based methods described. For example, the proposal in [43] essentially would use coefficient thresholding, with positive (negative) coefficients moved to +1 (−1).

Object recognition can be given a clear Bayesian flavour. With the notation of Section 3, we would base recognition on the posterior $p(y_i|x) \propto p(x|y_i)p(y_i)$, typically, but not necessarily, through the posterior mode. However, rather than using the complete data x, we could use only part of the wavelet transform, possibly after filtering, and similarly for the candidate y_i. For example, if the transform, according to levels, is described as $(c_1, c_2, ..., c_k)$, we could start by classifying based on coarser coefficients c_1, i.e. classify based on $p(y_i|c_1)$; if classification is not conclusive, say because several objects still have high probability of being the candidate, we may go further down in detail and base classification on $p(y_i|c_1, c_2)$, and so on. This sequential approach has the clear advantage of dealing with lower dimensionality data. Note that, under assumptions of conditional independence, we have

$$p(y_i|c_1, c_2) \propto p(c_2|y_i)p(y_i|c_1),$$

so updating would be relatively fast.

For a fixed level, classification has been based conventionally on classic methods such as k-nearest neighbors or feedforward neural networks. For methods with a Bayesian flavour, one possibility would be to use Lavine and West's [50] approach based on mixtures of normals, with the thresholded c_i's as data. Alternatively, we may also use Bayesian versions of a neural network model, see e.g. [51], [52].

One issue that remains to be addressed is when to stop the procedure, the central goal being to limit the computational effort. Sometimes experience will dictate which levels to use for classification. More heuristic approaches could stop the procedure when there is little change in discrimination from one level to the next one, or when the energy saturated by the retained coefficients attains a certain level. Vemuri and Radisajlevic [27] introduce a supervised learning procedure that may be used to determine the stopping level.

A more flexible approach may be undertaken using deformable templates. An interesting introduction to these concepts, with applications to face recognition, may be seen in [53]. Specifically related to wavelets is the approach described in [27], to deal with 3D shapes of cortical and subcortical structures from brain MRI data. To account for normalization issues, a 3D-shape described by its coordinates is represented as

$$x = q_c + R(q_\theta)(s(q_s) + d),$$

where q_c are the coordinates of the center of the model with respect to a global coordinate system, R is the rotational matrix (depending on parameters q_θ), s is a reference shape, e.g. a superquadric dependent on parameters q_s, and d is the displacement, which we may transform into the wavelet coefficients $q_d = (q_{dgl}, q_{dl})$. Hence a shape may be characterized as $q = (q_c, q_\theta, q_s, q_d)$ or $q' = (q_c, q_\theta, q_s, q_{dgl})$, if we retain only global coefficients q_{dgl}, and inference may be based on the posterior $p(q|D)$ or $p(q'|D)$.

17.5   Conclusions

Like Fourier transforms, wavelet transforms provide domains where certain operations can be performed either more effectively or more efficiently. For object modeling and recognition, wavelets have proven useful at least to achieve compact representations and extract shape features to be used for discrimination purposes.

Perhaps as a consequence of their novelty, wavelets have not been applied to object modeling and recognition in a systematic way, but rather following some researchers' intuition to confront specific problems.

Much work has still to be done; in particular we can mention:

- Achieve efficient methods to cope with translation, rotation and scale coefficient variability.

- Develop robust techniques for selecting which coefficients, and at which decomposition level are more appropriate for characterizing objects' shapes.

The development of techniques which work directly in the wavelet transform domain is still to come, although it is the authors' belief that wavelet transforms have as much potential for solving specific imaging and vision problems as Fourier techniques. As we have mentioned, Bayesian approaches may aid in these endeavours.

Acknowledgments: This work has been partially funded by the Spanish Commission for Science and Technology (grants CICYT TAP94-0305-C03-02 and CICYT TIC98-0272-C02-01) and Madrid's Autonomous Community (grant 06T/020/96). The authors gratefully acknowledge the help provided by Belén Moreno (3D skull data), and Soledad Herrero and José Miguel Espadero (figures 17.1–17.5).

17.6   References

[1] D. Marr and E. Hildreth: Theory of Edge Detection, *Proceedings of the Royal Society*, (London), ser. B, vol. 207, pp. 187–217, 1980.

[2] A. Rosenfeld, Ed.: *Multiresolution Image Processing and Analysis*, Springer Verlag, 1984.

[3] E. L. Grimson: Introduction of the Special Issue on Interpretation of 3D Scenes. *IEEE Trans. on PAMI*, **14**, 2, (1992), 97–98.

[4] S. J. Dickinson, R. Bergevin, I. Biederman, J. Eklundh, R. Munck-Fairwood and A. Pentland: The Use of Geons for Generic 3D Object Recognition. *Proc. 13th Int. Conf. on Artificial Intelligence*, Chambery, France,(1993) pp. 1693-1699.

[5] J. Wood: Invariant Pattern Recognition: A Review. *Pattern Recognition*, vol. 29 (1), pp. 1–17, 1996.

[6] A. Laine, S. Schuler y V. Girish: Orthonormal wavelet representations for recognizing complex annotations, *Machine Vision and Applications*, vol. 6, pp. 110–123, 1993.

[7] P. Wunsch and A. F. Laine: Wavelet Descriptors for Multiresolution Recognition of Handprinted Characters. *Pattern Recognition*, vol. 28 (8), pp. 1237–1249, 1995.

[8] Gene C.H. Chuang y C.C. Jay Kuo: Wavelet Descriptor of Planar Curves: Theory and Applications. *IEEE Trans. on Image Processing*, vol. 5 (1), pp. 56–70, Jan. 1996.

[9] W. Wang, Z. Bao, Q. Meng, G.M: Flachs y J.B. Jordan: Hand Recognition by Wavelet Transform and Neural Network. *Proc. of the SPIE*, vol. 2484, pp. 346–353, 1995.

[10] S.G. Mallat: Zero-crossing of a Wavelet Transform. *IEEE Trans. on Information Theory*, vol. 37 (14), pp. 1019–1033, 1991.

[11] Z. Berman y J. S. Baras: Properties of the Multiscale Maxima and Zerocrossings Representations. *IEEE Trans. Signal Processing*, vol. 41 (12), pp. 3216–3230, 1993.

[12] R. Hummel and R. Moniot: Reconstruction from zero-crossings in scale-space, *IEEE Trans. on Acoustic Speech Signal Processing*, vol. 37 (12), Dec. 1989.

[13] W. W. Boles: A Security System Based on Human Iris Identification Using Wavelet Transform. *First International Conference on Knowledge-Based Intelligent Electronic Systems*, Ed. L.C. Jain, pp. 533–541, 21–23 May., 1997.

[14] E.P. Simoncelli, W.T. Freeman, E.H. Adelson y D.J. Heeger: Shiftable Multiscale Transforms. *IEEE Trans on Information Theory*, vol. 38 (2), pp. 587–607, Mar. 1992.

[15] Q. M. Tieng and W. W. Boles: Complex Daubechies Wavelet Based Affine Invariant Representation for Object Recognition. *ICIP'94*, pp. 198–202, 1994.

[16] J.C. Pesquet, H. Krim and H. Carfantan: Time-Invariant Orthonormal Wavelet Representations. *IEEE Trans. on Signal Processing*, vol. 44 (8), pp. 1964–1970, Aug. 1996.

[17] W. Yu-Ping and S.L. Lee: Scale-space derived from B-splines. *IEEE Trans. on PAMI*, vol. 20 (10), October, 1998,

[18] W. Yu-Ping: Image Representations Using Multiscale Differential Operators. Submitted to *IEEE Trans. on Image Processing* in 1998. Avaliable as a preprint in:
http://wavelets.math.nus.edu.sg/download_papers/Preprints.html

[19] L. M. Reissell: Wavelet multiresolution representation of curves and surfaces. *Graphical Models and Image Processing*, vol. 58(3), pp. 198–217, 1996.

[20] M. Lades, J.C. Vorbrüggen, J. Buhmann, J. Lange, C.v.d. Malsburg, R.P. Würtz y W. Konen: Distortion Invariant Object Recognition in the Dynamic Link Architecture. *IEEE Trans. on Computers*, vol. 42 (3), pp. 300–311, Mar. 1993.

[21] L.Wiskott, J.M. Fellous, N. Krüger y C. von der Malsburg: Face Recognition by Elastic Bunch Graph Matching. *IEEE Trans. on PAMI*, vol. 19 (7), pp. 775–779, Jul. 1997.

[22] R. P. Würtz: Object Recognition Robust Under Translations, Deformations, and Changes in Background. *IEEE Trans. on PAMI*, vol. 19 (7), pp. 769–775, Jul. 1997.

[23] J.G. Daugman: High Confidence Visual Recognition of Persons by a Test of Statistical Independence. *IEEE Trans. on PAMI*, vol. 15 (11), pp. 1148–1161, Nov. 1993.

[24] M. Eck, T. DeRose, T. Duchamp, H. Hoppe, M. Lounsbery and W. Stuetzle: Multiresolution analysis of arbitrary meshes. *Computer Graphics Proceedings (SIGGRAPH 95)*, pp. 173–182, 1995.

[25] S. Campagna and H.-P. Seidel: Parametrerizing Meshes with Arbitrary Topology, in H. Niemann, H.-P. Seidel and B. Girod (eds.), *Image and Multidimensional Signal Processing'98*, pp. 287-290, 1998.

[26] L. Pastor, A. Rodríguez: Aproximación de superficies utilizando técnicas multirresolución. *Technical report, Dept. Tecnología Fotónica, Universidad Politécnica de Madrid*, 1997.
http://www.dtf.fi.upm.es/~arodri/apsumul.ps.gz

[27] B. C. Vemuri y A. Radisavljevic: Multiresolution Stochastic Hybrid Shape Models with Fractal Priors, *ACM Transactions on Graphics*, vol. 13 (2), pp. 177–207, ACM, 1994.

[28] M.H. Gross y R. Gatti: Efficient Triangular Surface Approximations using Wavelets and Quadtree Data Structures. *IEEE Trans. on Visualization and Computer Graphics*, vol. 2 (2), pp. 1–13, Jun. 1996.

[29] M. Lounsbery, T. DeRose y J. Warren: Multiresolution analysis for surfaces of arbitrary topological type. *ACM Trans. on Graphics*, vol 16 (1), pp. 34–73, Jan. 1997.

[30] Ikeuchi K. and Hebert M.: Spherical Representations: From EGI to SAI. *Proc. Intl, NSI–ARPA Workshop*, pp. 327–345, 1995.

[31] P. Schröeder and W. Sweldens: Spherical wavelets: Efficiently representing functions on a sphere. *Computer Graphics Proceedings (SIGGRAPH 95)*, pp. 161–172, 1995.

[32] P. Schröeder: Wavelets in Computers Graphics. *Proc. of the IEEE*, vol. 84 (4), Abr., pp. 615–625, 1996.

[33] E. J. Stollnitz, T. D. DeRose and D. H. Salesin: *Wavelets for Computer Graphics*. Morgan Kauffman Publishers (1996).

[34] D. Zorin, P. Schröeder and W. Sweldens: Interactive Multiresolution Mesh Editing. *Computer Graphics Proceedings (SIGGRAPH 97)*, pp. 259–269, 1997.

[35] I. Biederman: Recognition by Components: A Theory of Human Image Understanding, *Psycological Review*, vol. 94 (2), pp. 115–147, American Psychological Association, 1987.

[36] Aggarwal, J. K., Ghosh, J., Nair, D. y Taha, I.: A Comparative Study of Three Paradigms for Object Recognition—Bayesian Statistics, Neural Networks and Expert Systems, Bowyer, K. and N. Ahuja, Ed., *Advances in Image Understanding. A Festschrift for Azriel Rosenfeld*, pp. 196–208, IEEE Comp. Society, 1996.

[37] R. Jain, R. Kasturi y B.G. Schunck: *Machine Vision*, McGraw-Hill, 1995.

[38] S. Ullman: *High-level Vision. Object Recognition and Visual Cognition*, MIT Press, 1996.

[39] S. J. Dickinson, A. P. Pentland and A. Rosenfeld: From Volumes to Views: An Approach to 3D Object Recognition, *CVGIP: Image Understanding*, vol. 55 (2), pp. 130–154, Academic Press, 1992.

[40] T. Chang and C.-C.J. Kuo: Texture analysis and classification with tree-structured wavelet trasform. *IEEE Trans. Im. Process.*, vol 2 (4), pp. 429–441, 1993.

[41] A. Laine and J. Fan: Texture classification and segmentation using wavelet frames. *IEEE Trans. Pattern Analysis and Machine Intelligence*, vol 15 (11), pp. 1186–1190, 1993.

[42] G. Van de Wouwer, S. Livens and D. van Dyck: Color Texture Classification by wavelet energy-correlation signatures. In *Proc. of the International Conference on Computer Analysis and Image Processing, 1997-Firenze,,* pp. 327-334, 1997.

[43] C. E. Jacobs, A. Finkelstein and D. H. Salesin: Fast Multiresolution Image Querying. *Proc. of SIGGRAPH'95*, pp. 277–286, ACM, New York, 1995.

[44] L. Pastor, A. Rodríguez, O. Robles and J.M. Espadero: Estudio de técnicas de búsqueda de objetos por contenido utilizando wavelets. *Technical report, Dpto. Tecnología Fotónica, Universidad Politécnica de Madrid,* 1998.
http://www.dtf.fi.upm.es/~arodri/recwav.ps.gz

[45] Q. M. Tieng and W. W. Boles: Recognition of 2D Object Contours Using the Wavelet Transform Zero-Crossing Representation. *IEEE Trans. on PAMI*, vol. 19 (8), pp. 910–916, 1997.

[46] X. Wu and B. Bhanu: Gabor Wavelet Representation for 3-D Object Recognition. *IEEE Trans. on Image Processing*, vol. 6 (1), pp. 47–64, Jan. 1997.

[47] L. Pastor, J. Durán and A. Rodríguez: Qualitative shape modeling of 3D data. *Proc of Second International Conference on Mathematics & Design 98*, San Sebastián, Spain, pp. 459–466, Jun. 1998.

[48] D. Knill and W. Richards: *Perception as Bayesian Inference*, Cambridge University Press, 1996.

[49] B. Vidakovic: *Statistical Modeling by Wavelets*, Wiley, 1999.

[50] M. Lavine, M. West: A Bayesian method for classification and discrimination, *Can. Jour. Stats.*, 20, 451-461, 1992.

[51] P. Muller, D. Rios Insua: Issues in Bayesian analysis of neural network models, *Neural Computation*, pp. 571–592, 1998.

[52] C. Bishop: *Neural Networks for Pattern Recognition*, Chapman Hall, 1996.

[53] D.B. Phillips and A.M.F. Smith: Bayesian faces via hierarchical template modeling, *Jour. Amer. Stat. Assoc.*, 89, pp. 1151–1163, 1994.

18

Bayesian Denoising of Visual Images in the Wavelet Domain

Eero P. Simoncelli

ABSTRACT The use of multi-scale decompositions has led to significant advances in representation, compression, restoration, analysis, and synthesis of signals. The fundamental reason for these advances is that the statistics of many natural signals, when decomposed in such bases, are substantially simplified. Choosing a basis that is adapted to statistical properties of the input signal is a classical problem. The traditional solution is principal components analysis (PCA), in which a linear decomposition is chosen to diagonalize the covariance structure of the input. The most well-known description of image statistics is that their Fourier spectra take the form of a power law [e.g., 1, 2, 3]. Coupled with a constraint of translation-invariance, this suggests that the Fourier transform is an appropriate PCA representation. Fourier and related representations are widely used in image processing applications. For example, the classical solution to the noise removal problem is the Wiener filter, which can be derived by assuming a signal model of decorrelated Gaussian-distributed coefficients in the Fourier domain.

Recently a number of authors have noted that statistics of order greater than two can be utilized in choosing a basis for images. Field [2, 4] noted that the coefficients of frequency subbands of natural scenes have much higher kurtosis than a Gaussian density. Recent work on so-called "independent components analysis" (ICA) has sought linear bases that optimize higher-order statistical measures [e.g., 5, 6]. Several authors have constructed optimal bases for images by optimizing such information-theoretic criterion [7, 8]. The resulting basis functions are oriented and have roughly octave bandwidth, similar to many of the most common multi-scale decompositions. A number of authors have explored the optimal choice of a basis from a library of functions based on entropy or other statistical criterion [e.g. 9, 10, 11, 12, 13].

In this chapter, we examine the empirical statistical properties of visual images within two fixed multi-scale bases, and describe two statistical models for the coefficients in these bases. The first is a non-Gaussian marginal model, previously described in [14]. The second is a joint non-Gaussian Markov model for wavelet subbands, previous versions of which have been described in [15, 16]. We demonstrate the use of each of these models in

Research supported by NSF CAREER grant MIP-9796040.

Bayesian estimation of an image contaminated by additive Gaussian white noise.

18.1   Marginal Statistical Model

A number of authors have observed that wavelet subband coefficients have highly non-Gaussian statistics [e.g., 17, 4, 14]. The intuitive explanation for this is that images typically have spatial structure consisting of smooth areas interspersed with occasional edges or other abrupt transitions. The smooth regions lead to near-zero coefficients, and the structures give occasional large-amplitude coefficients.

As an example, histograms for subbands of separable wavelet decompositions of several images are plotted in figure 1[1]. These densities may be accurately modeled with a two-parameter density function of the form [17, 14]:

$$P_c(c; s, p) = \frac{e^{-|c/s|^p}}{Z(s,p)}, \tag{1}$$

where the normalization constant is $Z(s,p) = 2\frac{s}{p}\Gamma(\frac{1}{p})$. Each graph in figure 1 includes a dashed curve corresponding to the best fitting instance of this density function, with the parameters $\{s, p\}$ estimated by maximizing the likelihood of the data under the model. For subbands of images in our collection, values of the exponent p typically lie in the range $[0.5, 0.8]$. The density model fits the histograms remarkably well, as indicated by the relative entropy measures given below each plot.

18.1.1   Bayesian denoising: Marginal model

Consider an image whose pixels are contaminated with i.i.d. samples of additive Gaussian noise. Because the wavelet transform is orthonormal, the noise is also Gaussian and white in the wavelet domain. Thus, each coefficient in the wavelet expansion of the noisy image is written as $y = c + n$, where c is drawn from the marginal density given in equation (1), and n is Gaussian.

A standard estimator for c given the corrupted observation y is the maximum a posteriori (MAP) estimator:

$$\hat{c}(y) = \arg\max_c P_{c|y}(c|y) \tag{2}$$

$$= \arg\max_c P_{y|c}(y|c)P_c(c) \tag{3}$$

$$= \arg\max_c P_n(y-c)P_c(c) \tag{4}$$

---

[1]The specific wavelet decomposition used for these examples is described in section 18.2.2.

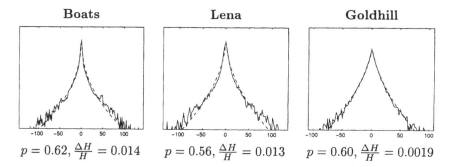

FIGURE 1. Examples of 256-bin coefficient histograms for a single vertical wavelet subband of three images, plotted in the log domain. All images are size 512x512. Also shown (dashed lines) are fitted model densities corresponding to equation (1). Below each histogram is the maximum-likelihood value of p used for the fitted model density, and the relative entropy (Kullback-Leibler divergence) of the model and histogram, as a fraction of the total entropy of the histogram.

where Bayes' rule allows us to write this in terms of the probability densities of the noise ($\mathcal{P}_n$) and the prior density of the signal coefficient ($\mathcal{P}_c$). In order to use this equation to estimate the original signal value c, we must know both density functions.

Figure 2 shows a set of (numerically computed) MAP estimators for the model in equation (1) with different values of the exponent p, assuming a Gaussian noise density. In the special case of $p = 2$ (i.e., Gaussian source density), the estimator assumes the well-known linear form:

$$\hat{c}(y) = \frac{\sigma_c^2 \, y}{\sigma_c^2 + \sigma_n^2},$$ (5)

estimators for other values of p are nonlinear: the $p = 0.5$ function resembles a hard thresholding operator, and $p = 1$ resembles a soft thresholding operator. Donoho has shown that these types of shrinkage operator are nearly minimax optimal for some classes of regular function (e.g., Besov) [18]. Other authors have established connections of these these results with statistical models [19, 20]. In addition, thresholding techniques are widely used in the television and video engineering community, where they are known as "coring" [e.g., 21, 22, 23]. For example, most consumer VCR's use a simple coring technique to remove magnetic tape noise.

If one wishes to minimize squared error, the mean of the posterior distribution provides an optimal estimate of the coefficient c, given a measurement of y:

$$\hat{c}(y) = \int dc \, \mathcal{P}_{c|y}(c|y) \, c = \frac{\int dc \, \mathcal{P}_{y|c}(y|c) \, \mathcal{P}_c(c) \, c}{\int dc \, \mathcal{P}_{y|c}(y|c) \, \mathcal{P}_c(c)} = \frac{\int dc \, \mathcal{P}_n(y - c) \, \mathcal{P}_c(c) \, c}{\int dc \, \mathcal{P}_n(y - c) \, \mathcal{P}_c(c)}.$$ (6)

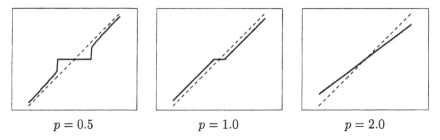

$p = 0.5$ $p = 1.0$ $p = 2.0$

FIGURE 2. MAP estimators for the model given in equation (1), with three different exponents. The noise is additive and Gaussian, with variance one third that of the signal. Dashed line indicates the identity function.

$p = 0.5$ $p = 1.0$ $p = 2.0$

FIGURE 3. Bayesian least-squares estimators for the model given in equation (1), with three different exponents, p. The noise is additive and Gaussian, with variance one third that of the signal. Dashed line indicates the identity function.

The denominator is the pdf of the noisy observation y, computed by marginalizing the convolution of the noise and signal pdf's.

Figure 3 shows (numerically computed) Bayesian least-squares estimators for the model of equation (1), with three different values of the exponent p. Again, for the special case of $p = 2$ the estimator is linear and of the form of equation (5). As with the MAP estimators, smaller values of p produce a nonlinear shrinkage operator, somewhat smoothed in comparison to those of figure 2. In particular, for $p = 0.5$ (which is well-matched to wavelet marginals such as those shown in figure 1), the estimator preserves large amplitude values and suppresses small amplitude values. This is intuitively sensible: given the substantial prior probability mass at $c = 0$, small values of y are assumed to have arisen from a value of $c = 0$.

The quality of a denoising algorithm will depend on the exponent p. To quantify this, figure 4 shows the (numerically computed) error variance for the Bayesian least-squares estimate (see figure 3), as a function of p. Note that the error variance drops significantly for values of p less than one.

In practice, one must estimate the parameters $\{s, p\}$ and σ_n from the

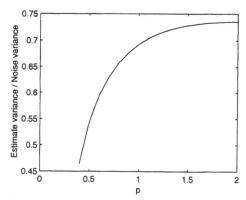

FIGURE 4. Error variance of the Bayes least-squares estimator relative to the noise variance, as a function of the model parameter p of equation (1). Noise variance σ_n was held constant at one third of the signal variance.

noisy collection of coefficients $\{y_k\}$. A simple solution is a maximum likelihood estimator:

$$\{\hat{s}, \hat{p}, \hat{\sigma}_n\} = \arg \max_{\{s,p,\sigma_n\}} \prod_k \mathcal{P}_y(y_k; s, p, \sigma_n)$$

$$= \arg \max_{\{s,p,\sigma_n\}} \prod_k \int dc \, e^{-|c/s|^p} \, e^{-(y_k-c)^2/2\sigma_n^2} \quad (7)$$

where the product is taken over all coefficients within the subband. In practice, both the integration and the optimization are performed numerically. Furthermore, in the examples shown in section 18.3, we assume σ_n is known, and optimize only over $\{s, p\}$. As a starting point for the optimization, we solve for the parameter pair $\{s, p\}$ corresponding to a density with kurtosis and variance matching those of the histogram [as in 14].

18.2 Joint Statistical Model

In the model of the previous section, we treated the wavelet coefficients as if they were independent. Empirically, orthonormal wavelet coefficients are found to be fairly well decorrelated. Nevertheless, it is quite evident that wavelet coefficients of images are *not* statistically independent. Figure 5 shows the magnitudes (absolute values) of coefficients in a four-level separable wavelet decomposition. In particular, previous work has shown that large-magnitude coefficients tend to occur near each other within subbands, and also occur at the same relative spatial locations in subbands at adjacent scales, and orientations [e.g., 15, 16].

As an example, consider two coefficients representing horizontal information at adjacent scales, but the same spatial location of the "Boats" image.

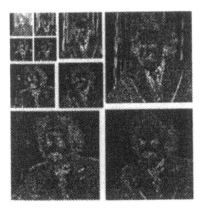

FIGURE 5. Coefficient magnitudes of a wavelet decomposition. Shown are absolute values of subband coefficients at three scales, and three orientations of a separable wavelet decomposition of the "Einstein" image.

Figure 6A shows the conditional histogram $\mathcal{H}(c|p)$ of the "child" coefficient conditioned on a coarser-scale "parent" coefficient. The histogram illustrates several important aspects of the relationship between the two coefficients. First, they are (second-order) decorrelated, since the expected value of c is approximately zero for all values of p. Second, the variance of the conditional histogram of c clearly depends the value of p. Thus, although c and p are uncorrelated, *they are still statistically dependent.* Furthermore, this dependency cannot be eliminated through further linear transformation.

The structure of the relationship between c and p becomes more apparent upon transforming to the log domain. Figure 6B shows the conditional histogram $\mathcal{H}\left(\log_2(c^2)|\log_2(p^2)\right)$ The right side of the distribution is unimodal and concentrated along a unit-slope line. This suggests that in this region, the conditional expectation, $\mathbf{E}(c^2|p^2)$, is approximately proportional to p^2. Furthermore, vertical cross sections (i.e., conditional histogram for a fixed value of p^2) have approximately the same shape for different values of p^2. Finally, the left side of the distribution is concentrated about a horizontal line, suggesting that c^2 is independent of p^2 in this region.

The form of the histograms shown in figure 6 is surprisingly robust across a wide range of images. Furthermore, the qualitative form of these statistical relationships also holds for pairs of coefficients at adjacent spatial locations ("siblings"), adjacent orientations ("cousins"), and adjacent orientations at a coarser scale ("aunts"). Given the linear relationship between the squares of large-amplitude coefficients and the difficulty of characterizing the full density of a coefficient conditioned on its neighbors, we've examined a linear predictor for the squared coefficient. Figure 6C shows a histogram of $\log_2(c^2)$ conditioned on a linear combination of the squares of eight adjacent coefficients in the same subband, two coefficients at other

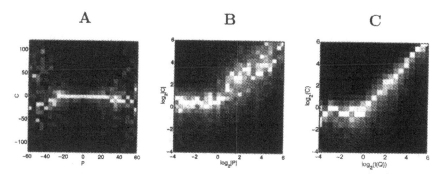

FIGURE 6. Conditional histograms for a fine scale horizontal coefficient. Brightness corresponds to probability, except that each column has been independently rescaled to fill the full range of display intensities. **A:** Conditioned on the parent (same location and orientation, coarser scale) coefficient. Data are for the "Boats" image. **B:** Same as **A**, but in the log domain. **C** Conditioned on a linear combination of neighboring coefficient magnitudes.

orientations, and a coefficient at a coarser scale. The linear combination is chosen to be least-squares optimal (see equation (9)). The histogram is similar to the single-band conditional histogram of figure 6B, but the linear region is extended and the conditional variance is significantly reduced.

The form of these observations suggests a simple Markov model, in which the density of a coefficient, c, is conditionally Gaussian with variance a linear function of the squared coefficients in a local neighborhood:

$$ \mathcal{P}\left(c \mid \vec{p}\right) = \mathcal{N}\left(0; \sum_k w_k p_k^2 + \alpha^2\right). \tag{8} $$

Here, the neighborhood $\{p_k\}$ consists of coefficients at other orientations and adjacent scales, as well as adjacent spatial locations. Note that although we utilize a normal distribution, this is not a jointly Gaussian density in the traditional sense, since the *variance* rather than the mean is dependent on the neighborhood. Figure 7 shows a set of conditional histograms, and the best-fitting instantiation of the model in equation (8). The fits are seen to be reasonably good, as indicated by the low relative entropy values. We have used variants of this model for applications of compression [15] and texture synthesis [24].

18.2.1   Bayesian denoising: Joint model

As in the previous section, assume a coefficient is contaminated with Gaussian white noise: $y = c + n$. If we assume the neighbor coefficients are known, the conditionally Gaussian form of equation (8) leads to a linear Bayesian

Boats   Lena   Goldhill

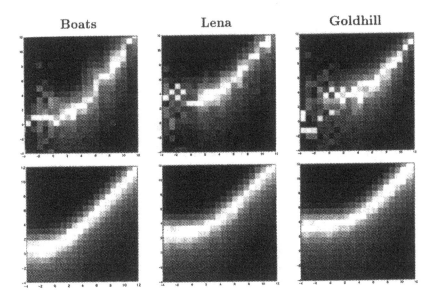

FIGURE 7. Top: Examples of log-domain conditional histograms for the second-level horizontal subband of different images, conditioned on an optimal linear combination of coefficient magnitudes from adjacent spatial positions, orientations, and scales. **Bottom:** Model of equation (8) fitted to the conditional histograms in the left column. Intensity corresponds to probability, except that each column has been independently rescaled to fill the full range of intensities.

estimator:

$$\hat{c}(y) = \frac{\sum_k w_k p_k^2 + \alpha^2}{\sum_k w_k p_k^2 + \alpha^2 + \sigma_n^2} y.$$

In a more realistic implementation, we must estimate c given the noisy observations of the neighbors. A complete solution for this problem is difficult, since the conditional density of the variance of the clean coefficient given the noisy neighbors cannot be computed in closed form. A numerical solution should be feasible, but for the purposes of the current paper, we instead choose to utilize the marginal model estimator from the previous section.

Specifically, we first compute estimates of the coefficients in a subband, $\hat{c}$, using equation (6). We then use these estimated coefficients to estimate the weight parameters, $\{w_k\}$, and the constant, α, by minimizing the squared error:

$$\{\hat{w}, \hat{\alpha}\} = \arg\min_{\{\vec{w}, \alpha\}} \mathbf{E} \left[\hat{c}^2 - \sum_k w_k \hat{p}_k^2 - \alpha^2 \right]^2. \tag{9}$$

Note that we are using the marginal estimates of both the coefficient *and* the neighbors. We have also implemented (numerically) a maximum likelihood solution, but it was found to be computationally expensive and did not yield any improvement in performance.

Given $\hat{w}$, the joint model estimate, $\hat{c}(y)$, is computed from the noisy observation, y, using the marginal estimates of the neighbors:

$$\hat{c}(y) = \frac{\sum_k \hat{w}_k \hat{p}_k^2 + \hat{\alpha}^2}{\sum_k \hat{w}_k \hat{p}_k^2 + \hat{\alpha}^2 + \sigma_n^2}\, y. \tag{10}$$

Although clearly sub-optimal, this estimate is easily computed and gives reasonable results.

18.2.2   Choice of Basis

As mentioned in the introduction, a number of recent researchers have derived wavelet-like bases for images using information-theoretic optimality criterion. Here, we compare the denoising abilities of two different types of discrete multi-scale basis.

The first is a separable critically-sampled 9-tap quadrature mirror filter (QMF) decomposition, based on filters designed in [25]. This is a linear-phase (symmetric) approximation to an orthonormal wavelet decomposition. The lowpass filter samples are:

$$l[n] \quad = \quad [0.028074, -0.060945, -0.073387, 0.41473, 0.79739,$$
$$0.41473, -0.073387, -0.060945, 0.028074].$$

The highpass filter is obtained via $h[n] = (-1)^n l[N-n+1]$, and the system diagram is shown in figure 8. Compared with orthonormal wavelets, this decomposition has the advantage that the basis functions are symmetric. The drawback is that the system does not give perfect reconstruction: the filters are designed to optimize a residual function. This is not a serious problem for applications such as the one discussed in this chapter, since reconstruction signal-to-noise ratios (SNRs) are typically about 55dB.

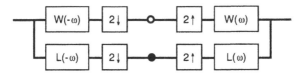

FIGURE 8. Single-scale system diagram for a critically-sampled QMF or wavelet decomposition, in one dimension. Boxes correspond to convolution, downsampling, and upsampling operations. Two-dimensional decomposition is achieved by applying the one-dimensional decomposition in the vertical direction, and then to both resulting subbands in the horizontal direction. Multi-scale decompositions are constructed by inserting the system into itself at the location of the filled circle.

The second decomposition is known as a *steerable pyramid* [26]. In this decomposition, the image is subdivided into subbands using filters that are polar-separable in the Fourier domain. In scale, the subbands have octave bandwidth with a functional form constrained by a recursive system diagram. In orientation, the functional form is chosen so that the set filters at a given scale span a rotation-invariant subspace. The decomposition can performed with any number of orientation bands, K, each of orientation bandwidth $2\pi/K$ radians. The full two-dimensional transform is overcomplete by a factor of $4K/3$, and is a tight frame (i.e., the matrix corresponding to the inverse transformation is equal to the transpose of the forward transformation matrix). Spatial subsampling of each subband respects the Nyquist criterion, and thus the representation is translation-invariant (free of aliasing). An idealized system diagram is shown in figure 9. The trans-

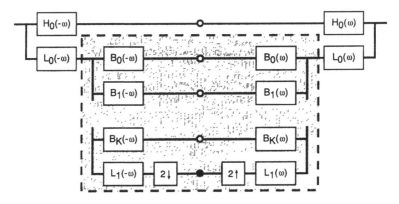

FIGURE 9. Single-scale system diagram for a steerable pyramid. Multiscale decompositions are constructed by inserting the portion of the system within the gray region at the location of the filled circle.

form is implemented using a set of oriented filters that are polar-separable when expressed in the Fourier domain:

$$F_{n,k}(r,\theta) = B_n(r)G_k(\theta), \quad n \in [0, M], k \in [0, K-1],$$

where $B_n(r) = \cos\left(\pi/2 \log_2\left(2^n r/\pi\right)\right)$ and

$$G_k(\theta) = \begin{cases} \left[\cos(\theta - \frac{\pi k}{K})\right]^{K-1}, & |\theta - \frac{\pi k}{K}| < \frac{\pi}{2} \\ 0, & \text{otherwise,} \end{cases}$$

where r, θ are polar frequency coordinates. Subbands are subsampled by a factor of 2^n along both axes. In addition, one must retain the (non-oriented) lowpass residual band, which is computed using the following filter:

$$L(r) = \begin{cases} \cos\left(\frac{\pi}{2} \log_2\left(\frac{2^{(M+1)}r}{\pi}\right)\right), & r \in [\frac{\pi}{2^{M+1}}, \frac{\pi}{2^M}] \\ 1 & r < \frac{\pi}{2^{M+1}} \\ 0 & r > \frac{\pi}{2^M}. \end{cases}$$

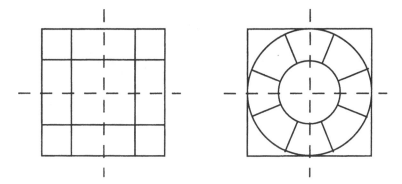

FIGURE 10. Idealized partition of frequency domain associated with each decomposition. Each axis covers the range $[-\pi, \pi]$ radians/pixel. Left: Separable QMF or wavelet. Right: Steerable pyramid, $K = 4$.

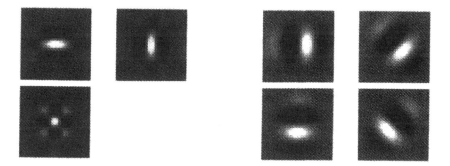

FIGURE 11. Basis functions at a single scale. Left: QMF decomposition. Right: 4-orientation steerable pyramid.

Figure 10 shows the idealized frequency partition of the two decompositions, figure 11 shows the basis functions at a single scale, and figure 12 shows the decomposition of the Einstein image using the two bases.

Figure 13 shows a scatter plot of estimated values of p for the images shown in figure 1. Note that the values for some of the separable QMF bands are quite high (particularly, band 3, which contains the mixed diagonals). On average, the steerable pyramid values less than those of the separable critically-sampled system. This is consistent with the preference for oriented basis functions, as mentioned in the introduction.

In addition to the smaller values of p, another advantage of the steerable pyramid in the context of denoising is the translation-invariance property. Previous work has emphasized the importance of translation-invariance for image processing tasks such as denoising [26, 11, 12, 14]. One drawback of this representation is that our assumption of orthonormality is violated. In particular, the representation is heavily overcomplete and thus there are

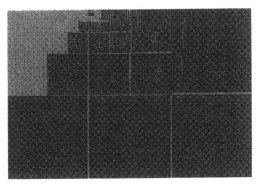

FIGURE 12. Example multi-scale decompositions. Left: separable QMF decomposition, a linear-phase approximation to an orthonormal wavelet decomposition. Right: steerable pyramid, with $K = 4$ orientation bands.

strong linear dependencies between the coefficients. The marginal model of section 18.1 assumes that the coefficients are statistically independent, and the joint model of section 18.2 assumes that they are decorrelated.

18.3 Results

In this section, we show examples of image denoising using the two models described in previous sections. In all cases, we assume the noise variance, σ_n^2, is known. Gaussian noise of this variance, truncated to a range of three standard deviations, is added to the original image. This contaminated image is transformed to the relevant basis, the appropriate estimator is applied to all coefficients within each subband, and then the transformation is inverted. The estimators are computed as follows:

- Linear estimator:

    1. Estimate $\sigma_c^2 \approx \max\{0, \mathbf{E}\left(c^2 - \sigma_n^2\right)\}$.
    2. $\hat{c}(y) = \frac{\sigma_c^2}{\sigma_c^2 + \sigma_n^2} y$.

- Threshold estimator:

    1. $t = 3\sigma_n$
    2. $\hat{c}(y) = \begin{cases} y, & |y| > t \\ 0, & \text{otherwise} \end{cases}$

- Bayesian marginal (coring) estimator:

    1. Compute parameter estimates $\{\hat{s}, \hat{p}\}$ by maximizing likelihood of the subband data (equation (7)).

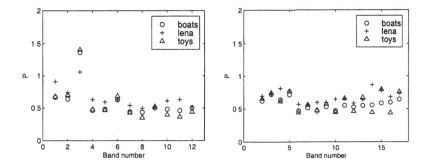

FIGURE 13. Values of p for subbands from the 512×512 images analyzed in figure 1. Left: Subbands of separable QMF pyramid. Right: Subbands of steerable pyramid ($K = 4$ orientations). Subband numbering runs from high to low frequency. Orientation bands of the separable decomposition are ordered (vertical,horizontal,mixed-diagonal), and orientation bands of the steerable pyramid start with vertical and proceed counterclockwise.

2. Compute the conditional mean estimator $f(y)$ numerically using equation (6).

3. $\hat{c}(y) = f(y)$

- Bayesian joint estimator:

1. Compute $\hat{c}(y)$ for all subbands using the Bayesian marginal estimator.

2. Estimate weights $\hat{w}$ and $\hat{\alpha}$ using equation (9).

3. $\hat{\hat{c}}(y) = \frac{\sum_k \hat{w}_k \hat{p}_k^2 + \hat{\alpha}^2}{\sum_k \hat{w}_k \hat{p}_k^2 + \hat{\alpha}^2 + \sigma_n^2} \, y.$

All expectations are estimated by summing spatially. For the joint estimator, we use a neighborhood consisting of the 12 nearest spatial neighbors (within the same subband), the 5 nearest cousin coefficients (in other orientation bands at the same scale), the 9 nearest parent coefficients (in the adjacent subband of coarser scale), a single aunt (from each orientation at the adjacent coarser scale), and a single grandparent.

Table 18.3 shows signal-to-noise ratios (SNRs) for all four algorithms, applied to all three decompositions, at four different contamination levels. Note that Bayesian algorithms outperform the other two techniques for all examples. Also note that the steerable pyramid decompositions significantly outperform the separable QMF decomposition, and the six-orientation decomposition shows a noticeable improvement over the four-orientation decomposition.

Finally, figures 14 and 15 show some example images. Figure 14 shows results using the separable decomposition. The Bayesian results appear both sharper (because high-amplitude coefficients are preserved) and less

decomposition type	noisy image	Estimator			
		Linear	Threshold	BayesCore	BayesJoint
QMF-9	1.59	8.26	8.09	9.66	10.58
	4.80	9.71	10.10	11.64	12.31
	8.99	12.02	12.46	14.21	14.49
	13.03	14.81	14.65	16.61	16.62
Spyr (4 ori)	1.59	9.28	10.12	10.35	10.96
	4.80	10.60	11.71	11.98	12.61
	9.02	12.58	14.04	14.19	14.81
	13.06	14.96	15.89	16.50	16.99
Spyr (6 ori)	1.59	9.34	10.37	10.55	11.12
	4.80	10.67	12.03	12.18	12.74
	9.02	12.66	14.23	14.43	14.90
	13.06	15.02	16.28	16.71	17.09

TABLE 1. Denoising results for four estimators, three different decompositions, and four different levels of additive Gaussian noise added to the "Einstein" image. All values indicate signal-to-noise ratio in decibels ($10 \log_{10}$(signal variance/error variance)).

noisy (because low-amplitude coefficients are suppressed) than the linear estimator. The aliasing artifacts of the critically sampled transform are most evident with the thresholding estimator, and least evident in the Bayes joint estimator.

Figure 15 shows results using the 4-orientation steerable pyramid. Note that although the use of this decomposition eliminates the aliasing artifacts and produces higher SNR results, the results now look more blurred. The results can be made more visually appealing by a subsequent sharpening operation, although this reduces the SNR.

18.4   Conclusion

We have described two non-Gaussian density models for visual images, and used them to develop nonlinear Bayesian estimators that outperform classical linear estimators and simple thresholding estimators.

We have implemented these estimators in the context of two different multi-scale decompositions. The results obtained with the overcomplete steerable pyramid bases are superior to the separable QMF basis, due to translation-invariance and more kurtotic statistics. This is true in spite of the fact that the statistical model is no longer correct: the use of an overcomplete basis creates linear dependencies between coefficients. It would be interesting to compare these results to other translation-invariant denoising schemes, such as the cycle-spinning approach of [27]. It would also be interesting to explore whether the results can be improved by adaptively choosing an optimal basis. Such optimization could be done over a collec-

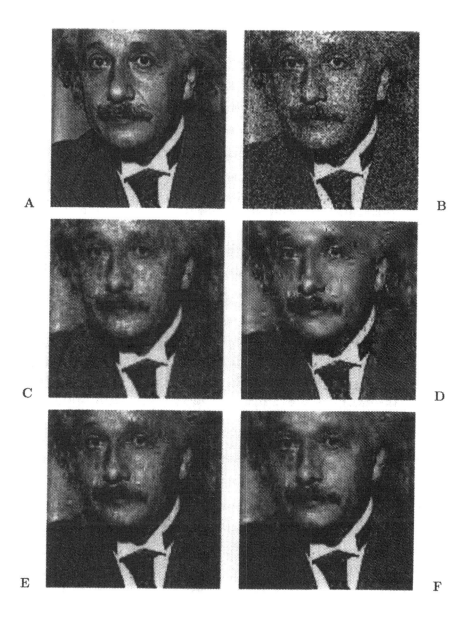

FIGURE 14. Cropped denoising results using an (approximately) orthonormal separable decomposition. **A**: Original "Einstein" image (cropped). **B**: Noisy image (SNR = 4.8dB). **C**: Linear least-squares estimator. **D**: Optimal thresholding. **E**: Bayes - marginal model. **F**: Bayes - joint model.

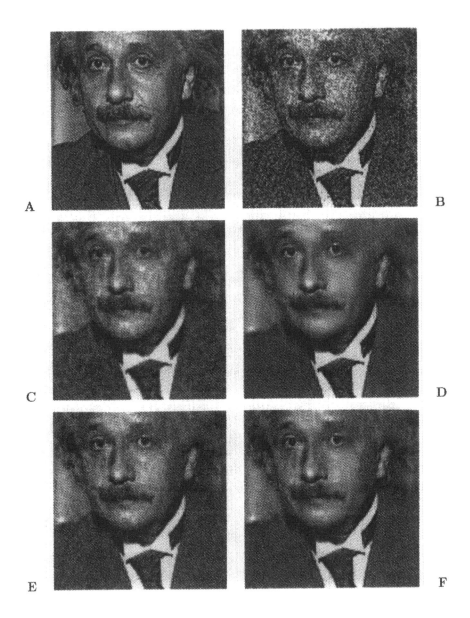

FIGURE 15. Cropped denoising results using a 4-orientation steer-able pyramid decomposition. **A**: Original "Einstein" image (cropped). **B**: Noisy image (SNR = 4.8dB). **C**: Linear least-squares estimator. **D**: Optimal thresholding. **E**: Bayes - marginal model. **F**: Bayes - joint model.

tion of images drawn from a particular class, or individually for each image (assuming spatial stationarity).

A number of improvements should be made to the statistical models, and the Bayesian estimators described in this paper. In particular, the marginal densities that come from integrating the joint model density are inconsistent with those of the generalized Gaussian marginal model of section 18.1. Although the empirical evidence for the linear dependency of coefficient variance on a single neighbor is quite strong, the linear estimate based on the full neighborhood needs to be validated. A full joint Bayesian estimator, which estimates both the density parameters and the clean coefficient based on the noisy observations, should be implemented numerically. The distortion model should be extended to include spatial blurring. A fully blind denoising technique (i.e., including estimation of σ_n) should also be explored.

Finally, we note the need for a measure of image distortion that adequately reflects human perceptual salience. Although the Bayesian denoising results in figure 15 are excellent according to a squared error measure, informal questioning suggests that most observers prefer a sharper image, even if it contains more noticeable artifacts.

18.5 REFERENCES

[1] Alex Pentland. Fractal based description of natural scenes. *IEEE Trans. PAMI*, 6(6):661–674, 1984.

[2] D J Field. Relations between the statistics of natural images and the response properties of cortical cells. *J. Opt. Soc. Am. A*, 4(12):2379–2394, 1987.

[3] D L Ruderman and W Bialek. Statistics of natural images: Scaling in the woods. *Phys. Rev. Letters*, 73(6), 1994.

[4] D J Field. What is the goal of sensory coding? *Neural Computation*, 6:559–601, 1994.

[5] J F Cardoso. Source separation using higer order moments. In *ICASSP*, pages 2109–2112, 1989.

[6] P Common. Independent component analysis, a new concept? *Signal Process.*, 36:387–314, 1994.

[7] B A Olshausen and D J Field. Natural image statistics and efficient coding. *Network: Computation in Neural Systems*, 7:333–339, 1996.

[8] A J Bell and T J Sejnowski. Learning the higher-order structure of a natural sound. *Network: Computation in Neural Systems*, 7:261–266, 1996.

[9] R R Coifman and M V Wickerhauser. Entropy-based algorithms for best basis selection. *IEEE Trans. Info. Theory*, IT-38:713–718, March 1992.

[10] Stephane Mallat and Zhifeng Zhang. Matching pursuits with time-frequency dictionaries. *IEEE Trans. Signal Proc.*, December 1993.

[11] D L Donoho and I M Johnstone. Ideal denoising in an orthogonal basis chosen from a library of bases. *C.R. Acad. Sci.*, 319:1317–1322, 1994.

[12] M V Wickerhauser. *Adapted Wavelet Analysis: From Theory to Software.* A K Peters, Wellesley, MA, 1994.

[13] J C Pesquet, H Krim, D Leporini, and E Hamman. Bayesian approach to best basis selection. In *Proc Int'l Conf Acoustics, Speech and Signal Proc*, pages 2634–2638, Atlanta, May 1996.

[14] E P Simoncelli and E H Adelson. Noise removal via Bayesian wavelet coring. In *Third Int'l Conf on Image Proc*, volume I, pages 379–382, Lausanne, September 1996. IEEE Sig Proc Society.

[15] R W Buccigrossi and E P Simoncelli. Image compression via joint statistical characterization in the wavelet domain. Technical Report 414, GRASP Laboratory, University of Pennsylvania, May 1997. Accepted (3/99) for publication in IEEE Trans Image Processing.

[16] E P Simoncelli. Statistical models for images: Compression, restoration and synthesis. In *31st Asilomar Conf on Signals, Systems and Computers*, pages 673–678, Pacific Grove, CA, November 1997. IEEE Computer Society. Available at: ftp://ftp.cns.nyu.edu/pub/eero/simoncelli97b.ps.gz.

[17] S G Mallat. A theory for multiresolution signal decomposition: The wavelet representation. *IEEE Pat. Anal. Mach. Intell.*, 11:674–693, July 1989.

[18] D Donoho and I Johnstone. Adapting to unknown smoothness via wavelet shrinkage. *J American Stat Assoc*, 90(432), December 1995.

[19] F Abramovich, T Sapatinas, and B W Silverman. Wavelet thresholding via a bayesian approach. *J R Stat Soc B*, 60:725–749, 1998.

[20] D Leporini and J C Pesquet. Multiscale regularization in besov spaces. In *31st Asilomar Conf on Signals, Systems and Computers*, Pacific Grove, CA, November 1998.

[21] J P Rossi. *JSMPTE*, 87:134–140, 1978.

[22] B. E. Bayer and P. G. Powell. A method for the digital enhancement of unsharp, grainy photographic images. *Adv in Computer Vision and Im Proc*, 2:31–88, 1986.

[23] J. M. Ogden and E. H. Adelson. Computer simulations of oriented multiple spatial frequency band coring. Technical Report PRRL-85-TR-012, RCA David Sarnoff Research Center, April 1985.

[24] E Simoncelli and J Portilla. Texture characterization via joint statistics of wavelet coefficient magnitudes. In *Fifth IEEE Int'l Conf on Image Proc*, volume I, Chicago, October 4-7 1998. IEEE Computer Society.

[25] E P Simoncelli and E H Adelson. Subband transforms. In John W Woods, editor, *Subband Image Coding*, chapter 4, pages 143–192. Kluwer Academic Publishers, Norwell, MA, 1990.

[26] E P Simoncelli, W T Freeman, E H Adelson, and D J Heeger. Shiftable multi-scale transforms. *IEEE Trans Information Theory*, 38(2):587–607, March 1992. Special Issue on Wavelets.

[27] R R Coifman and D L Donoho. Translation-invariant de-noising. Technical Report 475, Statistics Department, Stanford University, May 1995.

19

Empirical Bayes Estimation in Wavelet Nonparametric Regression

Merlise A. Clyde and Edward I. George

ABSTRACT Bayesian methods based on hierarchical mixture models have demonstrated excellent mean squared error properties in constructing data dependent shrinkage estimators in wavelets, however, subjective elicitation of the hyperparameters is challenging. In this chapter we use an Empirical Bayes approach to estimate the hyperparameters for each level of the wavelet decomposition, bypassing the usual difficulty of hyperparameter specification in the hierarchical model. The EB approach is computationally competitive with standard methods and offers improved MSE performance over several Bayes and classical estimators in a wide variety of examples.

19.1 Introduction

Wavelet shrinkage has become an increasingly popular method for compression and denoising of data in the context of signal and image processing as well as nonparametric regression (Donoho and Johnstone (1994, 1995). The nonparametric regression model can be specified as

$$Y_i = f_i + \epsilon_i$$

where f_i represents the underlying unknown mean function and ϵ_i are independent $N(0, \sigma^2)$ random errors, representing additive white noise. In the wavelet domain, this can be equivalently expressed as

$$D_{jk} = \beta_{jk} + \epsilon_{jk} \tag{1}$$

where D_{jk} represent the elements of the data after applying the discrete wavelet transformation (DWT) and β_{jk} represent the wavelet coefficients of the function f; the double indices reflect the multiresolution decomposition in the wavelet domain. Wavelet shrinkage estimation proceeds by estimating the β_{jk} by some shrinkage procedure, and then transforming the estimated coefficients back to the original domain by applying the inverse discrete wavelet transformation to obtain an estimate of the function f.

Bayesian methods, which offer coherent data-dependent shrinkage, have exhibited excellent integrated mean squared error properties in several studies (Abramovich et al. 1998, Chipman et al. 1997, Clyde et al. 1998) for estimation of f. The above Bayesian methods involve taking the standard linear model (1) with independent normal errors and embedding it in a conjugate hierarchical mixture model that takes into account that some wavelet coefficients will be zero or close to zero. The multiresolution decomposition suggests a natural grouping of wavelet coefficients by level which is reflected in specifying the distribution for β_{jk} conditional on the level j. Clyde et al. (1998) use a hierarchical model that expresses the belief that some of the wavelet coefficients β_{jk} are zero

$$\beta_{jk}|\gamma_{jk} \quad \sim \quad N(0, c_j \gamma_{jk} \sigma^2) \tag{2}$$

$$\gamma_{jk} \quad \sim \quad Bernoulli(\omega_j) \tag{3}$$

through the indicator variable γ_{jk} that determines if the coefficient is non-zero ($\gamma_{jk} = 1$), arising from a normal distribution with variance $c_j \sigma^2$, or degenerate at zero ($\gamma_{jk} = 0$). In the next stage of the hierarchy, the indicator variables γ_{jk} have independent Bernoulli distributions with $P(\gamma_{jk} = 1) = \omega_j$, for some fixed hyperparameter ω_j. The hyperparameter ω_j reflects the expected fraction of non-zero wavelet coefficients at level j. By collapsing these two stages, the prior distribution for β_{jk} can be equivalently represented as a two point mixture distribution,

$$\beta_{jk} \sim (1 - \omega_j)\delta(0) + \omega_j N(0, c_j \sigma^2)$$

where $\delta(0)$ represents a point-mass at 0. Chipman et al. (1997) consider a similar prior, but replace the point-mass at zero by a normal distribution that is tightly distributed around zero as in George and McCulloch (1993).

Because of the conditional independence structure in the prior distributions, the β_{jk}s are *a posteriori* conditionally independent,

$$p(\beta_{jk}|\gamma_{jk}, Y) \sim N\left(\gamma_{jk}\frac{c_j}{1+c_j}D_{jk}, \ \gamma_{jk}\sigma^2\frac{c_j}{1+c_j}\right).$$

Threshold estimators, where some of the coefficients are set to zero, can be obtained by selecting the highest posterior probability model, $\hat{\gamma}$, and using the posterior mean conditional on $\hat{\gamma}$,

$$E(\beta_{jk}|\hat{\gamma}, Y) = \hat{\gamma}_{jk}\frac{c_j}{1+c_j}D_{jk}. \tag{4}$$

(Clyde and George 1998). The posterior median (Abramovich et al. 1998) is another thresholding estimator.

An alternative shrinkage estimator is based on the posterior mean under Bayesian model averaging which takes into account uncertainty about γ_{jk},

$$E(\beta_{jk}|Y) = \pi(\gamma_{jk} = 1|Y)\frac{c_j}{1+c_j}D_{jk}, \tag{5}$$

(Clyde et al. 1998). Given Y, the γ_{jk} are independently distributed as Bernoulli random variables with

$$\pi(\gamma_{jk} = 1|Y) = \frac{O_{jk}}{1 + O_{jk}}, \tag{6}$$

where O_{jk} is the posterior odds that $\gamma_{jk} = 1$,

$$O_{jk} = (1 + c_j)^{-1/2} \left(\frac{\omega_j}{1 - \omega_j} \right) \exp\left\{ \frac{1}{2} \left(\frac{D_{jk}}{\sigma} \right)^2 \left(\frac{c_j}{1 + c_j} \right) \right\}. \tag{7}$$

While Bayesian methods are very flexible in the range of shrinkage patterns they can produce, subjective elicitation of the hyperparameters ω_j and c_j at each level j, is a difficult task. Clyde et al. (1998) used ideas of George and Foster (1997) to specify the prior hyperparameters so that the highest posterior model corresponds to the model selected using a classical model selection criterion. This, in effect, requires that one either elicit utilities/losses for model selection, which can be as difficult as specifying the prior hyperparameters, or use default choices such as AIC (Akaike 1973), BIC (Schwartz 1978), or RIC (Foster and George 1994). Abramovich et al. (1998) establish a relationship between the prior hyperparameters and Besov space parameters (α, β) which allows them to take into account the likely smoothness and regularity properties of the function. They assume the hyperparameters have the following structure

$$\begin{aligned} c_j &= C_1(2^{-j})^\alpha \\ \omega_j &= \min(1, C_2(2^j)^\beta) \end{aligned}$$

where C_1 and C_2 are additional hyperparameters (see also the chapter by Abramovich and Sapatinas in this volume). Noting that it is often difficult to elicit prior information about the smoothness of the function, they suggest default choices for α and β and use method of moments estimators for C_1 and C_2.

Because of the difficulties of subjective elicitation, lack of knowledge about the function, and concern that a default prior may be at odds with the data, many of the proposed Bayesian methods use some form of data-dependent prior combined with assumptions about how the hyperparameters are related by level (Abramovich et al. 1998, Chipman et al. 1997, Yau and Kohn 1999). Rather than imposing any structure on the hyperparameters, Clyde and George (1998) and Johnstone and Silverman (1998) take an Empirical Bayes (EB) approach and estimate the hyperparameters in the prior distribution based on the marginal distribution of the data. These EB procedures not only bypass the difficulty of specifying the hyperparameters in the prior distributions, but are also very competitive with other wavelet shrinkage methods on computational grounds. In this chapter, we review these Empirical Bayes approaches and show how they can be used to construct both thresholding and shrinkage estimators for wavelet nonparametric regression.

19.2   Empirical Bayes

In an Empirical Bayes analysis, one would estimate the hyperparameters of the hierarchical model by some estimation procedure, commonly method of moments or maximum likelihood, and then proceed with the posterior analysis for the parameters of interest by treating the estimated hyperparameters as if they were known *a priori*. In the hierarchical model given by (1), (2), and (3), the unknown hyperparameters are σ^2, c_j and ω_j. Many papers have considered estimating σ^2 using the MAD estimate,

$$\hat{\sigma} = \text{Median}(|D_{1k}|)/0.6745$$

(Donoho et al. 1995) using the wavelet coefficients at the finest level of resolution. We will first consider estimation of c_j and ω_j conditional on using the MAD estimate of σ, and then later proceed with joint estimation of σ in addition to c_j and ω_j by maximum likelihood estimation. Maximum likelihood estimates can be found by either direct maximization of the marginal likelihood (Clyde and George 1998) or by using the EM algorithm with an augmented likelihood (Johnstone and Silverman 1998).

19.2.1   Direct Maximum Likelihood Estimation of c_j and ω_j

Given an estimate for σ, such as the MAD estimate, c_j and ω_j can be estimated via maximum likelihood estimation using the marginal distribution of the data at level j. Marginalizing over β_{jk} and γ_{jk}, and conditioning on c_j, ω_j, and σ, the observations D_{jk} are independently distributed as a mixture of two normal components. The log likelihood $\mathcal{L}$ for c_j and ω_j is

$$\mathcal{L}(c_j, \omega_j) = \sum_k \log \left[\omega_j \phi \left(D_{jk}; 0, \sqrt{1 + c_j}\sigma \right) + (1 - \omega_j)\phi \left(D_{jk}; 0, \sigma \right) \right]$$

$$= \text{constant} + \sum_k \log \left[1 + \omega_j \left((1 + c_j)^{-\frac{1}{2}} e^{\frac{1}{2}\frac{D_{jk}^2}{\sigma^2}\frac{c_j}{1+c_j}} - 1 \right) \right] \quad (8)$$

where $\phi(x; \mu, \sigma)$ denotes the normal density evaluated at the point x with mean μ and standard deviation σ. This form does not lead to closed form solutions for the maximum likelihood estimates $\hat{c}_j$ and $\hat{\omega}_j$, and numerical methods must be used to obtain the MLEs. Clyde and George (1998) used nonlinear Gauss-Seidel iteration (see Thisted 1988, pp. 187-188). This involves solving the single variable optimization problem to first find $\hat{c}_j$ as function of ω_j and then finding $\hat{\omega}_j$ using the estimate of $\hat{c}_j$. One cycles through these two optimization problems, successively substituting the current estimate until convergence is achieved. Any popular root finding algorithm may be used to solve the single variable equations. If the Hessian is positive definite for all values of c_j and ω_j (excluding the boundaries), then if the algorithm converges the solution is the global maximum.

19.2.2 Maximum Likelihood Estimation using the EM Algorithm

Johnstone and Silverman (1998) use an EM algorithm to find the MLE, based on a derivation that introduces an entropy function to create a modified likelihood, where the global maximum of the modified likelihood function is the global MLE of the marginal likelihood. This approach is equivalent to the general EM algorithm given by Neal and Hinton (1998). The EM algorithm in exponential family problems is particularly simple to implement (Dempster, Laird and Rubin 1977, Tanner 1996), and we present this alternative derivation. To implement the EM algorithm, we consider the likelihood given D and the latent variable γ, rather than the marginal likelihood (8). The log likelihood for the "augmented" or "complete" data, $X = (D, \gamma)$,

$$
\mathcal{L}(c_j, \omega_j | D, \gamma) = \left[\log \left(\frac{\omega_j}{1 - \omega_j} \right) - \frac{1}{2} \log(1 + c_j) \right] \sum_k \gamma_{jk}
$$
$$
- \frac{1}{2}(1 + c_j)^{-1} \sum_k \gamma_{jk} D_{jk}^2 / \sigma^2 \tag{9}
$$
$$
+ n_j \log(1 - \omega_j) - \frac{1}{2} \sum_k (1 - \gamma_{jk}) \frac{D_{jk}^2}{\sigma^2}
$$

belongs to a regular exponential family of the form $a(\theta)^T b(X) + c(\theta) + d(X)$ where $\theta = (c_j, \omega_j)$, $a(\theta)$ is the vector of natural parameters and $b(X)$ is the vector of sufficient statistics, $b(X) = (\sum_k \gamma_{jk}, \sum_k (\gamma_{jk} D_{jk}^2 / \sigma^2))^T$.

Because of the exponential family form, the E-step of the EM algorithm consists of computing the expectation of the sufficient statistics with respect to the distribution of γ of given D

$$
E[b(X) \mid D, c_j^{(i)}, \omega_j^{(i)}] = \left(\sum_k \hat{\gamma}_{jk}^{(i)}, \sum_k \hat{\gamma}_{jk}^{(i)} D_{jk}^2 / \sigma^2 \right)^T = \hat{b}^{(i)}(X)
$$

where $\hat{\gamma}_{jk}^{(i)} = O_{jk}^{(i)} / (1 + O_{jk}^{(i)})$ is the posterior mean of γ_{jk} and $O_{jk}^{(i)}$ is the posterior odds (7) evaluated using the current estimates $\hat{c}_j^{(i)}$ and $\hat{\omega}_j^{(i)}$.

The M-step consists of maximizing $c(\theta) + a(\theta)^T \hat{b}^{(i)}(X)$, resulting in the solution

$$
\hat{c}_j^{(i+1)} = \max \left(0, \frac{\sum_k \hat{\gamma}_{jk}^{(i)} D_{jk}^2}{\sum_k \hat{\gamma}_{jk}^{(i)} \sigma^2} - 1 \right) \tag{10}
$$
$$
\hat{\omega}_j^{(i+1)} = \frac{\sum_k \hat{\gamma}_{jk}^{(i)}}{n_j}. \tag{11}
$$

If the estimates are in the interior of the parameter space, because the augmented likelihood belongs to a regular exponential family, the solutions for c_j and ω_j are the unique global solutions (conditional on $\hat{\gamma}_{jk}$) which follows from standard exponential family theory. The E and M steps are repeated until the estimates converge, and yield a stationary point of the marginal likelihood (8). Because the convergence rate of the EM algorithm is linear (Dempster, Laird and Rubin 1977), the Gauss-Seidel algorithm applied to (8) may be faster. As in the Gauss-Seidel algorithm, this results in a global solution if and only if the marginal likelihood is unimodal. In practice, however, we have noticed little difference in performance between the two approaches or difficulties with convergence.

19.2.3   Maximum Likelihood Estimation of σ^2

Rather than using the MAD estimate for σ, the augmented data likelihoods (9) at each level j can be combined to construct a complete data likelihood for estimating σ^2 through the EM algorithm. This complete data likelihood is still in a regular exponential family. The sufficient statistics for σ involve the same terms as in estimating c_j, so the E-step only involves the expectation of γ_{jk}. The M-step for estimating σ^2 has solution

$$\sigma^{2\,(i+1)} = \frac{1}{N} \sum_{j,k} \left(D_{jk}^2 - \frac{c_j}{1+c_j} \hat{\gamma}_{jk}^{(i)} D_{jk}^2 \right)$$

while the M-steps for c_j and ω_j are the same as before. The M-steps for σ^2 and c_j now involve iterative solutions. This approach takes full advantage of the data at all levels to construct an estimate of σ^2, unlike the MAD estimate.

19.2.4   Conditional Likelihood Estimates

Clyde and George (1998) also consider a conditional likelihood approximation to the full likelihood, which yields rapidly computable analytic expressions for $\hat{c}_j$ and $\hat{\omega}_j$. This can be viewed as taking the augmented likelihood (9) and evaluating it at the mode for γ_{jk}, rather than using the posterior mean, as in the EM algorithm. At level j, consider models γ where $q_j = \sum_k \gamma_{jk}$ is the number of nonzero wavelet coefficients. For fixed j, let $D_{j(k)}^2$ denote the sorted values (in decreasing order) of D_{jk}^2. Then the most likely model with q_j nonzero components, $\gamma^*(q_j)$, corresponds to $\gamma_{j(k)}^* = 1$ if $k \leq q_j$, and 0 otherwise, based on assigning the q_j largest D_{jk}^2 values to the mixture component representing signal, and the remaining to the noise component. For each value of q_j, the values of c_j and ω_j that maximize the conditional log likelihood are

$$c_j(q_j) = \max\left\{0, \frac{\sum_k \gamma^*_{j(k)}(q_j)D^2_{j(k)}}{\sum_k \gamma^*_{j(k)}(q_j)\sigma^2} - 1\right\} \tag{12}$$

$$= \max\left\{0, \frac{\sum_{k \leq q_j} D^2_{j(k)}}{q_j\sigma^2} - 1\right\}$$

$$\omega_j(q_j) = \frac{\sum_k \gamma^*_{j(k)}}{n_j} = \frac{q_j}{n_j}. \tag{13}$$

It is straightforward to find the $\hat{q}_j$ that maximizes the conditional likelihood, and the corresponding $\hat{c}_j$ and $\hat{\omega}_j$ that yield the largest mode. Note that $L(c_j, \omega_j \mid \hat{\gamma}(\hat{q}_j))$ may be thought of as a profile likelihood approximation to $L(c_j, \omega_j)$. The conditional maximum likelihood estimates $\hat{c}_j$ and $\hat{\omega}_j$ are alternative EB estimates which can be rapidly computed.

19.2.5   Comparing the Hyperparameter Estimators

Figure 1 illustrates profile likelihood plots of the marginal log likelihood, $\mathcal{L}(\hat{c}_j(\omega_j), \omega_j)$ based on (8), (left column) and the conditional log likelihood, $\mathcal{L}(\hat{c}_j(\omega_j), \omega_j \mid \hat{\gamma}(q_j(\omega_j)))$ based on (9), (right column) as a function of ω for a wavelet decomposition with 7 levels. To construct the profile likelihoods, $\hat{c}_j(\omega_j)$ is the MLE of c_j obtained by fixing ω_j in the marginal and conditional log likelihoods respectively. The corresponding marginal and conditional maximum likelihood estimates are given in Table 1. Although comparison of the marginal and conditional maximum likelihood estimates in Table 1 shows relatively close agreement, there is a suggestion of systematic bias in the conditional estimates, with a slight underestimation of ω_j and overestimation of c_j. Also, for the finest level of resolution there is a bimodality in the conditional loglikelihood, in which case we cannot distinguish between noise and signal. For cases like this in practice we find that the likelihood is extremely flat with estimates near the boundary with $\hat{\omega}_j \approx 0$ or $\hat{c}_j \approx 0$. As the posterior mean is approximately the same under both cases, this has not resulted in any serious bias for estimation in our experience. By comparing the EM and conditional MLE estimators (10) to (12) and (11) to (13), one sees that the estimators have the exact same form, but the EM estimates are evaluated with γ_{jk} at the posterior mean while the conditional estimates are evaluated with γ_{jk} at the posterior mode. One can see that in general the conditional and marginal maximum likelihood estimates will not agree, unless the posterior distribution of γ_{jk} is degenerate at 1 or 0, in which case the expected values and the modes for γ_{jk} will coincide. This difference will not disappear, even as the number of coefficients grows asymptotically, as posterior model probabilities will not necessarily converge to 0 or 1 asymptotically. For the coarser levels, with predominantly large coefficients (in absolute value) the posterior mean of

Level	MMLE ω	CMLE ω	MMLE c	CMLE c
s	0.85	0.81	1477.5	1557.7
6	0.71	0.63	660.5	760.5
5	0.62	0.53	313.6	367.4
4	0.39	0.36	391.8	431.4
3	0.21	0.18	197.7	238.7
2	0.08	0.07	72.5	93.9
1	0.04	0.03	21.5	35.5

TABLE 1. Maximum likelihood estimates of c_j and ω_j from the marginal (MMLE) and conditional likelihoods (CMLE).

γ_{jk} is often close to 1, resulting in less bias. The difference between the conditional and marginal estimators will be the most extreme if the posterior means of all the γ_{jk} equal one half. Fortunately in wavelets, a good basis should result in posterior model probabilities being close to zero or one, reducing the potential for bias. To understand the effect of the bias on shrinkage, note that the posterior model probabilities are nonlinear functions of c_j and ω_j, and it is the linear shrinkage in the form $c_j/(1 + c_j)$ and the multiple shrinkage through the posterior model probabilities that is critical in determining the posterior mean. As we will see later in the simulation study, these two errors appear to cancel each other for estimating the posterior mean.

19.2.6 Empirical Bayes Estimators

The EB estimates of σ^2, c_j and ω_j are now used in the hierarchical model as if they had been fixed in advanced and are used to construct Bayesian estimators of the wavelet coefficients. A threshold estimator is obtained by first selecting the highest posterior probability model $\hat{\gamma}$, where $\hat{\gamma}_{jk} = 1$ if $\hat{\pi}(\gamma_{jk} = 1|Y) \geq 0.5$ and is zero otherwise, and $\hat{\pi}(\gamma_{jk} = 1|Y)$ is obtained by inserting the EB estimates into (6). The conditional posterior mean $E(\beta_{jk}|\hat{\gamma}, Y)$ in (4) is given by

$$\hat{E}(\beta_{jk}|\hat{\gamma}, Y) = \hat{\gamma}_{jk}\frac{\hat{c}_j}{1 + \hat{c}_j}D_{jk}. \tag{14}$$

This model selection shrinkage estimator thresholds the data by setting $\hat{\beta}_{jk} = 0$ whenever $\hat{\gamma}_{jk} = 0$. This is useful for compression problems where dimension reduction and elimination of negligible coefficients is important.

Alternatively, one might use the EB estimates to estimate the overall posterior mean, $E(\beta_{jk}|Y)$ in (5), which yields

$$\hat{E}(\beta_{jk}|Y) = \hat{\pi}(\gamma_{jk} = 1|Y)\frac{\hat{c}_j}{1 + \hat{c}_j}D_{jk}. \tag{15}$$

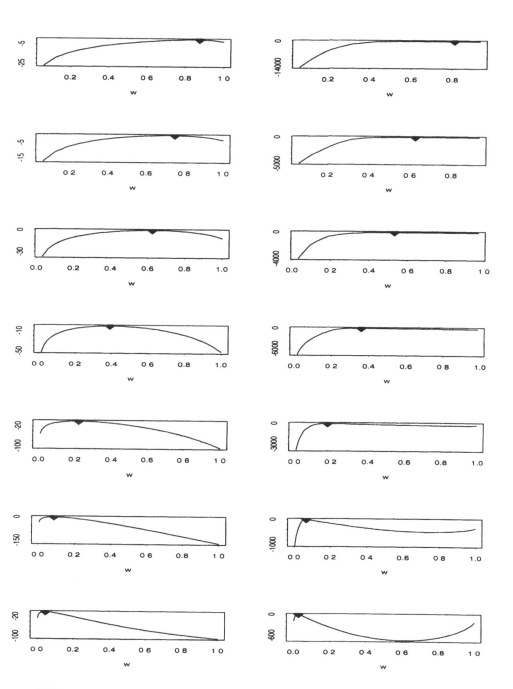

FIGURE 1. Marginal loglikelihood, $\log L(\hat{c}_j(\omega_j), \omega_j)$, (left column) and conditional loglikelihood $\log L(\hat{c}_j(\omega_j), \omega_j \mid \hat{\gamma}(q_j(\omega_j)))$, (right column) as a function of ω_j, for $j = s, 6, \ldots, 1$. The triangles represent the location of the maximum.

This multiple shrinkage estimator (George 1986, Clyde et al. 1998) corresponds to Bayesian model averaging. In contrast to the thresholding behavior induced by $\hat{\gamma}_{jk}$ in (14), (15) includes an additional shrinkage factor $\hat{\pi}(\gamma_{jk} = 1|Y)$ to compensate for model uncertainty and appears to offer improved performance (Clyde and George 1998). Finally, note that both of the EB estimators (14) and (15) are fully automatic, as opposed to (4) and (5) which require hyperparameter specification.

19.3 Simulations

We compared the EB estimators to several existing shrinkage strategies: HARD: Hard thresholding with the universal rule (Donoho and Johnstone 1994) and SURE: SureShrink adaptive shrinkage rule as implemented in S+Wavelets, based on Donoho and Johnstone's (1995) Sureshrink procedure, and RIC, which fixes the hyperparameters so that $c_j \equiv 1048561$ and $\omega_j \equiv 0.50$ corresponding to the Risk Inflation Criterion of Foster and George (1994). We used the four test functions "blocks", "bumps", "doppler", "heavisine", proposed by Donoho and Johnstone, and generated 100 samples of each function with $N = 1024$ and $\sigma = 1$. The signal-to-noise ratio SNR = 7 and the wavelet bases are chosen to match Donoho and Johnstone (1995). We evaluated the performance based on the average mean squared error (MSE) from the 100 simulations as

$$\text{MSE} = \frac{1}{100} \sum_{l=1}^{100} \sum_{i=1}^{N} \frac{(f_i - \hat{f}_i^l)^2}{N}$$

where f_i is the true signal and $\hat{f}_i^l$ is the estimate of the function from simulation l.

Table 2 presents the average MSEs and standard deviations from the simulation study. We compared the EB model averaging estimator (15) with the marginal MLE of c_j and ω_j and the conditional MLE estimates using the MAD estimate of σ (the first two columns respectively) to the joint MLE of σ^2, c_j and ω_j (column 3). The results indicate that all three EB estimators are superior to HARD, SURE, and RIC in this setting. Interestingly, performance is hardly affected, if at all, by using the conditional EB estimates instead of the marginal EB estimates. Apparently, the individual biases of the estimates of c and ω discussed in Section 2.5 have little effect. Using the MLE EB estimate of σ^2 is generally more efficient than the robust MAD estimate, as one would expect since it is a function of all of the data. However, the EM algorithm for estimation of σ^2, c_j, and ω_j often took much longer to converge (sometimes more than 50 iterations), than the EM algorithm for c_j and ω_j with the MAD estimate of σ^2.

Figure 2 shows the distribution of the maximum likelihood estimates of c_j and ω_j for the four test functions from the 100 simulations using

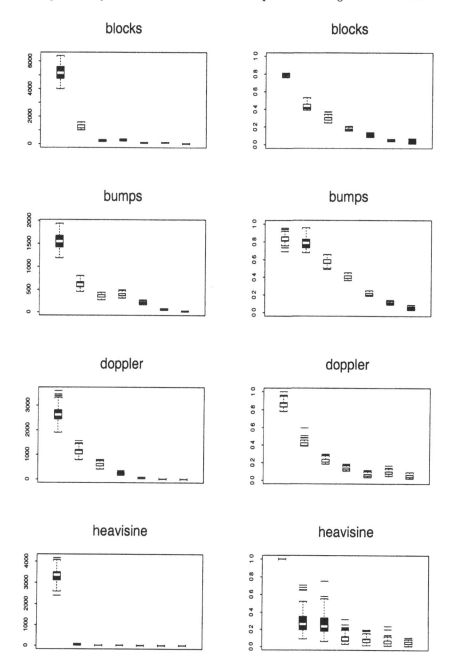

FIGURE 2. Distribution of the EB estimates of c_j (left column) and ω_j (right column) by level from the 100 simulations for the four functions.

Function	EB CML	EB MML	EB MML (all)	RIC	HARD	SURE
blocks	**0.100**	0.102	0.101	0.106	0.142	0.213
	(0.002)	(0.002)	(0.002)	(0.002)	(0.002)	(0.002)
bumps	0.323	0.310	**0.308**	0.342	0.433	0.382
	(0.003)	(0.003)	(0.003)	(0.003)	(0.0040)	(0.003)
doppler	0.147	0.144	**0.143**	0.152	0.208	0.204
	(0.002)	(0.002)	(0.002)	(0.002)	(0.003)	(0.002)
heavisine	0.084	**0.082**	0.083	0.096	0.109	0.118
	(0.002)	(0.001)	(0.001)	(0.002)	(0.002)	(0.002)

TABLE 2. Average mean squared error and (standard deviations) from 100 simulations for each function based on SNR = 7. The EB estimates are based on the posterior mean using the conditional MLE (CML) and the marginal MLE (MML) of c_j and ω_j using the MAD estimate of σ^2, and the joint MLE of σ^2, c_j, and ω_j (MML all). Values in bold indicate the estimator with the minimum average MSE.

the estimates from the joint estimation of c_j, ω_j and σ^2. The variation of the estimates across the different levels is striking, revealing strong decay in both ω_j and c_j from top to bottom, but very different rates across functions. Although such decay might be roughly anticipated using a fixed hyperparameter Bayes setup with subjective prior inputs, it is very difficult to pre-specify the appropriate magnitude and rate of decay. Indeed, such fixed Bayes estimators did not perform as well as the EB estimators in Clyde and George (1998).

19.4   Discussion

In this chapter, we have discussed Empirical Bayes methods for wavelet estimation. Embedding the wavelet setup in a hierarchical normal mixture model, we considered conditional and marginal likelihood estimates of the unknown hyperparameters for each wavelet level. We then obtained shrinkage and threshold estimators based on posterior means under the estimated prior distributions. When applied to a variety of simulated examples, these shrinkage estimators performed better than current methods including fixed hyperparameter Bayes estimators. Johnstone and Silverman (1998) obtain similar results using the posterior median as an estimator.

Clyde and George (1998) consider extensions of the normal hierarchical model to include scale mixtures of normals. This allows for robustness to outliers through the use of heavy tailed error distributions such as the Student-t or power exponential distribution (Box and Tiao 1973). The EB approach yields robust estimators that are computationally competitive with classical methods (order N). The hierarchical Student-t EB estimates are superior across a wide variety of situations.

An explanation for the improved performance of the EB estimators is that they allow for wide variation of hyperparameter values across different wavelet levels, yielding flexible shrinkage patterns. One could also achieve this by elaborating our hierarchical setup to include prior distributions on all the hyperparameters. If the computational issues for this approach could be simplified, this would be a promising competitor to our methods, and would provide improved estimates of the posterior variances over the naive EB approach that ignores uncertainty in the hyperparameter estimates.

Clyde and George (1998) found that the EB estimates can be very sensitive to the choice of σ. When prior information or data are available, the EB approach can easily incorporate prior information about the noise level σ. Some additional improvement could be made by placing a prior distribution on σ^2 and using MCMC methods, but with additional computational cost. The EB methods can also be adapted to the case of correlated noise, by replacing σ with a level dependent estimate σ_j as in Johnstone and Silverman (1997).

Finally, another avenue for future research in this area is performance evaluation. Any simulation evaluation such as ours is necessarily limited to one part of the overall parameter space. Although our Bayesian estimators do not appear to offer oracle or risk inflation like minimax guarantees (Donoho and Johnstone 1994; Foster and George 1994), it would be worthwhile to investigate regions of worst performance. In this vein, we would expect the EB estimators offer more robustness than fixed hyperparameter Bayes estimators. Another interesting, but difficult, direction would be asymptotic evaluation of the EB procedures. This is complicated by the fact that the model dimension is always increasing with the sample size. While the marginal EB estimates of c_j and ω_j appear to be asymptotically consistent (as n_j goes to infinity), this is not necessarily the case with the conditional EB estimates (Johnstone and Silverman 1998). But, even if the hyperparameter estimates are consistent, the posterior model probabilities do not necessarily converge to 0 or 1 asymptotically (particularly when c_j is small), thus model selection will not generally be asymptotically consistent in the wavelet context.

References

Abramovich, F., Sapatinas, T., and Silverman, B.W. (1998). "Wavelet Thresholding via a Bayesian Approach," *Journal of the Royal Statistics Society, Series B*, 60, 725-749.

Akaike, H. (1973). Information theory and an extension of the maximum likelihood principle. In *2nd International Symposium on Information Theory*, Eds B.N. Petrov and F. Csaki, pp. 267-81. Budapest: Akademia Kiado.

Box, G.E.P. and Tiao, G.C. (1973). *Bayesian Inference in Statistical Analysis*,

Wiley, NY.

Chipman, H., Kolaczyk, E., and McCulloch, R. (1997). "Adaptive Bayesian Wavelet Shrinkage", Journal of the American Statistical Association, 92, 1413-1421.

Clyde, M. and George, E.I. (1998). "Robust Empirical Bayes Estimation in Wavelets", ISDS Discussion Paper 98-21/ http://www.isds.duke.edu/

Clyde, M., Parmigiani, G., Vidakovic, B. (1998). "Multiple Shrinkage and Subset Selection in Wavelets," *Biometrika*, 85, 391-402.

Dempster, A.P. Laird, N.M. and Rubin, D.B. (1977). Maximum likelihood estimation from incomplete data via the EM algorithm (with discussion). *Journal of the Royal Statistical Society, Series B*, 39, 1-38.

Donoho, D.L., and Johnstone, I.M., (1994). "Ideal spatial adaptation by wavelet shrinkage," *Biometrika*, 81, 425-256.

Donoho, D. and Johnstone, I. (1995). "Adapting to Unknown Smoothness via Wavelet Shrinkage," *Journal of the American Statistical Association*, 90, 1200-1224.

Donoho, D., Johnstone, I., Kerkyacharian, G., and Picard, D. (1995). "Wavelet shrinkage: Asymptopia?" *Journal of the Royal Statistical Society, Series B*, 57, 301-369.

Foster, D. and George, E. (1994). The risk inflation criterion for multiple regression. *Annals of Statistics* 22, 1947–75.

George, E.I. (1986). "Minimax multiple shrinkage estimation," *Annals of Statistics*, 14, 188-205.

George, E.I. and Foster, D.P. (1997). "Empirical Bayes Variable Selection", Tech Report, University of Texas at Austin.

George, E.I. and McCulloch, R. (1993). Variable selection via Gibbs sampling. *Journal of the American Statistical Association* 88, 881–89.

Johnstone, I.M. and Silverman, B.W. (1997). "Wavelet Threshold Estimators for Data with Correlated Noise," *Journal of the Royal Statistical Society, Series B*, 59, 319-351.

Johnstone, I.M. and Silverman, B.W. (1998). "Empirical Bayes approaches to mixture problems and wavelet regression", Technical report. Department of Mathematics, University of Bristol.

Neal, R. M. and Hinton, G. E. (1998) "A view of the EM algorithm that justifies incremental, sparse, and other variants", in M. I. Jordan (editor) *Learning in Graphical Models*, Dordrecht: Kluwer Academic Publishers, pages 355-368.

Schwarz, G. (1978). Estimating the dimension of a model. *Annals of Statistics* 6, 461-464.

Tanner, M.A. (1996). Tools for statistical inference: methods for the exploration of posterior distributions and likelihood functions. New York: Springer, 3rd Edition. Chapter 4, pages 64-89.

Yau, P. and Kohn, R. (1999). "Wavelet Nonparametric Regression Using Basis Averaging". To appear in *Bayesian Inference in Wavelet Based Models* eds P. Müller and B. Vidakovic. Springer-Verlag.

20

Nonparametric Empirical Bayes Estimation via Wavelets

Marianna Pensky

ABSTRACT   We consider a traditional nonparametric empirical Bayes (EB) setting with a parameter θ being a location or a scale parameter. EB estimators are constructed and are shown to provide adaptation to the unknown degree of smoothness of $g(\theta)$. The case when the conditional distributions belongs to a one-parameter exponential family is also studied. Examples are given for familiar families of conditional distributions. The advantages of using wavelets are discussed.

20.1   Introduction

Statistical estimation based on wavelet approximation has become more and more popular in the last decade. One of the reasons for this popularity is that wavelets provide an accurate and a rapidly converging approximation for a function of an unknown degree of smoothness. The present chapter develops wavelet techniques in an empirical Bayes (EB) estimation problem.

The traditional EB model is as follows. One observes independent two-dimensional random vectors (X_1, Θ_1), ,..., (X_n, Θ_n), where each Θ_i is distributed according to some unknown prior density function g and, given $\Theta_i = \theta$, X_i has a known conditional density function $q(x|\theta)$, $\theta \in (a, b) \subseteq (-\infty, \infty)$, $x \in (c, d) \subseteq (-\infty, \infty)$. In each pair the first component is observable, but the second is not. After the $(n+1)$-th observation, $y \equiv X_{n+1}$, is made, the objective is to estimate $t \equiv \Theta_{n+1}$.

In the case of continuous random variables X and θ the majority of practical applications of the EB model deal with the situation where θ is a location parameter or a scale parameter. The present chapter consideres both of these cases, as well as the situation when $q(x|\theta)$ belongs to a one-parameter exponential family.

Denote by $\mathbf{E}_g$, $\mathbf{E}_q$, $\mathbf{E}_p$ and $\mathbf{E}_{p^n}$ the expectations with respect to the densities $g(\theta)$, $q(x|\theta)$,

$$p(x) = \int_a^b q(x|\theta)g(\theta)d\theta \tag{1}$$

and $\prod_{i=1}^{n} p(x_i)$, respectively, and

$$\Psi(x) = \int_a^b \theta q(x|\theta) g(\theta) d\theta. \tag{2}$$

If we knew the prior density $g(\theta)$, then the Bayes estimator of t under the squared loss would have the form

$$\beta(y) = \Psi(y)/p(y). \tag{3}$$

As the prior density $g(\theta)$ is unknown, we construct an EB estimator $\beta_n(y) = \beta_n(y; X_1, X_2, ..., X_n)$ as the estimator of (3) based on observations $X_1, X_2, ..., X_n$.

An EB estimator $\beta_n(y)$ may be characterized by its posterior risk

$$R(y; \beta_n) = (p(y))^{-1} \mathbf{E}_{p^n} \int_a^b (\beta_n(y) - \theta)^2 q(y|\theta) g(\theta) d\theta,$$

or by its prior risk

$$R(\beta_n) = \mathbf{E}_p R_n(y) = \int_c^d R(y; \beta_n) p(y) dy.$$

Note that both $R(y; \beta_n)$ and $R(\beta_n)$ can be partitioned into sums of two components. The first components of these sums are, respectively, the posterior risk $R(y; \beta) = (p(y))^{-1} \int_a^b (\beta(y) - \theta)^2 q(y|\theta) g(\theta) d\theta$ and the prior risk $R(\beta) = \int_c^d R(y; \beta) p(y) dy$ of the Bayes estimator (3), and they both are independent of $\beta_n(y)$. Thus, the quality of EB estimators is measured by the second components

$$R_n(y) = \mathbf{E}_{p^n} (\beta_n(y) - \beta(y))^2 \tag{4}$$

and

$$R_n = \int_c^d R_n(y) p(y) dy. \tag{5}$$

In this paper we shall use mainly the posterior risk to characterize the asymptotic behavior of an EB estimator and will measure its quality by the risk function (4). Note that the risk function (4) has at least two advantages compared with (5). Firstly, $R_n(y)$ enables one to calculate the mean squared error for the **given** observation y which is the quantity of interest. Secondly, by using the risk function (4) we eliminate the influence on the risk function of the observations having very low probabilities. However, since one may also be interested in a risk function which does not depend on y, we provide upper bounds for $R_n(y)$ over any interval $[A, B] \subset (c, d)$ such that $\inf_{x \in [A, B]} p(x) > 0$. The risk function (5) will be used when $q(x|\theta)$

belongs to a one-parameter exponential family so that the rate of descent of $p(x)$ can be predicted.

Note that sometimes EB estimation is easier to perform if $\beta(y)$ is rearranged and presented in the form

$$\beta(y) = \beta_0(y) + \beta^*(y), \quad \text{with} \quad \beta^*(y) = \Phi(y)/p(y), \tag{6}$$

and $\Phi(y) = \int_a^b K(y, \theta)g(\theta)d\theta$. Here the functions $\beta_0(y)$ and $K(y, \theta)$ are known. Clearly, (3) is a particular case of (6) with $\beta_0(y) = 0$ and $K(y, \theta) = \theta q(y|\theta)$.

Example 1. Location-parameter family. Let θ be a location parameter, i.e.

$$q(x|\theta) = q(x - \theta), \quad x, \theta \in (-\infty; \infty). \tag{7}$$

Observe that the numerator $\Psi(y)$ of the Bayes estimator (3) can be written as $\Psi(y) = yp(y) - \int_{-\infty}^{\infty}(y-\theta)q(y-\theta)g(\theta)d\theta$. Therefore, the Bayes estimator of θ has the form

$$\beta(y) = y - \Phi(y)/p(y), \tag{8}$$

where

$$\Phi(y) = \int_{-\infty}^{\infty}(y - \theta)q(y - \theta)g(\theta)d\theta. \tag{9}$$

Example 2. Scale-parameter family. If θ is a scale parameter,

$$q(x|\theta) = \theta^{-1}q(x\theta^{-1}), \quad x, \theta \in (0, \infty). \tag{10}$$

The Bayes estimator of θ in this case is given by (6) with $\beta_0(y) = 0$ and

$$\Phi(y) \equiv \Psi(y) = \int_0^{\infty} q(\theta^{-1}y)g(\theta)d\theta. \tag{11}$$

Example 3. One-parameter exponential family. Consider the case when $q(x|\theta)$ belongs to a one-parameter exponential family

$$q(x|\theta) = C(\theta)h(x)\exp(x\theta). \tag{12}$$

Then $p(y) = h(y)\int C(\theta)e^{y\theta}g(\theta)d\theta$, $\Psi(y) = h(y)\int \theta C(\theta)e^{y\theta}g(\theta)d\theta$, and thus, there is the obvious relation $\Psi(y)/h(y) = [p(y)/h(y)]'$ between $\Psi(y)$ and $p(y)$. Therefore, $\beta(y)$ has the form (6) with $\Phi(y) = p'(y)$ and $\beta_0(y) = -h'(y)/h(y)$.

In what follows, we will make no parametric assumptions on $g(\theta)$, which means that we restrict ourselves to nonparametric empirical Bayes estimation. The nonparametric EB model was previously studied extensively and EB estimators were constructed for different classes of conditional densities $q(x|\theta)$. These include: a one-parameter exponential family (Singh(1979)), a location-parameter family (Pensky (1997a)),a scale-parameter family (Singh and Wei (1992), Pensky (1996)), a family of uniform density functions (Nogami (1988)), a family of Pareto densities (Tiwari and Zalkikar (1990)), etc. Some attempts were also made to work out an approach to the problem which is suitable for various classes of conditional densities (see, for example, Walter (1981), Robbins (1983), Pensky (1997b), also, Maritz and Lwin (1989) and Carlin and Louis (1996)).

The wavelet approach to curve estimation has become more and more popular after wavelet methods were proved to provide adaptation to erratic behavior of an unknown function (see, for example, Kerkyacharian and Picard (1992), Antoniadies *et al* (1994), Donoho (1994), Masry (1994), Donoho and Johnstone (1995), Hall and Patil (1995), Donoho *et al* (1996), Hall *et al* (1998)). Since an EB estimator is presented via the ratio of two unknown functions, it seems natural to use methods based on wavelet expansions in nonparametric EB estimation. The present chapter examines the application of wavelet techniques in this area. EB estimators are obtained by using wavelet approximations. Convergence rates of the EB estimators are investigated and the advantages and the disadvantages of using wavelets for EB estimation are discussed.

20.2   EB estimation for the location parameter family

We remind the reader that, in the location parameter case, $q(x|\theta)$ has the form (7) and the Bayes estimator of θ can be presented as (8). Throughout the paper we use the notation $\tilde{f}(\omega) = \int_{-\infty}^{\infty} e^{-i\omega x} f(x)dx$ for the Fourier transform of a function $f(x)$, and $\check{f} = \int_{0}^{\infty} z^{x-1} f(z)dz$ for the Mellin transform of a function f (see Erdélyi(1954)). Denote the sets of integer numbers and the set of nonnegative integer numbers by **Z** and **N**, respectively, and let $\|f\|_c = \sup_y |f(y)|$ and $\|f\|_{L_k} = \left\{ \int_{-\infty}^{\infty} |f(x)|^k dx \right\}^{1/k}$. Assume that $g(\theta)$ is square integrable, and that $q(x)$ satisfies the following conditions

C1. $\tilde{q}(\omega)$ does not vanish for real ω; $\|q\|_c < \infty$;
C2. $\int_{-\infty}^{\infty} |y|^2 q(y)dy < \infty$;
C3. $\| \exp(\gamma|\omega|^\nu) (w^2 + 1)^{\varrho/2} \tilde{q}(\omega)\|_c < \infty$;
C4. $|\tilde{q}'(\omega)| \le C_q (|\omega| + 1)^\lambda |\tilde{q}(\omega)|$ with $\lambda \le 0$ whenever $\gamma = 0$ in **C3**.

For the sake of construction of wavelet approximations of $\Phi(y)$ and $p(y)$, consider a scaling function $\varphi(x)$ and a corresponding wavelet $\psi(x)$

in $L^2(-\infty, \infty)$. Then, $\Phi(y)$ and $p(y)$ can be written as

$$\Phi(y) = \sum_{k \in \mathbf{Z}} a_{m,k} \varphi_{m,k}(y) + \sum_{j=0}^{\infty} \sum_{k \in \mathbf{Z}} \alpha_{j,k} \psi_{j,k}(y), \tag{1}$$

$$p(y) = \sum_{k \in \mathbf{Z}} b_{m,k} \varphi_{m,k}(y) + \sum_{j=0}^{\infty} \sum_{k \in \mathbf{Z}} \beta_{j,k} \psi_{j,k}(y), \tag{2}$$

respectively. Here

$$\varphi_{m,k}(x) = 2^{m/2} \varphi(2^m x - k), \quad \psi_{j,k}(x) = 2^{(m+j)/2} \psi\left(2^{m+j} x - k\right), \tag{3}$$

with $k \in \mathbf{Z}$, $j \in \mathbf{N}$, and the coefficients in (1) and (2) have the forms
$a_{m,k} = \int_{-\infty}^{\infty} \Phi(x) \varphi_{m,k}(x) dx$, $b_{m,k} = \int_{-\infty}^{\infty} p(x) \varphi_{m,k}(x) dx$,
$\alpha_{j,k} = \int_{-\infty}^{\infty} \Phi(x) \psi_{j,k}(x) dx$, and $\beta_{j,k} = \int_{-\infty}^{\infty} p(x) \psi_{j,k}(x) dx$.

20.2.1   Linear wavelet EB estimation

To construct the estimators of $\Phi(y)$ and $p(y)$, we will use a special class of band-limited wavelets, Meyer-type wavelets (see Walter (1994), Zayed and Walter (1996)). In the case of Meyer-type wavelets, $\tilde{\varphi}(\omega)$ and $\tilde{\psi}(\omega)$ have bounded supports: supp $\tilde{\varphi} \subseteq [-4\pi/3, 4\pi/3]$ and supp $\tilde{\psi} \subseteq [-8\pi/3, -2\pi/3] \cup [2\pi/3, 8\pi/3]$. Moreover, $\tilde{\varphi}(\omega) = 1$ if $|\omega| < 2\pi/3$. In order to ensure that $\varphi(x)$ and $\psi(x)$ have sufficient rate of descent as $|x| \to \infty$, we choose the functions $\tilde{\varphi}(\omega)$ and $\tilde{\psi}(\omega)$ to be at least twice continuously differentiable on $(-\infty, \infty)$.

It is easy to see that $b_{m,k} = \mathbf{E}_p \varphi_{m,k}(X)$, $\beta_{j,k} = \mathbf{E}_p \psi_{j,k}(X)$, and, thus, unbiased estimators of $b_{m,k}$ and $\beta_{j,k}$ have, respectively, the forms

$$\hat{b}_{m,k} = n^{-1} \sum_{l=1}^{n} \varphi_{m,k}(X_l), \quad \hat{\beta}_{j,k} = n^{-1} \sum_{l=1}^{n} \psi_{j,k}(X_l). \tag{4}$$

For the sake of estimation of $a_{m,k}$, notice that if we found a function $u_{m,k}(x)$ such that it satisfies the equation

$$\int_{-\infty}^{\infty} q(x - \theta) u_{m,k}(x) dx = \int_{-\infty}^{\infty} (x - \theta) q(x - \theta) \varphi_{m,k}(x) dx, \tag{5}$$

then $a_{m,k} = \mathbf{E}_p u_{m,k}(X)$ and we would estimate $a_{m,k}$ by

$$\hat{a}_{m,k} = n^{-1} \sum_{l=1}^{n} u_{m,k}(X_l). \tag{6}$$

Applying the Fourier transform to both sides of the equation (5), we obtain

$$u_{m,k}(x) = 2^{m/2} U_m(2^m x - k), \tag{7}$$

where $U_m(x)$ is the inverse Fourier transform of the function

$$\tilde{U}_m(\omega) = i\,\tilde{q}'(-2^m\omega)\tilde{\varphi}(\omega)\,[\tilde{q}(-2^m\omega)]^{-1}. \tag{8}$$

The inverse Fourier transform of the function (8) exists since $\tilde{\varphi}(\omega)$ has a bounded support and $\tilde{q}(\omega)$ does not vanish. Similarly, we can estimate $\alpha_{j,k}$ by

$$\hat{\alpha}_{j,k} = n^{-1}\sum_{l=1}^{n} 2^{(j+m)/2}V_j(2^{j+m}X_l - k). \tag{9}$$

where $V_j(x)$ is the inverse Fourier transform of the function

$$\tilde{V}_j(\omega) = i\,\tilde{q}'(-2^{j+m}\omega)\tilde{\psi}(\omega)\,[\tilde{q}(-2^{j+m}\omega)]^{-1}. \tag{10}$$

Combining (1), (2), (4) and (6) – (10), and choosing sufficiently large $m = m_n$, we obtain the linear wavelet estimators of $\Phi(y)$ and $p(y)$

$$\hat{\Phi}_m(y) = \sum_{k\in\mathbf{Z}} \hat{a}_{m,k}\,\varphi_{m,k}(y), \qquad \hat{p}_m(y) = \sum_{k\in\mathbf{Z}} \hat{b}_{m,k}\,\varphi_{m,k}(y). \tag{11}$$

After the estimators of $\Phi(y)$ and $p(y)$ are constructed, following Penskaya (1995), we estimate the ratio $\Phi(y)/p(y)$ by $H(\hat{\Phi}_m(y)/\hat{p}_m(y), \delta_n)$, where

$$H(x,\delta) = x(1 + \delta x^{2L})^{-1}, \quad L = 1, 2, 3, \ldots \tag{12}$$

Thus,

$$\beta_n(y) = y - H(\hat{\Phi}_m(y)/\hat{p}_m(y), \delta_n). \tag{13}$$

To calculate the risk $R_n(y)$ of the EB estimator (13) assume that $g(\theta)$ belongs to the Sobolev space H_α, that is

$$\int_{-\infty}^{\infty} |\tilde{g}(\omega)|^2(\omega^2 + 1)^\alpha d\omega < \infty. \tag{14}$$

Choose

$$m_n = \begin{cases} (2\alpha + 2\varrho)^{-1}\log_2 n, & \text{if } \gamma = 0, \\[2mm] \log_2\frac{3}{2\pi} + \frac{1}{\nu}\log_2\left[\frac{1}{2\gamma}\left(\ln n - \frac{2\varrho}{\nu}\ln\ln n\right)\right], & \text{if } \gamma > 0. \end{cases} \tag{15}$$

and

$$\delta_n \sim \begin{cases} n^{-\frac{2\alpha+2\varrho-1}{4\alpha+4\varrho}}, & \text{if } \gamma = 0, \\[2mm] n^{-1/2}(\ln n)^{\frac{\max(\lambda,0)+1/2}{\nu}}, & \text{if } \gamma > 0, \end{cases} \tag{16}$$

Then the following Theorem is true (see Pensky (1998b))

Theorem 1. *Let the EB estimator be defined by (13) with m_n and δ_n given by formulae (15) and (16), respectively. Then, under the conditions C1 – C4 and (14),*

$$
\sup_{y \in [A,B]} R_n(y) =
\begin{cases}
O\left(n^{-\frac{2\alpha+2\varrho-1}{2\alpha+2\varrho}}\right), & \text{if } \gamma = 0, \\[2mm]
O\left(n^{-1}(\ln n)^{\frac{2\lambda^*+1}{\nu}}\right), & \text{if } \gamma > 0.
\end{cases}
\tag{17}
$$

for any interval $[A, B]$ such that $\inf_{y \in [A,B]} p(y) > 0$. Here $\lambda^* = \max(\lambda, 0)$.

The EB estimator (13) is optimal in a sense that the rate of convergence cannot be improved (see Pensky (1997a)). If the conditional density $q(x)$ is supermooth ($\gamma > 0$), then convergence rate is close to $O(n^{-1})$, deviating from this by only a logarithmic factor. Moreover, in the case of $\gamma > 0$ the estimator is adaptive: the choice of m_n and δ_n is independent of the unknown value of α in (14).

20.2.2   *Nonlinear wavelet EB estimation*

Although the estimator (13) with $\hat{\Phi}_m(y)$ and $\hat{p}_m(y)$ given by (11) has outstanding properties in the case when $\tilde{q}(\omega)$ has an exponential descent, it also has at least two shortcomings when $\tilde{q}(\omega)$ decreases at a polynomial rate. The first is that, in this situation, the estimator (13) is not adaptive, that is, the choice of parameters m_n and δ_n depends on the unknown degree of smoothness α of the density $g(\theta)$. The second shortcoming is that we obtain the estimator adjusted to the lowest possible degree of smoothness of $\Phi(y)$ and $p(y)$. Consider a simple example of this situation. Let $q(x)$ and $g(\theta)$ be the densities of gamma distributions with shape parameters ϱ_1 and ϱ_2, respectively. Then it is easy to see that $p(x)$ is the gamma p.d.f. with the shape parameter $(\varrho_1 + \varrho_2)$. Hence, $p(x)$ is infinitely differentiable everywhere except for $x = 0$. Nevertheless, using the Meyer-type wavelets we cannot obtain a rate of convergence higher than $O\left(n^{-(2\varrho_1+2\varrho_2-2)/(2\varrho_1+2\varrho_2-1)}\right)$ since the discontinuity of the $(\varrho_1 + \varrho_2)$-th derivative of $p(x)$ at $x = 0$ affects the estimator of $p(x)$ at any point. This deficiency is due to the fact that the Meyer-type wavelets have infinite supports, and can be avoided by the use of wavelets with bounded supports, say, Daubechies wavelets or B-spline wavelets (see, for example, Daubechies (1992) or Hernández and Weiss (1996)).

Consider EB estimation of θ in the case when $\tilde{q}(\omega)$ has a polynomial rate of descent, i.e. $\|(w^2 + 1)^{\varrho/2}\tilde{q}(\omega)\|_c < \infty$, $\varrho > 1/2$. Choose a real number $r > \max(\varrho, 2)$, a real r-regular scale function $\varphi(x)$ with a bounded support $[-A_\varphi, A_\varphi]$ and a corresponding wavelet $\psi(x)$ with a bounded support $[-A_\psi, A_\psi]$. We estimate coefficients $a_{m,k}$, $\alpha_{j,k}$, $b_{m,k}$ and $\beta_{j,k}$ by $\hat{a}_{m,k}$, $\hat{\alpha}_{j,k}$,

$\hat{b}_{m,k}$ and $\hat{\beta}_{j,k}$ using formulae (4) and (6) – (10). Then, we construct the estimators $\hat{\Phi}_m(y)$ and $\hat{p}_m(y)$ of $\Phi(y)$ and $p(y)$ of the form

$$\hat{\Phi}_m(y) = \sum_{k \in \mathbf{Z}} \hat{a}_{m,k} \varphi_{m,k}(y) + \sum_{j=0}^{M} \sum_{k \in \mathbf{Z}} \hat{\alpha}^*_{j,k} \psi_{j,k}(y), \qquad (18)$$

$$\hat{p}_m(y) = \sum_{k \in \mathbf{Z}} \hat{b}_{m,k} \varphi_{m,k}(y) + \sum_{j=0}^{M} \sum_{k \in \mathbf{Z}} \hat{\beta}^*_{j,k} \psi_{j,k}(y), \qquad (19)$$

where

$$\hat{\alpha}^*_{j,k} = \hat{\alpha}_{j,k} \, I(|\hat{\alpha}_{j,k}| > \Delta_n), \quad \hat{\beta}^*_{j,k} = \hat{\beta}_{j,k} \, I(|\hat{\beta}_{j,k}| > \Delta_n). \qquad (20)$$

Here, $I(A)$ is the indicator of a set A. Observe that, according to the condition **C4**, $Q = \|\tilde{q}'(\omega)/\tilde{q}(\omega)\|_c < \infty$, so that the inverse Fourier transforms (8) and (10) exist. Also, note that since $\varphi(x)$ and $\psi(x)$ have bounded supports, the sums in (18) and (19) contain only a finite number of nonzero terms. Now, as the estimators $\hat{\Phi}_m(y)$ and $\hat{p}_m(y)$ of $\Phi(y)$ and $p(y)$ are constructed, we estimate $\beta(y)$ by (13).

To calculate the risks (4) of the EB estimator (13), observe that since $\varrho > 1/2$, it is always possible to choose ε such that $2\varrho - 1 - \varepsilon > 0$. For this choice of ε the following statement is valid (see Pensky (1997c)).

Theorem 2. *Let conditions* **C1** – **C4** *hold with* $\gamma = 0$ *and assume that* $p(x)$ *and* $\Phi(x)$ *are* s *times continuously differentiable in the neighbourhood* $(y - 2^{1-m} A_\psi, \, y + 2^{1-m} A_\psi)$ *of the point* y. *If in* (12), (13) *and* (18) – (20) *we have* $m = m_n = \log_2(\ln n)$, $M = M_n = \mathbf{Ent}\left((2\varrho - 1 - \varepsilon)^{-1} \log_2 n\right)$, *and* $\Delta_n = \Delta_0 \sqrt{\ln n / n}$, *where*

$$\Delta_0 > 4 \left[\left(Q^2 \|q\|_c + Q\|\psi\|_L\right) \, \max\left(2, (2\varrho + 2 - \varepsilon)(2\varrho - 1 - \varepsilon)^{-1}\right)\right]^{\frac{1}{2}},$$

$L = L_n > 0.5 + (4\varrho - 2 - 2\varepsilon)^{-1}$ *and* $\delta_n = \sqrt{\ln n / n}$, *then*

$$R_n(y) = O\left(n^{-\frac{2\tau}{2\tau+1}} [\ln n]^{\frac{2\tau}{2\tau+1}}\right). \qquad (21)$$

Here $\tau = \min(r, s)$, $\mathbf{Ent}(x)$ *is the entire part of* x.

If $p(x)$ and $\Phi(x)$ are s times continuously differentiable at the point $x = y$, then the best rates of convergence that can be achieved by an EB estimator at the point y is $O\left(n^{-\frac{2s}{2s+1}}\right)$. Note that a large value of r implies that $r \geq s$, i.e. $\tau = s$. The latter will guarantee that the estimator (13) achives the rate of convergence which is only $(\ln n)^{\frac{2\tau}{2\tau+1}}$ times greater than the optimal convergence rate at the point y. However, the convergence rate in (21) cannot be improved unless α is known, since an adaptive estimator cannot have convergence rate higher than (21) (see Brown and Low (1996)). Therefore, in the case when $q(x)$ is not supersmooth, it seems more advantageous to use wavelets with bounded supports rather than the Meyer-type wavelets.

20.3   EB estimation in the scale parameter case

20.3.1   Construction of the estimator

If θ is a scale parameter, the conditional density has the form (10), and the EB estimator is given by (6) with $\Psi(y) \equiv \Phi(y)$ given by (11). Recall that $\hat{f}$ and $\check{f}$ denote the Fourier and the Mellin transforms of a function f, respectively, and assume that the following conditions hold

C1*. The Fourier transform of the function $e^{3x/2}q(e^x)$ does not vanish for real ω.
C2*. $\int_0^\infty (y^2+1)q^2(y)dy < \infty$, $\int_0^\infty (\theta^2+1)g^2(\theta)d\theta < \infty$, $\|x^3 q(x)\|_c < \infty$;
C3*. $\|\exp(\gamma|\omega|^\nu)(w^2+1)^{\varrho/2}\check{q}(0.5-i\omega)\|_c < \infty$;
C4*. $|\check{q}(0.5-i\omega)| \le C_q^*(|\omega|+1)^\lambda |\check{q}(1.5-i\omega)|$ with $\lambda \le 0$ whenever $\gamma = 0$ in **C3***.

We will consider the case of $x > 0$ keeping in mind that, if $q(x)$ is an even function on $(-\infty, \infty)$, then $p(y)$ and $\Psi(y)$ are also even, and, thus, we are interested in $p(|y|)$ and $\Psi(|y|)$ only. Therefore, we need an orthogonal system on $(0, \infty)$ rather than on $(-\infty, \infty)$. Consider two systems of functions

$$\{x^{-1/2}\varphi_{m,k}(\ln x)\}_{k\in\mathbf{Z}}, \quad \{x^{-1/2}\psi_{j,k}(\ln x)\}_{k\in\mathbf{Z}, j\in\mathbf{N}}, \tag{1}$$

where $\varphi_{m,k}(y)$ and $\psi_{j,k}(y)$ are defined in (3) and $\varphi(x)$ and $\psi(x)$ are a scaling function and a corresponding wavelet, respectively. It is easy to see that the first system in (1) forms an "approximation" of $L^2(0, \infty)$, while the combination of the first and the second is the orthonormal basis of $L^2(0, \infty)$. Observe that, according to the condition **C2***, both $\Phi(y)$ and $p(y)$ are square integrable. Hence, we approximate $\Psi(y)$ and $p(y)$ by

$$\Psi(y) = \sum_{k\in\mathbf{Z}} a_{m,k}\, y^{-1/2}\, \varphi_{m,k}(\ln y) + \sum_{j=0}^\infty \sum_{k\in\mathbf{Z}} \alpha_{j,k}\, y^{-1/2}\, \psi_{j,k}(\ln y),$$

$$p(y) = \sum_{k\in\mathbf{Z}} b_{m,k}\, y^{-1/2}\, \varphi_{m,k}(\ln y) + \sum_{j=0}^\infty \sum_{k\in\mathbf{Z}} \beta_{j,k}\, y^{-1/2}\, \psi_{j,k}(\ln y).$$

Here $b_{m,k} = \mathbf{E}_p\left[X^{-1/2}\varphi_{m,k}(\ln X)\right]$ and $\beta_{j,k} = \mathbf{E}_p\left[X^{-1/2}\psi_{j,k}(\ln X)\right]$, so that the coefficients $b_{m,k}$ and $\beta_{j,k}$ can be estimated by sample means $\hat{b}_{m,k}$

and $\hat{\beta}_{j,k}$. The coefficients $a_{m,k}$ and $\alpha_{j,k}$ have the forms

$$a_{m,k} = \int x^{-1/2} \varphi_{m,k}(\ln x) \, \Psi(x) dx, \quad \alpha_{j,k} = \int x^{-1/2} \psi_{j,k}(\ln x) \, \Psi(x) dx.$$

To present $a_{m,k}$ and $\alpha_{j,k}$ as mathematical expectations of some functions $u_{m,k}(x)$ and $v_{j,k}(x)$, respectively, we need to solve integral equations, similar to the location parameter case (see (5)). For example, $u_{m,k}(x)$ is given by the equation

$$\int_0^\infty \theta^{-1} q(x\theta^{-1}) u_{m,k}(x) dx = \int_0^\infty q(x\theta^{-1}) x^{-1/2} \varphi_{m,k}(\ln x) dx. \quad (2)$$

Change the variables $x = e^z$, $\theta = e^t$ in both sides of the formula (2) and denote $q_1(x) = e^{x/2} q(e^x)$, $q_2(x) = e^{3x/2} q(e^x)$ and $U_{m,k}(x) = e^{-x/2} u_{m,k}(e^x)$. Then the equation (2) can be written as

$$\int_{-\infty}^\infty q_2(z-t) U_{m,k}(z) dz = \int_{-\infty}^\infty q_1(z-t) \varphi_{m,k}(z) dz,$$

where $q_1(x)$ and $q_2(x)$ are square integrable according to **C2***. A similar equation can be obtained to derived $v_{j,k}(x)$. Solving both equations using the Fourier transform, as for the case of the location parameter, and expressing $\tilde{q}_1(\omega)$ and $\tilde{q}_2(\omega)$ via the Mellin transform $\check{q}$ of q, we finally obtain that

$$u_{m,k}(x) = \sqrt{x}\, 2^{m/2} U_m(2^m \ln x - k), \quad v_{j,k}(x) = \sqrt{x}\, 2^{\frac{m+j}{2}} V_j(2^{m+j} \ln x - k),$$

where $U_m(z)$ and $V_j(z)$ are the inverse Fourier transforms of

$$\tilde{U}_m(\omega) = \frac{\tilde{\varphi}(\omega)\check{q}(0.5 - i2^m\omega)}{\check{q}(1.5 - i2^m\omega)}, \quad \tilde{V}_j(\omega) = \frac{\tilde{\psi}(\omega)\check{q}(0.5 - i2^{m+j}\omega)}{\check{q}(1.5 - i2^{m+j}\omega)}.$$

The estimators $\hat{a}_{m,k}$ and $\hat{\alpha}_{j,k}$ of $a_{m,k}$ and $\alpha_{j,k}$ are the sample averages of the functions $u_{m,k}(x)$ and $v_{j,k}(x)$, respectively.

After the coefficients are estimated, as for the location parameter case, we can construct linear wavelet estimators of $\Psi(y)$ and $p(y)$

$$\hat{\Psi}_m(y) = \sum_{k \in \mathbf{Z}} \frac{\hat{a}_{m,k}}{\sqrt{y}} \varphi_{m,k}(\ln y), \quad \hat{p}_m(y) = \sum_{k \in \mathbf{Z}} \frac{\hat{b}_{m,k}}{\sqrt{y}} \varphi_{m,k}(\ln y), \quad (3)$$

or, nonlinear wavelet estimators of $\Psi(y)$ and $p(y)$

$$\hat{\Psi}_m(y) = \sum_{k \in \mathbf{Z}} y^{-1/2} \hat{a}_{m,k} \varphi_{m,k}(\ln y) + \sum_{j=0}^M \sum_{k \in \mathbf{Z}} \hat{\alpha}_{j,k}^* \, y^{-1/2} \, \psi_{j,k}(\ln y),$$

$$(4)$$

$$\hat{p}_m(y) = \sum_{k \in \mathbf{Z}} \hat{b}_{m,k} y^{-1/2} \varphi_{m,k}(\ln y) + \sum_{j=0}^M \sum_{k \in \mathbf{Z}} \hat{\beta}_{j,k}^* \, y^{-1/2} \, \psi_{j,k}(\ln y).$$

Here the coefficients $\hat{\alpha}^*_{j,k}$ and $\hat{\beta}^*_{j,k}$ are determined by (20). Now, the EB estimator of t has the form

$$\beta_n(y) = H(\hat{\Psi}_m(y)/\hat{p}_m(y), \delta_n) \qquad (5)$$

where the function H is defined in (12).

20.3.2   The risk of the EB estimator

To find the risk of the estimator (5), let us introduce the functions $\Psi^*(x) = e^{x/2}\Psi(e^x)$ and $p^*(x) = e^{x/2}p(e^x)$, $x \in R$. Therefore, $\Psi(y) = y^{-1/2}\Psi^*(\ln y)$, $p(y) = y^{-1/2}p^*(\ln y)$, and, $\beta(y) = \Psi^*(\ln y)/p^*(\ln y)$. Furthermore, observe that $a_{m,k}$ and $b_{m,k}$ are the coefficients of the wavelet expansion of the functions $\Psi^*(x)$ and $p^*(x)$, respectively. Hence, $\sqrt{y}\,\hat{\Psi}_m(y)$ and $\sqrt{y}\,\hat{p}_m(y)$ are the estimators of $\Psi^*(\ln y)$ and $p^*(\ln y)$. Also notice that functions $\Psi^*(y)$ and $p^*(y)$ have the convolution representations

$$p^*(y) = \int_{-\infty}^{\infty} q_1(y-t)g_1(t)dt, \quad \Psi^*(y) = \int_{-\infty}^{\infty} q_1(y-t)g_2(t)dt, \qquad (6)$$

where $g_1(t) = e^{t/2}g(e^t)$ and $g_2(t) = e^{3t/2}g(e^t)$, $q_1(x) = e^{x/2}q(e^x)$. The identities (6) allow us to reformulate the results of the Section 2. However, we draw attention to the fact that $p^*(y)$ is not the p.d.f. of $\exp(X_1)$, and neither $g_1(t)$ nor $g_2(t)$ is the p.d.f. of $\exp(\Theta_1)$. Therefore, the estimators (3) and (4) cannot be obtained by reparametrization of the initial model. The following theorem, for example, derives the risk of the estimator (5) based on linear wavelet estimators (3).

Theorem 3. *Let $\beta_n(y)$ have the form (5) with $\hat{\Psi}_m(y)$ and $\hat{p}_m(y)$ given by (3). Let m_n and δ_n be defined by (15) and (16), respectively. If the assumptions* **C1*** *–* **C4*** *are valid and*

$$\int_{-\infty}^{\infty} |\tilde{g}_i(\omega)|^2(\omega^2+1)^\alpha d\omega < \infty, \quad i = 1,2,$$

then the asymptotic expression (17) is valid.

Theorem 3 has a very useful Corollary. Observe that, for familiar distribution families, $\check{q}(\omega)$ can be expressed via a gamma-function.

Corollary 1. *Let*

$$\check{q}(\omega) = q_0\,a_1^\omega\,\Gamma(b_1\omega + b_2), \quad a_1 > 0, \ b_1 > 0, \ 0.5b_1 + b_2 > 0. \qquad (7)$$

Then, under the conditions **C1*** *and* **C2***,

$$\sup_{[A,B]} R_n(y) = O\left(n^{-1}\ln n\right) \qquad (8)$$

for any interval $[A, B] \subset (0, \infty)$ *such that* $\inf_{y \in [A, B]} p(y) > 0$.

It is easy to see that the class (7) contains the normal distribution $q(x) = (\sqrt{2\pi})^{-1} \exp(-x^2/2)$ for which $\check{q}(\omega) = (2\pi)^{-1/2}(\sqrt{2})^{\omega-2}\Gamma(\omega/2)$, the gamma-distribution $q(x) = x^\varrho e^{-x}[\Gamma(\varrho+1)]^{-1}$ with $\check{q}(\omega) = \Gamma(\omega + \varrho)/\Gamma(\varrho + 1)$, and also the double-exponential distribution $q(x) = 0.5 \exp(-|x|)$ with $\check{q}(\omega) = 0.5\,\Gamma(\omega)$.

20.3.3   EB estimation for the uniform distribution

In the case when $q(x)$ is the uniform distribution $q(x) = I(0 < x < 1)$, a wavelet based EB estimation was constructed by Huang (1997). Observe that in the case of the uniform distribution,

$$\beta(y) = x + (1 - P(x))/p(x),$$

where $p(x)$ is the marginal p.d.f. of X defined by (1) and $P(x)$ is the marginal c.d.f. of X. Huang (1997) considered an r-regular scaling function $\varphi(x)$ with a bounded support, and constructed the estimator $\hat{p}_m(y)$ of $p(y)$ in the form (11) with the coefficient $\hat{b}_{m,k}$ given by (4). Subsequently, the wavelet based EB estimator

$$\beta_n(y) = y \vee \{[y + (1 - P_n(y))/\hat{p}_m(y)] \wedge M\}, \tag{9}$$

was obtained, where $P_n(y)$ is the empirical c.d.f. of X.

To derive the risk of the estimator (9), Huang (1997) introduced the Sobolev space

$$H_2^s = \left\{ f : f^{(l)} \in L^2(-\infty, \infty),\ l = 0, ..., N,\ \text{and}\ J_{s,2}(u^{(N)}) < \infty \right\},$$

where N is a nonnegative integer, $0 < \sigma < 1$, $s = N + \sigma$ and

$$J_{s,2}(u^{(N)}) = \int_{-\infty}^{\infty} z^{-(2\sigma+1)} \int_{-\infty}^{\infty} \left[u^{(N)}(x+z) - u^{(N)}(x) \right]^2 dx\, dz.$$

The conditions

C5. supp $g \in (0, A)$ with A known.
C6. $0 < p(0) < \infty$.
C7. $\mathbf{E}_p\left(1/p^2(X)\right) < \infty$.
C8. $p \in H_2^s$.

lead to:

Theorem 4. *Let assumptions* **C5** - **C8** *be valid and* $r > s$. *Then as* $n \to \infty$

$$R_n = O\left(n^{-2s/(2s+1)}\right).$$

20.4   EB estimation for a one-parameter exponential family

In the case of a one-parameter exponential family, $q(x|\theta)$ has the expression (12) and $\beta(y) = p'(y)/p(y) - h'(y)/h(y)$. Let us construct an EB estimator assuming that $h(x)$ is $(r+1)$ times continuously differentiable, $r \geq 1$. We choose an r-regular scale function $\varphi(x)$ with a bounded support (see, for example, Daubechies (1992) or Hernández and Weiss (1996)) and derive the estimator $\hat{p}_m(y)$ of $p_m(y)$ (see (11)) and the estimator

$$\hat{p}_{m,1}(y) = \sum_{k \in \mathbf{Z}} \hat{a}_{m,k} \, \varphi_{m,k}(y)$$

of $p'(y)$. Here $\hat{b}_{m,k}$ and $\hat{a}_{m,k}$ are defined by (4) and (6), respectively, with $u_{m,k}(x) = -2^{3m/2}\varphi'(2^m x - k)$. Then the EB estimator has the form

$$\beta_n(y) = H(\hat{p}_{m,1}(y)/\hat{p}_m(y), \delta_n) - h'(y)/h(y), \tag{1}$$

where the function H is defined by (12).

Since in the case of the one-parameter exponential family the rate of descent of $p(x)$ is predictable, an upper bound for the prior risk R_n (see (4)) can be derived. For a function f and a point $x \in R$ denote

$$L(f, y, \delta) = \sup_{|x-y|<\delta} |f(x)|.$$

Assume that the following conditions are satisfied:

C9. $\int_{-\infty}^{\infty} [p(x)]^\mu dx < \infty$, $\mu < 1$.
C10. $\|p\|_c < \infty$.
C11. For some $\varepsilon_1, \varepsilon_2, 0 \leq \varepsilon_1, \varepsilon_2 \leq 1$ and any $\delta > 0$

$$\left\| \frac{L^2(\Phi^{(r_1)}, x, \delta) + L(p, x, \delta)}{[p(x)]^{1-\varepsilon_1}} \right\|_c < \infty, \quad \left\| \frac{L^2(p^{(r_2)}, x, \delta) + L(p, x, \delta)}{[p(x)]^{1-\varepsilon_2}} \right\|_c < \infty.$$

C12. $\int_{-\infty}^{\infty} |\Phi(y)/p(y)|^{2K} dy < \infty$, $K > 1$.

It is easy to see that the sufficient condition for **C12** is

$$\int_a^b |\theta|^{2\tau} g(\theta) d\theta < \infty, \quad \tau > 1. \tag{2}$$

Theorem 5. *Let* $q(x|\theta)$ *belong to the one-parameter exponential family and the conditions* **C9** $-$ **C12** *and* (2) *hold. Then*

$$R_n = O\left(n^{-\frac{2r\lambda}{2r+3}}\right), \tag{3}$$

with $\lambda = [\tau(1 + \varepsilon)]^{-1}(1 - \mu)(K - 1) - (2s + 1)^{-1}$ *provided* $s \leq 0.5\,\tau(1 + \varepsilon)/(1 - \mu)$ *and* $\delta_n \sim n^{-2sr[(2r+3)(2s+1)]^{-1}}$. *Here* ε *is an arbitrary small positive number.*

20.5   Examples and discussion.

20.5.1   Examples.

Example 1. The family of normal distributions. Let us first consider the case when

$$q(x|\theta) = q(x - \theta) = (\sqrt{2\pi})^{-1} \exp\left\{-(x - \theta)^2/2\right\}, \tag{1}$$

that is (1) is both a location parameter family with $\tilde{q}(\omega) = \exp\left\{-\omega^2/2\right\}$, and a one-parameter exponential family with $h(x) = \exp\left\{-x^2/2\right\}$ and $C(\theta) = (\sqrt{2\pi})^{-1} \exp\left\{-\theta^2/2\right\}$. Since $q(x)$ is supersmooth with $\gamma = 1$, $\nu = 2$ and $\lambda = 1$ in **C3**, the linear wavelet EB estimator (13), based on the Meyer-type wavelets, is adaptive and

$$\sup_{[A,B]} R_n(y) = O\left(n^{-1}(\ln n)^{3/2}\right)$$

for any $[A, B] \subset (-\infty, \infty)$. Moreover, the function $U_m(x)$ involved in the derivation of $\hat{a}_{m,k}$ (see (6) – (8)) can be explicitly expressed via $\varphi(x)$ as $U_m(x) = 2^m \varphi'(2^m x)$.

In the case when we are interested in the prior risk R_n rather than in $R_n(y)$ (see (4)), we construct a linear wavelet EB estimator based on an r-regular wavelet with a bounded support. It is easy to show that in the case of the normal distribution, **C9** is satisfied with $\mu > (2\tau + 1)^{-1}$, **C10** holds, and **C11** and **C12** are valid for any $\varepsilon_1, \varepsilon_2 \in (0, 1)$ and any $K > 0$. Thus, the following statement holds (see Pensky (1998a))

Corollary 2. *If the conditional density is normal (1), then, under the assumption (2), R_n has the form (3) with δ_n given in Theorem 5, $\lambda = [K(1 + \Delta_1)]^{-1}(1 - \mu)(K - 1) - (2S + 1)^{-1}$, $\mu = (2\tau + 1)^{-1} + \Delta_2$ and $S = 0.5 K(1 + \Delta_1)(1 + 2\tau)/[2\tau(1 - \Delta_2) - \Delta_2]$. Here r and K are arbitrary large numbers, and Δ_1 and Δ_2 are arbitrary small numbers.*

Let us compare the convergence rate (3) of the estimator (1) with the convergence rate of the estimators constructed by Singh (1979). Under the assumption (2), the best rate of convergence that can be attained by his estimator is $O\left(n^{-\frac{\tau-1}{\tau+1}}\right)$. Calculating the limit of $(2r + 3)^{-1}2r\lambda$ in (3) as $K \to \infty$, $r \to \infty$, $\Delta_1 \to 0$ and $\Delta_2 \to 0$, we obtain $2\tau(2\tau + 1)^{-1}$ which is greater than $(\tau - 1)(\tau + 1)^{-1}$. Thus, R_n can be made arbitrarily close to $O\left(n^{-\frac{2\tau}{2\tau+1}}\right)$ which is better than the convergence rate of the estimators obtained by Singh (1979). The reason for the improvement lies in the employment of the infinite smoothness of $p(x)$ and $\Phi(x)$. Singh (1979) in his construction used kernels of the order (2τ), whereas wavelets of higher orders can give better approximations.

Now let us consider the case when $q(x|\theta)$ is a family of normal distributions with a scale parameter, i.e. $q(x|\theta) = (\sqrt{2\pi}\,\theta)^{-1} \exp\{-x^2/(2\theta^2)\}$. In this situation, we may use the estimator (5) with $\hat{\Psi}_m(y)$ and $\hat{p}_m(y)$ defined in (3). The risk $R_n(y)$ of the estimator is given by (8).

Example 2. The family of gamma distributions. Consider the case when $q(x|\theta)$ belongs to the family of gamma-distributions with the location parameter, i.e. $q(x|\theta) = q(x - \theta)$ with

$$q(x) = x^\varrho e^{-x}[\Gamma(\varrho + 1)]^{-1}. \tag{2}$$

In this situation, $\tilde{q}(\omega) = (1-i\omega)^{-(\varrho+1)}$. Since $\tilde{q}(\omega)$ decreases at a polynomial rate, we choose $r > \varrho+1$, an r-regular scaling function $\varphi(x)$ with a bounded support and a corresponding wavelet $\psi(x)$.

To obtain $U_m(x)$ and $V_j(x)$, notice that $\tilde{q}'(\omega) = i(\varrho + 1)(1 - i\omega)^{-(\varrho+2)}$ which implies

$$\tilde{U}_m(\omega) = -(\varrho+1)(1+i2^m\omega)^{-1}\tilde{\varphi}(\omega), \quad \tilde{V}_j(\omega) = -(\varrho+1)(1+i2^{j+m}\omega)^{-1}\tilde{\psi}(\omega).$$

Multiplying both sides of the previous equations by $(1 + i2^m\omega)$ and $(1 + i2^{j+m}\omega)$, respectively, using the fact that $\mathcal{F}^{-1}[i\omega\tilde{f}(\omega)] = f'(x)$ for any function f with $\omega\tilde{f}(\omega) \in L_1(-\infty, \infty)$, and solving a simple linear differential equation, we obtain the explicit expressions for $U_m(x)$ and $V_j(x)$:

$$U_m(x) = (\varrho + 1) \int_0^\infty \varphi(x - 2^m z) e^{-z} dz,$$

$$\tag{3}$$

$$V_j(x) = (\varrho + 1) \int_0^\infty \psi(x - 2^{j+m} z) e^{-z} dz.$$

Under the condition (14), the error $R_n(y)$ of the estimator (13) has the form (21) with $\tau = \min(r, \alpha + \varrho - 0.5)$ if y lies in the neighborhood of zero, and $\tau = r$ otherwise. Since r is an arbitrary large number, this means that the convergence rate of the estimator (13) can be made as close to $O(n^{-1})$ as one wishes, everywhere except in the neighborhood of zero.

Now let us consider the case when θ is a scale parameter, i.e. $q(x|\theta)$ is defined by (10) with $q(x)$ given by (2). In this case, $\check{q}(\omega) = \Gamma(\omega+\varrho)/\Gamma(\varrho+1)$, and we implement estimator (5) based on (3). To find an explicit expression for $U_m(x)$, observe that $\tilde{U}_m(\omega) = (\varrho + 0.5 - i2^m\omega)^{-1}\tilde{\varphi}(\omega)$. Repeating the steps performed in the course of derivation of (3), we obtain

$$U_m(x) = \int_0^\infty \varphi(2^m z + x) \exp\{-(\varrho + 0.5)\} dz.$$

The error of the estimator (5) is determined by (8).

20.5.2 *Discussion*

In the present paper, the wavelet technique of EB estimation has been discussed, and several examples of construction of the estimators have been demonstrated. This technique has several nontrivial advantages in comparison with the other methods of nonparametric EB estimation.

Since the coefficients $a_{m,k}$, $b_{m,k}$, $\alpha_{j,k}$ and $\beta_{j,k}$ are based only on the "past" observations $X_1, ..., X_n$, the method described above enables one to pre-estimate $a_{m,k}$, $b_{m,k}$, $\alpha_{j,k}$ and $\beta_{j,k}$ in advance, and then use $\hat{a}_{m,k}$, $\hat{b}_{m,k}$, $\hat{\alpha}_{j,k}$ and $\hat{\beta}_{j,k}$ for every new observation y. This is impossible to do if EB estimation is based on kernels (see, for example, Singh (1979), Singh and Wei (1992), Tiwari and Zalkikar (1990)). This property becomes very important if the same pool of data $X_1, ..., X_n$ is used to estimate the values of the parameter corresponding to several new observations y.

The second major benefit is that EB estimators are obtained through a general technique which is suitable for various distribution families.

The other significant advantage of the wavelet method is that it provides high convergence rates which are optimal or very close to optimal (see Pensky (1997a)). The last property distinguishes the wavelet approach from the linear EB estimation (see Maritz and Lwin (1989)) and the technique based on estimation of the prior density $g(\theta)$ (see, for example, Walter (1981)). Also, the wavelet technique provides adaptation to the unknown degree of smoothness of a prior density $g(\theta)$.

The wavelet method does not possess another defect which is typical of a "naive" EB estimators: since $\varphi(x)$, $\psi(x)$ and $H(x, \delta)$ are infinitely differentiable (see (12)), the estimator $\beta_n(y)$ is also infinitely smooth. This is a very significant advantage of the estimator (13) in comparison with the estimators of Singh (1979) and Nogami (1988) which are not differentiable.

Another advantage of the wavelet method is that if we wish to try different values of m in order to determine the best one, we can use recursive relations between wavelet coefficients for recalculation of $\hat{a}_{m,k}$, $\hat{b}_{m,k}$, $\hat{\alpha}_{j,k}$ and $\hat{\beta}_{j,k}$ (see, for example, Walter (1994)).

However, together with a number of major advantages, the wavelet method also has certain shortcomings. The major one is that,for almost all interesting cases, the expressions for the scale function $\varphi(x)$ and the wavelet $\psi(x)$ cannot be obtained in a closed form and should be calculated numerically.

To summarize the preceding discussion, the wavelet approach to the EB estimation has major advantages in comparison with the other methods of nonparametric EB estimation and also some disadvantages. However, with the progress of computers the advantages of the wavelet technique (good precision, flexibility and computational convenience) become more and more attractive, while the shortcomings seem less and less serious.

References

Antoniadies, A., Grégoire, G., and McKeague, I.W. (1994) Wavelet method for curve estimation. *J. Amer. Statist. Assoc.*, 89, 1340-1353.

Brown, L.D. and Low M.G. (1996) A constrained risk inequality with applications to nonparametric functional estimation. *Ann. Stat.*, 24, 2524-2535.

Carlin, B.P. and Louis, T.A. (1996) *Bayes and Empirical Bayes Methods for Data Analysis*, Chapman & Hall, London.

Daubechies, I. (1992) *Ten Lectures in Wavelets*, SIAM, Philadelphia.

Donoho, D.L. (1994) Smooth wavelet decompositions with blocky coefficient kernels. In: *Recent Advances in Wavelet Analysis*, ed. by Schumaker, L.L. and Webb, G., Academic Press, London, 259- 308.

Donoho, D.L. and Johnstone, I.M. (1995) Adapting to unknown smoothness via wavelet shrinkage. *J. Amer. Statist. Assoc.*, 90, 1200 - 1224.

Donoho, D.L., Johnstone, I.M., Kerkyacharian, G., and Picard, D. (1996) Density estimation by wavelet thresholding. *Ann. Stat.*, 24, 508 - 539.

Erdélyi, A. et al (1954) *Tables of Integral Transforms*. Mc Graw-Hill Book Company, New York.

Hall, P., and Patil, P. (1995) Formulae for mean integrated squared error of nonlinear wavelet-based density estimators. *Ann. Stat.*, 23, 905-928.

Hall, P., Kerkyacharian, G., and Picard, D. (1998) Block thresholding rules for curve estimation using kernel and wavelet methods. *Ann. Stat.*, 26, 922-942.

Hernández, E., and Weiss, G. (1996) *A First Course on Wavelets*, CRC Press, Boca Raton.

Huang, Su-Yun (1997) Wavelet based empirical Bayes estimation for the uniform distribution. *Statist. Probab. Lett.*, 32, 141-146.

Kerkyacharian, G., and Picard, D. (1992) Density estimation in Besov spaces. *Statist. Probab. Lett.*, 13, 15-24.

Lahiri, P., and Park, D.H. (1991) Nonparametric Bayes and empirical Bayes estimators of mean residual life at age *t*, *J. Stat. Plan. Inference*, 29, 125-136.

Maritz, J., and Lwin, T. (1989) *Empirical Bayes Methods*, Chapman & Hall, London.

Masry, E. (1994) Probability density estimation from dependent observations using wavelet orthonormal bases. *Statist. Probab. Lett.*, 21, 181-194.

Meyer, I. (1993) *Wavelets. Algorithms and Applications*, SIAM, Philadelphia.

Nogami, Y. (1988) Convergence rates for empirical Bayes estimation in the uniform $U(0, \theta)$ distribution. *Ann. Stat.*, 16, 1335-1341.

Penskaya, M. (1995) On mean square consistent estimation of a ratio. *Scandinavian Journal of Statistics*, 22, 129-137.

Pensky, M. (1996) Empirical Bayes estimation of a scale parameter. *Mathematical Methods of Statistics*, 5, 316-331.

Pensky, M. (1997a) Empirical Bayes estimation of a location parameter. *Statist. Decisions*, 15, 1-16.

Pensky, M. (1997b) A general approach to nonparametric empirical Bayes estimation. *Statistics*, 29, 61-80.

Pensky, M. (1997c) Wavelet estimation in empirical Bayes model. *Proceedings of the Section on Bayesian Statistical Science*, American Statistical Association, 78-82.

Pensky, M. (1998a) Empirical Bayes estimation based on wavelets. *Sankhyā*, A60, 214 - 231.

Pensky, M. (1998b) Wavelet empirical Bayes estimation of a location or a scale parameter. *Journal of Statistical Planning and Inference*, submitted.

Robbins, H. (1983) Jerzy Neyman memorial lecture. Some thoughts on empirical Bayes estimation. *Ann. Statist.*, 11, 713-723.

Singh, R.S. (1979) Empirical Bayes estimation in Lebesgue -exponential families with rates near the best possible rate. *Ann.Statist.*, 7, 890-902.

Singh, R.S., and Wei, L. (1992) Empirical Bayes with rates and best rates of convergence in $u(x)c(\theta)\exp\left(-\frac{x}{\theta}\right)$ family estimation case. *Ann. Inst. Statist. Math.*, 44, 435-449.

Tiwari, R.C., and Zalkikar, J.N. (1990) Empirical Bayes estimation of a scale parameter in a Pareto distribution. *Comput. Statist. & Data Anal.*, 10, 261-270.

Walter, G.G. (1981) Orthogonal series estimators of the prior distribution. *Sankhya*, A43, 228-245.

Walter, G.G. (1994) *Wavelets and Other Orthogonal Systems with Applications*, CRC Press, Boca Raton.

Zayed, A.I., and Walter, G.G. (1996) Characterization of analytic functions in terms of their wavelet coefficients. *Complex Variables*, 29, 265-276.

21

Multiresolution Wavelet Analyses in Hierarchical Bayesian Turbulence Models

L. M. Berliner, C. K. Wikle, and R. F. Milliff

ABSTRACT Stochastic modeling plays a foundational role in the analysis of turbulent motion of fluids. We amplify on traditional stochastic models for turbulent fluids by adopting the Bayesian paradigm for stochastic modeling in the presence of uncertainty. This provides a mechanism for constructing analyses which incorporate physical reasoning and the effective use of observational data. A critical outcome of this modeling approach is the opportunity to reflect the complexities of turbulent motion at various spatial scales via statistical models based on wavelets. We present an example involving the prediction of near-surface ocean winds using spatially fine resolution satellite data in combination with coarse resolution information available from operational weather center "analysis fields."

21.1 Introduction

The mathematical formulation of fluid dynamics problems considered in this article asserts a coordinate system, say $\mathbf{x}$, describing the space in which the fluid exists and flows. The coordinates are either fixed (e.g., fluid contained in a laboratory setting) or rotating as a solid body with respect to a frame of reference (e.g., $\mathbf{x}$ represents a 3-dimensional coordinate system for the Earth's atmosphere or ocean). Fluid dynamical models consider the behavior of an idealized "fluid parcel." The idealization requires that the parcel is very small, yet large enough to ignore behaviors at the molecular level.

The velocity field of the fluid is denoted typically by the vector $\mathbf{u}(\mathbf{x}, t)$. This is a space-time field, indicating the velocity vector of a parcel at location $\mathbf{x}$ at time t. Similarly, other variables (vector or scalar) associated with the fluid (e.g., temperature, pressure) are represented as space-time fields, amenable to the mathematical analysis outlined here. Appropriate principles of physics (e.g., Newton's Laws, conservation laws) are applied, leading to an initial-value partial differential equation

$$\frac{D\mathbf{u}(\mathbf{x}, t)}{Dt} = \mathcal{L}(\mathbf{u}(\mathbf{x}, t)), \tag{1}$$

with initial condition

$$\mathbf{u}(\mathbf{x}, 0) = \mathbf{u}_o(\mathbf{x}).$$

(We suppress the dependence of the initial condition on $\mathbf{x}$ for the balance of this article.) The operator $\frac{D}{Dt}$ represents the "total derivative." It is applied to a space-time field such as $\mathbf{u}(\mathbf{x}, t)$ and represents the time derivative for a parcel following the flow. (Mathematically, to clarify this derivative, we could write the field as $\mathbf{u}(\mathbf{x}(t), t)$, though this is not common in the literature.)

The functional $\mathcal{L}$ is nonlinear and complicated. In general, solutions to the *Navier-Stokes equations* (1) are (i) hard to obtain and (ii) "turbulent." For our purposes it is sufficient to equate turbulence with unpredictability and complexity of solutions at a large variety of spatial scales. Unpredictability here refers to sensitivity to initial conditions, as in *chaos theory*. The form of nonlinearity present in a typical $\mathcal{L}$ means that even the slightest uncertainty in the initial field $\mathbf{u}_o$ eventually leads to major uncertainty as time evolves. (In most problems, uncertain boundary conditions also arise. We do not treat such cases here.) The complexity in spatial scales means intuitively that significant variation or fluctuations occur at many spatial resolutions.

This is not a paper about the theory of turbulence. Rather, we find motivation in the uncertainty alluded to above, as well as other sources to be discussed as the paper unfolds, to suggest a stochastic approach to modeling fluid flows. This is hardly a new suggestion. However, our viewpoint is not simply one of the traditional stochastic strategies. Rather, we suggest a Bayesian hierarchical modeling view to include both stochastic fluid modeling and a powerful method for efficiently learning about a flow based on both physics and observational data.

To explain the basic formulations of fluid dynamics needed to motivate our suggestions, we review common notions and uses of *averaging* of space-time fields in Section 2. The interplay between stochastic models and Bayesian learning from data are reviewed at the outset of Section 3. We then turn to general development of Bayesian hierarchical models for use in the analysis and prediction of turbulent fluids based on observational data. The critical suggestion here is to use wavelet transforms to represent small scale spatial behavior of turbulent fluids. Implementation requires thoughtful development of priors on parameters. Such developments are outlined in Section 3.4. An example implementation of the methodology is presented in Section 3.5. Section 4 is devoted to brief suggestions for generalizations of the models on Section 3. Finally, selected points for discussion are given in Section 5.

21.2 Averaging

In the last section we alluded to the imposition of uncertainty in initial conditions u_o into the realm of deterministic fluid models. In response, Batchelor (1953, p. 6) motivated a stochastic view of turbulence: "It is clear that if the initial conditions of the turbulent motion are known in probability only, we cannot hope to do more than determine the velocity field at later instants in the same way." Indeed, Batchelor *defined* the goal of analysis to be the derivation of the probability distribution of the fluid at future times using two inputs: (i) a distribution for the initial state and (ii) physically-based dynamics reflected in (1). From this viewpoint, Batchelor (1953, p. 14) suggested "Since we are supposing, as a matter of definition, that the velocity in a turbulent flow takes random values, and since, as a consequence, we are interested only in the *average* values of quantities, it is necessary to lay down rules about the method of taking averages."

21.2.1 Types of Averaging

Ensemble Averaging. Ensemble averaging is based on a standard probabilistic formulation. We view the initial condition to be a realization of a random process. Let $\mathbf{U}_o$ denote the random initial field; $\mathbf{u}_o$ denotes a realization of that process. Also, let P represent the probability specification (measure) that generates $\mathbf{U}_o$.

With this formalism, $\mathbf{u}(\mathbf{x}, t; \mathbf{U}_o)$ is a random process, even though the evolution dictated by the Navier-Stokes equations is deterministic. (A deterministic function of a random quantity is random.) An ensemble average is an expectation,

$$\bar{\mathbf{u}}(\mathbf{x}, t; \cdot) = E[\mathbf{u}(\mathbf{x}, t; \mathbf{U}_o)].$$

Alternatively, we have

$$\bar{\mathbf{u}}(\mathbf{x}, t; \cdot) = \int \mathbf{u}(\mathbf{x}, t; \mathbf{u}_o) dP(\mathbf{u}_o). \tag{2}$$

The place holder   is used to indicate what variable has been averaged out. (Since $\mathbf{U}_o$ is a random process, an equation like (2) actually requires careful mathematical clarification. We assume such structure is available and proceed viewing the operation in (2) as an abstract integral. We make no further allusions to measure theory here.)

The usual intuition behind ensemble averaging is the following frequentist thought experiment. Suppose we generate a collection of n initial conditions, $\mathbf{u}_o^{(1)}, \ldots, \mathbf{u}_o^{(n)}$, from the random initial process. We expect that for n very large

$$\frac{1}{n} \sum_{i=1}^{n} \mathbf{u}(\mathbf{x}, t; \mathbf{u}_o^{(i)}) \approx \bar{\mathbf{u}}(\mathbf{x}, t; \cdot).$$

Temporal Averaging.

Temporal "Ergodic" Averaging. Ergodic averaging involves long-run behavior of $\mathbf{u}(\mathbf{x}, t; \mathbf{u}_o)$. Consider the quantity

$$\lim_{T\to\infty} \frac{1}{T} \int_t^{t+T} \mathbf{u}(\mathbf{x}, t'; \mathbf{u}_o) dt'.$$

Let $\bar{\mathbf{u}}(\mathbf{x}, \cdot; \mathbf{u}_o)$ denote the result. The formal ergodic result is that the limiting value is independent of $\mathbf{u}_o$. The final result, denoted by $\bar{\mathbf{u}}(\mathbf{x}, \cdot)$, is a steady-state, purely spatial field.

Temporal Smoothing. By temporal smoothing, we mean averaging, or smoothing, or low-pass filtering of the process. The averaging is often over finite time intervals. Let δ be some positive constant. Define the average,

$$\bar{\mathbf{u}}_\delta(\mathbf{x}, t; \mathbf{u}_o) = \int_{t-\delta}^{t+\delta} \mathbf{u}(\mathbf{x}, t'; \mathbf{u}_o) d\mu_\delta(t'),$$

where the $d\mu$ notation allows for selection of a weight function (kernel) for smoothing; in simple averaging $d\mu_\delta(t') = \frac{1}{2\delta} dt'$.

Spatial Averaging

Spatial "Ergodic" Averaging. Let $B_\Delta(\mathbf{x})$ represent a box of width Δ in the spatial domain of $\mathbf{u}$, centered at a point $\mathbf{x}$. Consider the quantity

$$\lim_{\Delta\to\infty} \frac{1}{|B_\Delta(\mathbf{x})|} \int_{B_\Delta(\mathbf{x})} \mathbf{u}(\mathbf{y}, t; \mathbf{u}_o) d\mathbf{y}. \tag{3}$$

If this limit exists, it is independent of the point $\mathbf{x}$. The resulting average is denoted by $\bar{\mathbf{u}}(\cdot, t; \mathbf{u}_o)$.

Spatial Smoothing.

By spatial smoothing, we mean application of averaging similar to that in (3), but without the limit: Define

$$\bar{\mathbf{u}}_\Delta(\mathbf{x}, t; \mathbf{u}_o) = \int_{B_\Delta(\mathbf{x})} \mathbf{u}(\mathbf{y}, t; \mathbf{u}_o) d\nu(\mathbf{y}), \tag{4}$$

where $d\nu$ is a weighting measure supported in $B_\Delta(\mathbf{x})$.

Generally, smoothing may be viewed as a *convolution*. However, we only use such averaging here and therefore do not introduce arbitrary averaging kernels.

Spatial Gridding. A crucial form of averaging arises in the development of computational approaches for solving the Navier-Stokes equations. Consider a specified gridding of the spatial domain of the flow, leading to N points, say $\mathbf{x}_1, \ldots, \mathbf{x}_N$. We then apply averaging as in (4) for "boxes" B indexed by the grid. These actions yield a matrix process of averaged *state variables*,

$$\bar{\mathbf{U}}(t; \mathbf{u}_o) = (\bar{\mathbf{u}}(\mathbf{x}_1, t; \mathbf{u}_o), \ldots, \bar{\mathbf{u}}(\mathbf{x}_N, t; \mathbf{u}_o))^\top. \tag{5}$$

In the balance of this article, we view $\bar{\mathbf{U}}$ as a vector; we can "string out" the $\bar{\mathbf{u}}(\mathbf{x}_i, t; \mathbf{u}_o)$ as vectors.

Finally, note that various combinations of spatial and temporal averaging can be considered.

21.2.2 Relationships among Averages

Perhaps the most celebrated relationship between forms of averages is the Ergodic Theorem. One of the requirements for this result is *invariance*: if $\mathbf{U}_o$ is generated according to P, then the random state $\mathbf{u}(\mathbf{x}, t; \mathbf{U}_o)$ is also distributed according to P for all t. This implies that $\bar{\mathbf{u}}(\mathbf{x}, t; \cdot)$ cannot actually depend on t. This also relates ergodic processes and temporally stationary stochastic processes. We then write the ensemble mean as a steady-state field, $\bar{\mathbf{u}}(\mathbf{x}, \cdot)$. The conclusion of the Ergodic Theorem is that (with probability one, under P),

$$\bar{\mathbf{u}}(\mathbf{x}, \cdot; \mathbf{u}_o) = \bar{\mathbf{u}}(\mathbf{x}, \cdot)$$

for every $\mathbf{u}_o$. That is, for every realization, the "ergodic" time average is equal to the ensemble average. Of course, the invariant distribution is typically unknown. Also, it is not unique. Tacitly, modelers argue away this issue by trusting the existence of a unique, natural or "physically realizable" ergodic distribution. If such a distribution exists, the notion of *statistical regularity* suggests that as long as $\mathbf{u}_o$ is generated according to any distribution that is absolutely continuous with respect to the natural ergodic distribution, ergodicity applies. That is, after a burn-in or relaxation period, ergodic means converge as expected.

A notion of spatial stationarity can also be formalized. The key is to consider the probabilistic behavior of $\mathbf{u}$ at any arbitrary collection of points, say $\mathbf{x}_1, \ldots, \mathbf{x}_k$, in the $\mathbf{x}$-domain. Stationarity means that the joint distribution of $\mathbf{u}$ at these points is invariant to a location shift; that is, if the k points are all relocated (by the same rule) anywhere in the $\mathbf{x}$-domain. Though the language/notation is cumbersome, the idea is simple: the probabilistic structure of the field $\mathbf{u}$ is the same in all regions of the $\mathbf{x}$-domain. In this case we first have that the ensemble average $\bar{\mathbf{u}}(\mathbf{x}, t; \cdot)$ cannot actually depend on $\mathbf{x}$. We then write the ensemble mean as a temporal function $\bar{\mathbf{u}}(\cdot, t)$. Further, under appropriate conditions, we expect that for every t, (with probability one, under P),

$$\bar{\mathbf{u}}(\cdot, t; \mathbf{u}_o) = \bar{\mathbf{u}}(\cdot, t),$$

for every $\mathbf{u}_o$. That is, for every realization, the "ergodic" spatial average is equal to the ensemble average.

Also, note that temporally and spatially stationary processes are possible. In such cases, it must be that $\bar{\mathbf{u}}(\cdot, t) = \bar{\mathbf{u}}(\mathbf{x}, \cdot)$ everywhere in space and time; hence, the ensemble average is a scalar.

In practice, we cannot actually evaluate infinite time averages. It is natural to hope that temporal filter averages over sufficiently long time intervals approximate temporal ergodic averages. Of course, this only makes sense if the process under study is temporally stationary.

21.2.3 Using Averaging

Once a particular form of averaging is selected, the state variable is written compactly as the average plus fluctuation or *anomaly*:

$$\mathbf{u} = \bar{\mathbf{u}} + \mathbf{u}'. \tag{6}$$

Note that if we average an anomaly, we get zero, as long as we average in the same way that was used to obtain (6).

A common goal is to develop of a dynamical model for the evolution of the averaged process $\bar{\mathbf{u}}$, assuming of course that the averaging is not temporal-ergodic (in which case $\bar{\mathbf{u}}$ would be constant in time). We first present an elementary, but illustrative, example. Suppose $u(t)$ is a scalar process satisfying the differential equation $\frac{du}{dt} = u^2$. Rewriting u as average-plus-anomaly, we have $\frac{d(\bar{u}+u')}{dt} = \bar{u}^2 + 2\bar{u}u' + u'^2$. Applying the averaging rule to *both* sides of this expression, we have, after arguing that the average of the time derivative of the anomaly is zero, $\frac{d\bar{u}}{dt} = \bar{u}^2 + \overline{u'^2}$. Use of this expression requires specification of the unknown term $\overline{u'^2}$. Specifications of such unknown averages of functions of anomalies are known as *closure rules*. The critical point is that if the original dynamics are nonlinear, such terms arise in averaged equations for averages. In most problems in fluid dynamics, if ensemble averaging or some related stochastic view of the anomalies is adopted, the terms requiring closure involve various moments of the joint distribution of $\mathbf{u}'$.

Returning to the general formulation, substituting (6) into (1), we have $\frac{D(\bar{\mathbf{u}}+\mathbf{u}')}{Dt} = \mathcal{L}(\bar{\mathbf{u}} + \mathbf{u}')$. Using mathematics and parameterizations of closure rules, (Lesieur, 1990), the equations are reduced to a new system for the average $\bar{\mathbf{u}}$:

$$\frac{D\bar{\mathbf{u}}(\mathbf{x}, t)}{Dt} = \mathcal{G}(\bar{\mathbf{u}}(\mathbf{x}, t); \theta_{\mathcal{G}}) + \bar{\mathcal{H}}(\mathbf{u}'(\mathbf{x}, t; \mathbf{U}_o); \theta_{\mathcal{H}}). \tag{7}$$

Here, $\mathcal{G}$ and $\mathcal{H}$ depend on the type of averaging and closure rules used in the derivation. Closure rules are not derived from first principles and can be rather ad hoc. Often, parameters $\theta_{\mathcal{G}}$, and $\theta_{\mathcal{H}}$ are introduced as well. Note that in ensemble averaging the second term in (7) is an expected value $E[\mathcal{H}(\mathbf{u}'(\mathbf{x}, t; \mathbf{U}_o); \theta_{\mathcal{H}})]$. Throughout such analyses, one assumes the validity of interchanges of order of differentiation and averaging.

For discrete spatial gridding, we obtain a vectorized dynamical system for the evolution of a state vector $\bar{\mathbf{U}}$ (see (5)). In numerical computations it is typical to also discretize time, leading to a discrete space-time dynamical system. There is no universal approach to this step, however. It seems natural to grid the time axis and then temporally average in gridded intervals. However, it is actually more common to use small temporal grids and replace temporal derivatives by their natural discrete approximations. In either case, we represent the resulting system as

$$\bar{\mathbf{U}}_{t+\delta} = \tilde{\mathcal{G}}(\bar{\mathbf{U}}_t; \theta_{\tilde{\mathcal{G}}}) + \tilde{\mathcal{H}}(\mathbf{u}'(\mathbf{x}, t; \mathbf{U}_o); \theta_{\tilde{\mathcal{H}}}), \tag{8}$$

where δ is the gridding time step. Note that the use of numerical time derivatives leads to another source of uncertainty. (In the next section, we assume that the variables have been scaled appropriately so that the time step $\delta = 1$.) For discussion of other sources of uncertainty, see Lesieur (1990, pp 319-320).

The following remarks set-up our approach. The appearance of the quantity $\bar{\mathcal{H}}$ in (7) is interpreted as a need to understand how activity at small scales affects large-scale mean flow. Analogously, $\tilde{\mathcal{H}}$ reflects the impact of *unresolved, subgrid-scale* processes on the gridded flow. However, the equations do not actually contain realized anomalies or subgrid-scale fluctuations, but rather information on their "statistical" behavior. In our analysis, we do not attack (7) or (8) directly. In particular, ensemble averaging, though often viewed as a foundation for turbulence analysis, plays little role here. Alternatively, we deviate from Bathelor's suggestion, as quoted above, that only averages are of interest. Rather, one intent of our approach is to use data to learn about the realized, though unobserved, *actual* anomalies for the process studied. This is a very different viewpoint from the mainstream, where ensemble averaging appears to be viewed as the fundamental notion, and other averagings are used merely as approximations. However, the form of statistical modeling and analysis suggested not only permits estimation of the current flow, but also includes learning about the distributions of both the large scale flow and subgrid-scale processes. That is, we can estimate particular flows based on data without abandoning the more traditional uses of data to estimate statistical properties.

Second, in principle, (7) has a special sort of uniqueness associated with it if the averaging used is ergodic. However, if the averaging is space-time filtering or gridding, as in (8), then the analyses depend critically on the averaging rules, i.e., averaging kernels, grid box size, time step, δ. In particular, use of finer (or coarser) gridding leads to different equations. One can imagine re-filtering anomalies to a finer grid yielding something like $\mathbf{u}' = \overline{\mathbf{u}'} + (\mathbf{u}')'$. This can be repeated over and over again. That is, the process displays *multiresolution* dynamics. This property is critical to turbulence and is one motivation for the use of wavelet models (Wornell, 1993).

21.3   An Operational Approach

We focus on discrete space-time gridding in this development. We allow for contexts in which observations reflecting finer (and/or coarser) scales will be available. Statistical modeling permits observations at the fine, subgrid scale to influence the modeled impacts on the gridded process.

After a brief overview of the Bayesian paradigm, we formulate a stochastic space-time wavelet model capable of capturing behavior at various scales. While guided by physical reasoning, this model does not employ specific physics, as in the "mean-flow plus subgrid processes" viewpoint presented in the previous section. We comment on enhancements of the wavelet modeling strategy that can also employ physics associated with mean-flow models in Section 4.

We note that there may be a philosophical hurdle to overcome in conveying this sort of view to the traditional turbulence community. For us, "going fully stochastic" and thinking like Bayesian statistical modelers means we are: (i) trying to find effective, useful models that describe statistically the state variables, (ii) incorporate our knowledge and observations, and (c) give good predictions. While that is a fine scientific goal, the approach differs from the recipe; obtain the "best" approximation to the true dynamics and use that approximation for prediction.

21.3.1   Bayesian Statistical Models

We now embed the modeling and data analysis problem into the Bayesian paradigm. For general reviews, see Berger (1985) and Bernardo and Smith (1994). For reviews and examples directly related to the modeling notions here, see Berliner (1996), Berliner et al. (1998), Royle et al. (1998), and Wikle et al. (1998a).

In the Bayesian approach all uncertain quantities, generically denoted by $\mathbf{Y}$ (state variables, model parameters, initial conditions, etc.), are viewed as random and endowed with probability model specifications. We think of these probability specifications as forming a joint *prior* distribution, say $\pi(\mathbf{y})$; i.e., they are based on our information and scientific modeling prior to observing a data set. Next, observational data, say $\mathbf{Z}$, are modeled via statistical distributions, conditional on the unknown targets of the inquiry. That is, we formulate a probability distribution $p(\mathbf{z}|\mathbf{y})$. Formal Bayesian learning-from-data is performed by finding the *posterior* distribution $\pi(\mathbf{y}|\mathbf{z})$; Bayes' Theorem is that $\pi(\mathbf{y}|\mathbf{z}) \propto p(\mathbf{z}|\mathbf{y})\pi(\mathbf{y})$. The posterior is the basis upon which inferences about $\mathbf{Y}$ are made. The posterior combines all prior information ("science") with information contained in the data in an efficient manner. In principle, the paradigm is unassailable; Bayes' Theorem is a theorem. In practice, the approach depends on the perhaps difficult specification of the inputs (π, p).

To develop models for complex space-time processes we rely on the hierarchical Bayesian viewpoint; e.g., see Wikle et al. (1998a). First, assume that primary interest is on inferences for the state vectors (see (8)) $\bar{\mathbf{U}}_o, \bar{\mathbf{U}}_1, \ldots, \bar{\mathbf{U}}_T$, where T is some fixed time horizon and $\bar{\mathbf{U}}_o$ is an initial condition. Data of the form $\mathbf{Z} = (\mathbf{Z}_{t_1}, \ldots, \mathbf{Z}_{t_d})$ will be available. Let $p(t_k)$ be the dimension of $\mathbf{Z}_{t_k}$. The extra notation in the time index of the data reflects the fact that data need not be available at every time step of the $\bar{\mathbf{U}}$-process. We do require that $0 \leq t_1$ and $t_d \leq T$.

Specifically, the modeling is done in three basic stages:

1. *Data step:* Formulate $p(\mathbf{Z}|\bar{\mathbf{U}}_o, \bar{\mathbf{U}}_1, \ldots, \bar{\mathbf{U}}_T; \theta_d)$, where θ_d is a collection of parameters associated with the statistics of the measurement process; e.g., variances of measurement errors.

2. *Primary model:* Formulate a statistical model for $\bar{\mathbf{U}}_o, \bar{\mathbf{U}}_1, \ldots, \bar{\mathbf{U}}_T$. In the context of (8), this involves specification of a model $\tilde{\mathcal{G}}$ and specification of a distribution for the η process. The function $\tilde{\mathcal{G}}$ may include unknown parameters, say $\theta_{\tilde{\mathcal{G}}}$. Also, we imagine the inclusion of statistical models parameterized via another collection of parameters, denoted by θ_η.

3. *Prior on parameters:* Finally, we "close" the Bayesian formulation by specifying prior distributions for unknown quantities. In particular, we formulate a prior $\pi(\theta_d, \theta_{\tilde{\mathcal{G}}}, \theta_\eta)$. Note that in general, these parameters can be modeled as time-varying.

21.3.2 Multiresolution Wavelets in Space-Time

We discuss implementations of the hierarchical Bayesian viewpoint outlined in Section 3.1. We begin with the *primary model* referred to as Step 2 above.

Assume that the state vector $\bar{\mathbf{U}}$ evolves dynamically and exhibits complicated and possibly non-stationary behavior in time and space. Further, if $\bar{\mathbf{U}}$ exhibits multiscale "turbulent" scaling behavior over space, it is natural to assume that $\bar{\mathbf{U}}$ can be modeled as a $1/f$ family of processes. That is, a fractal, self-similar process that is generally nonstationary, but, appears stationary when considered through certain spatially-invariant filters. A flexible collection of models to handle these behaviors are *wavelet* models. For discussions in the context of turbulence and general references, see Chin et al. (1998), Katul and Vidakovic (1998), and Wikle et al. (1998b). In this analysis, we use wavelets to capture all important scales; that is, we allow our wavelet model to include phenomena represented in both the $\tilde{\mathcal{G}}$ and $\tilde{\mathcal{H}}$ terms in (8).

Our first task is to specify a probability model

$$[\bar{\mathbf{U}}_o, \bar{\mathbf{U}}_1, \ldots, \bar{\mathbf{U}}_T | \boldsymbol{\theta}_U].$$

There are two ways to develop a simple wavelet model. We present both because they each provide intuition needed in the development of priors.

Model 1. Consider a one-step, Markov vector autoregression model for the evolution of the $\bar{U}$-process (Berliner 1996; Wikle et al. 1998a). Namely, first assume that

$$[\bar{U}_o, \bar{U}_1, \dots, \bar{U}_T | \theta_U] = [\bar{U}_o | \theta_o] \prod_{t=0}^{T-1} [\bar{U}_{t+1} | \bar{U}_t, \theta_{t+1}]. \tag{9}$$

Next, assume that the conditional distributions are based on linear models: for $t = 0, \dots, T - 1$,

$$\bar{U}_{t+1} = \tilde{G}(t+1)\bar{U}_t + \eta_{t+1}, \tag{10}$$

where $\tilde{G}(t)$ is an $n \times n$ matrix of regression coefficients and $\{\eta_t\}$ is a sequence of independent, zero-mean errors with variance-covariance matrices $\Sigma_\eta(t)$.

Next, to enable the projection of the $\bar{U}$-process onto a wavelet basis, assume that for each $0 \leq t \leq T$,

$$\bar{U}_t = \Psi a_t, \tag{11}$$

where Ψ is a *fixed*, $n \times p$ $(p \leq n)$ matrix of wavelet basis functions, and a_t is a $p \times 1$ vector of random wavelet coefficients. This step represents an inverse discrete wavelet transform.

Note that (10) and (11) are not contradictory (under a minor assumption). Suppose that the $\bar{U}_t$ evolve stochastically according to (10). For each sequence, we can solve for the corresponding coefficients in (11), since

$$a_t = J\bar{U}_t,$$

where $J = (\Psi'\Psi)^{-1}\Psi'$. This requires Ψ to be of full rank (i.e., the required inverse exists).

Model 2. Assume first that (11) holds for each t. In a sense then there are "no $\bar{U}$." Rather, we must derive a stochastic model for the coefficients a_t. We again assume a first order vector autoregression:

$$a_{t+1} = \tilde{G}_a(t+1)a_t + \eta_{t+1}^a,$$

where each $\tilde{G}_a(t)$ is an $p \times p$ propagator matrix and the η_t^a are zero-mean, independent error vectors with covariance matrices $\Sigma_{\eta^a}(t)$.

To interrelate these models, first note that under Model 2, (11) implies that for each t, the variance-covariance matrix of $\bar{U}_t$, say $\Sigma_U(t)$, is given by

$$\Sigma_U(t) = \Psi \Sigma_a(t) \Psi'. \tag{12}$$

Further, under Model 1, we can derive the following recursive relationship for the $\Sigma_U(t)$:

$$\Sigma_U(t+1) = \tilde{G}(t+1)\Sigma_U(t)\tilde{G}(t+1)' + \Sigma_\eta(t+1).$$

Similarly, under Model 2, we have that

$$\Sigma_a(t+1) = \tilde{G}_a(t+1)\Sigma_a(t)\tilde{G}_a'(t+1) + \Sigma_{\eta^a}(t+1).$$

If one adds a stationarity-in-time assumptions (e.g., neither propagator matrices nor noise covariance matrices depend on time), the implied stationary covariances are the "fixed points" of these recursions. For example, the fixed point of $\Sigma_a = \tilde{G}_a\Sigma_a\tilde{G}_a' + \Sigma_{\eta^a}$ is representable as

$$vec(\Sigma_a) = [I_{p^2} - \tilde{G}_a \otimes \tilde{G}_a]^{-1}vec(\Sigma_{\eta^a}),$$

if $I_{p^2} - \tilde{G}_a \otimes \tilde{G}_a$ is positive definite.

In some sense Models 1 and 2 ought to be equivalent (i.e., $\eta_t^a = J\eta_t$). Specifically, substituting (11) into (10) and performing some algebra, we obtain

$$\mathbf{a}_{t+1} = \tilde{G}_a(t+1)\mathbf{a}_t + J\eta_{t+1},$$

where $\tilde{G}_a(t) = J\tilde{G}(t)\Psi$. These versions, along with the covariance formulas in the previous paragraph offer relationships and constraints among the required inputs (i.e., models for error covariances and propagator matrices).

An additional enhancement of the modeling is possible. It seems natural to question the assumption that $\bar{U}_t = \Psi\mathbf{a}_t$ exactly, especially if $p < n$. To relax this assumption, we can generalize the model by recasting (11) stochastically: for $0 \le t \le T$,

$$\bar{U}_t = \Psi\mathbf{a}_t + \omega_t,$$

where the ω_t are zero-mean, independent error vectors with covariance matrices $\Sigma_\omega(t)$. For brevity, we do not pursue this model here.

21.3.3   Data Models

A critical point in motivating the Bayesian approach is that we can incorporate observational data at arbitrary scales in the estimation of the $\bar{U}$-process in space and time. To that end, we formulate a probability model $[Z_{t_1}, \ldots, Z_{t_d}|\bar{U}_o, \bar{U}_1, \ldots, \bar{U}_T; \theta_d]$. To indicate the main issues, we assume that

$$[Z_{t_1}, \ldots, Z_{t_d}|\bar{U}_o, \bar{U}_1, \ldots, \bar{U}_T; \theta_d] = \prod_{k=1}^{d}[Z_{t_k}|\bar{U}_{t_k}; \theta_d]. \qquad (13)$$

That is, observations from time period to time period are conditionally independent and only depend on the "current" value of the $\bar{U}$-process. To formulate the terms on the left hand side of (13), we must answer questions like "What do our observations measure?" "What biases are present?" "What is the precision of the measurement device?"

A popular and plausible model is to assume that measurement errors are Gaussian. In tune with the averaging notions discussed in this paper, two approaches suggest themselves:

1. Assume that

$$\mathbf{Z}_{t_k} | \bar{\mathbf{U}}_{t_k} ; \theta_d \sim \text{Gau} \left(\mathbf{K}_{t_k} \bar{\mathbf{U}}_{t_k}, \Sigma_d(t) \right),$$

where $\mathbf{K}_{t_k}$ is a $p(t_k) \times n$ matrix which "maps" the gridded $\bar{\mathbf{U}}$-process to the observations, and $\Sigma_d(t)$ is the covariance matrix of the measurement errors.

2. Assume that

$$\mathcal{K}_{t_k} \mathbf{Z}_{t_k} | \bar{\mathbf{U}}_{t_k} ; \theta_d \sim \text{Gau} \left(\bar{\mathbf{U}}_{t_k}, \Xi_d(t) \right),$$

where $\mathcal{K}_{t_k}$ is an $n \times p(t_k)$ matrix which "maps" the observations to the $\bar{\mathbf{U}}$ grid.

In either strategy, both the mapping matrices and the measurement error covariances vary in time and can be parameterized as functions of unknown quantities (which in turn are endowed with priors and updated in the Bayesian analysis). Neither strategy imposes severe limitations in implementation; hence the choice should be based on the analyst's understanding of the measurement process.

21.3.4   Priors

Prior Information about Covariances. First, we briefly mention a standard idea in the turbulence literature. It is common in the physical sciences to examine the behavior of the *kinetic energy* (KE) of the system under study. Of course, KE is physically important. Further, since KE is essentially the square of velocity, when an "average KE" is defined appropriately, it can be related to second moment (and, hence, variance-covariance) structures of velocity fields. There is substantial experience in doing this in the spectral domain. After a Fourier transform, information about the spatial structure of KE is represented in a (spatial) spectrum. This object is used in the next argument.

Recall that we seek to model multiscale, "turbulent" scaling behavior over space. Pursuing the notion of a $1/f$ family of processes, such processes are typically described as following a power law relationship

$$S_U(k) \propto \frac{\sigma_U^2}{|k|^\kappa}, \tag{14}$$

where $S_U(k)$ is the (spatial) power spectral density of $\bar{\mathbf{U}}$ at wave number k, σ_U^2 is the variance of the $\bar{\mathbf{U}}$-process, and κ is the spectral parameter. Such relationships are prevalent in the natural world (e.g., Van Vliet, 1987) and have been associated with turbulence theory for many years (Kolmogorov,

1941 a,b). Typically, these processes exhibit both scale invariance and, in some sense, spatial invariance. It has been shown (Wornell, 1993) that multiresolution bases such as wavelets are ideally suited for modeling these processes.

Hence, we must rectify the spatial information reflected in (14) with the temporal evolution structure reflected in (12). To that end, we note that Chin et al. (1997) argued that a simple diagonal structure for $\Sigma_a(t)$ with variances decreasing proportionally to scale according to

$$\sigma_a^2(l) \propto 2^{-l(1+\kappa)-1},$$

where l is the level of the multiresolution decomposition, corresponds to a 2-dimensional fractal process, i.e., a spatial fractal process. Hence, we have a suggestion that diagonal $\Sigma_a(t)$ are plausible and further, we can control the variability in the the elements of $\Sigma_a(t)$ to match roughly the turbulence scaling law.

21.3.5   Example

Surface winds over the oceans are critical to climate and weather processes since they affect exchanges of heat, moisture, and momentum between the atmosphere and ocean over many scales of spatial and temporal variability. However, there are no spatially and temporally complete observations of such winds. In recent years, satellite-derived wind observations have become available. These estimates have high spatial resolution but cover only a limited area at any one instant in time. In contrast, the major weather centers produce wind fields over the ocean that are largely determined by deterministic models (guided by sparse observations). These *analysis fields* have global coverage but at low spatial resolution. Our goal is to combine these data sets via the spatio-temporal statistical model described in Section 3.2 such that the blended fields contain realistic physical signals from the smallest to largest scales.

We consider satellite data from the NASA scatterometer (NSCAT) instrument during a period from 28 October 1996 to 10 November 1996. We also consider analyzed wind fields from the National Centers for Environmental Prediction (NCEP) for the same period. Because it is an important region for many weather and climate phenomena, we consider these winds over the equatorial western Pacific ocean. Specifically, we develop spatio-temporal predictions on a 64 by 32 dimensional grid at one degree resolution in latitude and longitude and at six hourly time intervals.

We selected wavelet basis functions (Ψ) from the Daubechies class with two vanishing moments, including a modification for closed domains (e.g., Cohen et al. 1993). We modified the model of Section 3.2 by including a purely spatial mean process in the model. This addition makes good sense physically; there are "climatological" differences among subareas of the region of studied. Furthermore, in developing prior distributions for the innovation covariance matrix Σ_η, we noted that the energy spectrum of tropical surface winds has been shown to behave like a self-similar random fractal process (Wikle et al. 1998c), in which the parameter κ was shown to be approximately -5/3. This spectral decay rate is consistent with theoretical results in turbulence theory as discussed in Wikle et al (1998c). Both our "data models" and Bayesian updating, via Markov chain Monte Carlo, are direct analogues of a similar analysis described in Wikle et al. (1998b).

Winds are vector quantities and can be reported as speed and direction or as east-west (u) and north-south (v) components. Figure 1a shows the u-component of the wind from the NCEP analysis wind field at 12:00 UTC on 7 November 1996. This field is known to be unrealistically smooth (e.g., Wikle et al 1998c). Figure 1b shows the posterior mean of the large-scale spatial mean of our blended wind field for the same time. The posterior mean of the multiresolution wavelet portion of our blended field is then shown in Figure 1c. Finally, the complete (spatial mean plus the wavelet contribution) blended u-component posterior mean field for this time is shown in Figure 1d. We note that this blended field contains much more information at intermediate and small scales than the analysis wind in Figure 1a.

Figure 2 demonstrates the validity of the posterior mean wind from our model by comparing surface convergence estimated from the posterior wind analysis with independent cloud top temperature data during the life-cycle of Tropical Cyclone Dale (12:00 UTC, 7 November 1996). The posterior mean winds in Figure 2(b) can be used to compute surface convergences that correspond roughly to intense convection regions (coldest cloud top temperatures) of the cyclone (Figure 2c). We expect such correspondence since convergence implies the sort of upward vertical motion necessary for vigorous convection. These correspondences are absent completely from the convergence of the weather-center wind field (Figure 2a).

21.4   More General Models

As before, we begin with a one-step Markovian assumption (see (9)). Further, we consider a stochastic dynamical system for the evolution $\bar{U}_t$ as a replacement for (8):

$$\bar{U}_{t+1} = \tilde{G}_{t+1}(\bar{U}_t; \theta_{\tilde{G}}) + \eta_{t+1}, \qquad (15)$$

Here $\{\eta_t\}$ is a stochastic process representing sub-grid scale noise. We consider $(\tilde{G}; \tilde{H})$ (or $(G; H)$) and $(\tilde{G}; \{\eta_t\})$ specifications to come in pairs. The η-process is an attempt to model the realized sub-grid scale anomalies. The statistical properties of the sub-grid depend on the form of the grid model, $\tilde{G}$.

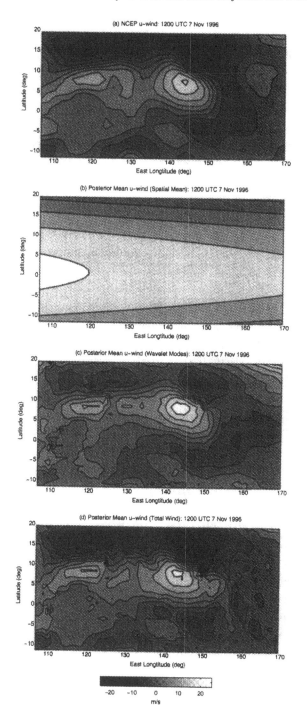

FIGURE 1. For 12:00 UTC on 7 November 1996: (a) u-component of wind from NCEP analysis wind field; (b) posterior mean of the spatial mean; (c) posterior mean of the multiresolution wavelet contribution; (d) u-component posterior mean field.

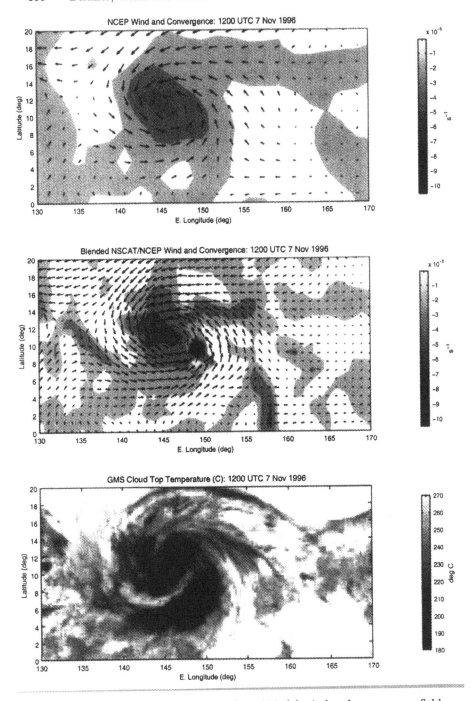

FIGURE 2. For 12:00 UTC on 7 November 1996: (a) wind and convergence fields based on NCEP analysis; (b) posterior estimates of wind and convergence fields from our model; (c) cloud top temperatures.

Substantial flexibility is possible in formulating models like (15), beyond the vector autoregression forms of the previous section. First, we see opportunities to let the η-process include model approximation error, external random forcing, etc. Second, one can include time-varying dynamics in the strategy by allowing the model $\tilde{\mathbf{G}}$ and/or the parameters $\theta_{\tilde{G}}$ to vary in time. Finally, we need not restrict to "additive" effects; i.e., we could consider models $\bar{\mathbf{U}}_{t+1} = \tilde{\mathbf{G}}(\bar{\mathbf{U}}_t, \eta_{t+1}; \theta)$.

Following the notions of modeling both large and small scale behavior, we suggest choosing the $\tilde{\mathbf{G}}_t$ to capture large scale dynamical physics. This can be accomplished in a variety of ways. For example, $\tilde{\mathbf{G}}_t$ might be specified to actually be the nonlinear, discretized evolution operators, $\tilde{\mathcal{G}}$ introduced in (8). Alternatively, $\tilde{\mathbf{G}}_t$ could be developed from numerical solutions of an approximate physical model. We have in mind a a model less complex than the discretized form in (8), yet one that may capture significant physics and be amenable to extensive exploratory computation. (See Wikle et al. (1998b) for an example.) The essential idea is to not project the original state variables onto a wavelet basis, but rather use quantitative information to explain large-scale features and then project the unexplained, yet spatially and temporally coherent, noise (η) onto a wavelet basis.

21.5   Discussion

In continuous-time, stochastically-driven models, the temporal evolution of the system is described probabilistically via solution of a Fokker-Planck equation. Specifically, we suppose that the process is a solution to a system of stochastic differential equations. That model incorporates both a dynamical system component and a stochastic term, used to describe uncertainty in the model and external noise. A probability distribution (say p_0) for the initial state is prescribed. Under appropriate conditions, a Fokker-Planck equation is developed to describe the evolution of the probability density function for the state vector of the system. This equation is a deterministic partial differential equation, and is solved with initial data (p_0). See Lasota and Mackey (1994) for more discussion. We note that a special case of the Fokker-Planck, the Liouville equation, describes the temporal evolution of the probability density function for the state vector of an uncorrupted system; namely, a deterministic system whose initial state is modeled as a random quantity (Ehrendorfer, 1994).

There are direct analogies with standard expressions for the continuity of mass in the equations of fluid motion. The term in the Fokker-Planck equation corresponding to the stochastic portion of the model takes the form of a Fickian diffusion operator balancing the continuity operator for the classical case of a white noise stochastic process. In this case, a direct analogy with the Brownian motion model for molecular diffusion pertains.

However, for practical problems of geophysical fluid flows, the dimensions of typical state vectors are enormous and explicit analytic solutions are not feasible.

These formulations can be incorporated in a Bayesian analysis (e.g., Berliner, 1996, and Miller et al., 1998). The essential idea is that the Fokker-Planck equation provides a primary model, as in our Section 3.1, for a continuous time process.

Acknowledgments: Support for this research was provided by the NCAR Geophysical Statistics Project, sponsored by NSF Grant DMS93-12686; the NCAR NSCAT Science Working Team cooperative agreement with NASA JPL; and the National Institute of Statistical Sciences.

References

Batchelor, G.K., 1953: *The Theory of Homogeneous Turbulence.* Cambridge University Press.

Batchelor, G.K., 1967: *An Introduction to Fluid Dynamics.* Cambridge University Press.

Berger, J. O., 1985: *Statistical Decision Theory and Bayesian Analysis.* New York: Springer-Verlag.

Berliner, L.M., 1996: Hierarchical Bayesian time series models. In *Maximum Entropy and Bayesian Methods*, K. Hanson and R. Silver (eds.), Kluwer Academic Publishers, Dordrecht, 15-22.

Berliner, L.M., J.A. Royle, C.K. Wikle, and R.F. Milliff, 1998: Bayesian Methods in the Atmospheric Sciences. In *Bayesian Statistics 6*, J.M. Bernardo, J.O. Berger, A.P. Dawid, and A.F.M. Smith (eds.), Oxford University Press, in press.

Bernardo, J. M. and Smith, A. F. M., 1994: *Bayesian Theory*, New York: Wiley.

Chin, T.M., Milliff, R.F., and W.G. Large, 1998: Multiresolution analysis of scatterometer winds and high-wavenumber effects on ocean circulation. *J. Atmospheric and Oceanic Technology*, **15**, 741-763.

Cohen, A., Daubechies, I., and Vial, P., 1993: Wavelets on the interval and fast wavelet transforms. *Applied and Computational Harmonic Analysis*, **1**, 54-81.

Ehrendorfer, M., 1994: The Liouville equation and its potential usefulness for the prediction of forecast skill. Part I: Theory. *Mon. Weather Rev.* **122**, 703-713.

Katul, G. and Vidakovic, B., 1998: Identification of low-dimensional energy containing/flux transporting eddy motion in the atmospheric surface layer using wavelet thresholding methods. *Journal of the Atmospheric Sciences* **55**, 377-389.

Kolmogorov, A. N., 1941a: The local structure of turbulence in incompressible viscous fluid for very large Reynolds numbers. *Dokl. Akad. Nauk. SSSR*, **30**, 301-305.

Kolmogorov, A. N., 1941b: On degeneration of isotropic turbulence in an incompressible viscous liquid. *Dokl. Akad. Nauk. SSSR*, **31**, 538-541.

Lasota, A. and Mackey, M.C., 1994: *Chaos, Fractals, and Noise: Stochastic Aspects of Dynamics.* Springer-Verlag, New York.

Lesieur, M., 1990: *Turbulence in Fluids: Stochastic and Numerical Modeling.* Boston: Kluwer Academic Publishers.

Miller, R.N., Carter, E.F., and Blue, S.T., 1998: Data assimilation into nonlinear stochastic models. *Tellus*, in press.

Royle, J.A., Berliner, L.M., Wikle, C.K., and R. Milliff, 1998: A hierarchical spatial model for constructing wind fields from scatterometer data in the Labrador Sea. *Case Studies in Bayesian Statistics, IV*, C. Gatsonis et al. (eds.), New York: Springer-Verlag.

Wikle, C.K., Berliner, L.M., and N. Cressie, 1998a: Hierarchical Bayesian space-time models. *Journal of Environmental and Ecological Statistics* **5**, 117–154.

Wikle, C.K., R.F. Milliff, D. Nychka, and L.M. Berliner, 1998b: Spatio-Temporal hierarchical Bayesian blending of tropical ocean surface wind data. submitted.

Wikle, C.K., R.F. Milliff, and W.G. Large, 1998c: Surface wind variability on spatial scales from 1 to 1000 km observed during TOGA COARE. *Journal of the Atmospheric Sciences*, in press.

Wornell, G. W., (1993): Wavelet-based representations for the $1/f$ family of fractals processes. *Proceedings of the IEEE* **81**, 1428-1450.

22

Low Dimensional Turbulent Transport Mechanics Near the Forest-Atmosphere Interface

Gabriel Katul and John Albertson

ABSTRACT Turbulent velocity, temperature, and water vapor concentration fluctuations time series were collected at the canopy-atmosphere interface of a 13 m tall uniform pine forest and a 33 m tall non-uniform hardwood forest. These measurements were used to investigate the existence of low- dimensional eddy motion (i.e. eddy motion described by few modes) responsible for much of the momentum and passive scalar transport mechanics. It is demonstrated that the origins of such "low dimensional" eddy-motion are Kelvin-Helmholtz like instabilities with well-defined characteristic length scale that can be computed via wavelet spectra of vertical velocity and cospectra of vertical velocity and other flow variables. A newly proposed filtering approach, which relies on Lorentz wavelet threshold of vertical velocity time series, provides simultaneous energy/covariance preserving characterization of vertical flux transporting eddies at the canopy-atmosphere interface.

22.1 Introduction

Understanding momentum, heat, and scalar mass exchanges between vegetation and the atmosphere is central to a wide range of atmospheric, ecological, and hydrological research thrusts. The estimation of (i) sensible heat for modeling the growth and decay of the atmospheric boundary layer, (ii) carbon dioxide uptake by forests for modeling carbon budgets, and (iii) evaporation in hydrologic budgets all require knowledge of momentum and scalar mass transfer from the land surface into the atmosphere. In the past two decades, many field experiments conducted above natural vegetation concluded that such exchange processes are governed by organized eddy motion that exhibits many universal characteristics (Thompson, 1979; Mulhearn and Finnigan, 1978; Amiro, 1990; Amiro and Davis, 1988; Moritz, 1989; Gao et al., 1992; Baldocchi and Meyers, 1988; 1989; Wilson, 1989; Paw U et al., 1992; Raupach et al., 1996). In turbulent flows, eddies are organized rotational structures embedded in the flow (e.g. swirls and whirls). Many of these studies primarily used temperature time series to identify self-organization in heat transporting eddy motion. Such time series commonly exhibits ramp-like signatures for unstable atmospheric conditions. The ramp-like structure identification is performed via conditional sampling analysis following a number of arbitrary conditioning criteria (see e.g. Shaw et al., 1989). In conditional analysis, ramps are identified by comparing rapid excursions in the temperature time series with a preset threshold. The resultant ramp-like structures are then used to identify their "mirror" events in the vertical velocity time series to describe types of vertical eddy motion (e.g. sequence of updrafts and downdrafts events). An argument for using temperature time series as a "tracer" for organized eddy motion (vis-á-vis vertical velocity) is that vertical velocity time series generally does not admit the same level of organization as the temperature time series upon visual inspection. Recently, Raupach et al. (1996) suggested that the eddy sizes responsible for the generation of coherency and spectral peaks in the vertical velocity time series are strongly coupled to eddy motion responsible for momentum transport near the canopy-atmosphere interface. Also, such coherency and spectral peaks in the vertical velocity time series exhibits many universal characteristics that can be well predicted from hydrodynamic stability theory (HST). In light of the recent findings by Raupach et al. (1996), it is possible that organized low dimensional eddy motion be completely embedded in the vertical velocity time series. Yet, such eddy-motion is not well revealed in the time domain as previously discussed. Whether appropriate transformation to another domain permits extracting such low dimensional eddy motion has not been explored and has motivated the present study.

The objective of this study is to investigate whether flux transporting eddies (hereafter referred to as active turbulence) for momentum and scalars can be directly inferred from vertical velocity time series without any reference to the scalar concentration time series. Here, eddy sizes responsible for co-spectral peaks of scalar fluxes and their relationship to active turbulence are considered. Active turbulence is identified from orthogonal wavelet decomposition that concentrates much of the vertical velocity energy in few wavelet coefficients. The remaining wavelet coefficients associated with "inactive", "wake", and "fine scale-turbulence" are thresholded using a Lorentz wavelet approach advanced by Vidakovic (1995) and Katul and Vidakovic (1996, 1998). Since canopy sublayer (CSL) turbulence is intermittent in the time domain with defined spectral properties in the Fourier domain, we argue that orthonormal wavelet decompositions permit a simultaneous time-frequency investigation of both flow characteristics. Much of the flow statistics derived by Raupach et al. (1996) in the time domain are extended to the wavelet domain. Specific attention is devoted to wavelet spectra and co-spectra used to investigate the characteristics of active turbulence for two morphologically distinct forest stands (i.e. different canopy heights, uniformity, leaf physiological properties, and leaf area distribution) and for a wide range of atmospheric stability conditions. Unless otherwise stated, the following notation is used throughout: U_i ($U_1=U$, $U_2=V$, and $U_3=W$) are the instantaneous longitudinal (U), lateral (V), and vertical (W) velocity components; C is the instantaneous concentration of a scalar entity ($C= T_a$, Q for air temperature and water vapor concentration); x_i ($x_1=x$, $x_2=y$,$x_3=z$) are the longitudinal (x), lateral (y), and vertical (z) directions, with z defined from the ground surface; $< . >$ is time averaging; u_i and c are the turbulent fluctuations around $< U_i >$ and $< C >$ so that $< u_i >=< c >=0$; and the x direction is aligned along the mean horizontal wind direction so that $< V >=0$.

22.2   Method of Analysis

Much of the turbulent transport at the canopy-atmosphere interface is dominated by the passage of transient unsteady gusts in time (Raupach et al., 1996; Paw U et al., 1992; 1995; Bergstrom and Hogstrom, 1989). Spectral measurements of many turbulent flow variables also suggest that turbulent energy is concentrated in low frequency (or larger-scale) eddies. Hence, it is natural to consider a wavelet transformation with sufficiently localized basis function in the time domain for analyzing turbulent flow measurements. In fact, Katul and Vidakovic (1996, 1998) demonstrated that orthonormal wavelets are better suited to concentrate much of the turbulent energy in few wavelet coefficients when compared to their Fourier counterparts. Much of the derivations below are described in Katul and Vidakovic (1996,1998) and Katul et al. (1998 a,b) but for completeness, key concepts are reviewed.

22.2.1   Fast Wavelet Transform (FWT)

The Haar wavelet basis is selected for (i) its differencing characteristics, (ii) its locality in the time domain, (iii) its short support which eliminates any edge effects in the transformed series, and (iv) its wide use in atmospheric turbulence research (Howell and Mahrt, 1994; Katul and Parlange, 1994; Katul and Vidakovic, 1996, 1998; Lu and Fitzjarald, 1994; Turner and Leclerc, 1994; Turner et al., 1994; Mahrt and Howell, 1994; Yee et al., 1996; Mahrt, 1991; Katul et al., 1998a,b). The Haar wavelet coefficients (WC) and coarse grained signal (or scaling function coefficients) ($S^{(m)}$) can be calculated using the usual multiresolution decomposition on a measurement vector ($f(x_j)$) originally stored in $S(0)$ for $m=0$ to $M-1$, where $< S(0) >=0$ and $M=log_2(N)$ is the total number of scales. The decomposition produces $N-1$ wavelet coefficients defining the Haar wavelet transform of $f(x_j)$ (see e.g. Katul et al., 1994;Katul and Parlange, 1994; and Katul and Vidakovic, 1996).

22.2.2   Wavelet Spectra and Cospectra

The N-1 discrete Haar wavelet coefficients satisfy the energy conservation

$$\sum_{j=0}^{j=N-1} f(j)^2 = \sum_{m=1}^{m=M} \sum_{i=0}^{i=2^{(M-m)}-1} (WC^{(m)}[i])^2,$$

which is analogous to Parseval's identity. The total energy T_E contained in a scale R_m is computed from the sum of the squared wavelet coefficients (recall $< S(0) >= 0$) at a scale index (m) using

$$T_E(R_m) = \frac{1}{N} \sum_{i=0}^{i=2^{M-m}-1} (WC^{(m)}[i])^2$$

where $R_m = 2^m dy$ (with $dy = f_s^{-1} < U >$) is typically estimated using Taylor's (1938) hypothesis and later adjusted by the ratio of convective to time-averaged velocities, and f_s is the sampling frequency (typically $10Hz$ for this experiment). From this identity, the Haar wavelet power spectrum (energy per unit wavenumber) for a given angular wavenumber K_m corresponding to scale R_m is (Meneveau, 1991)

$$K_m = \frac{2\pi}{R_m}$$

Hence, an eddy of wavenumber K_m may be thought of as some disturbance containing energy in the vicinity of K_m. Hereafter, K refers to wavenumber for representation simplicity unless the discrete K_m definition is required in a derivation. Hence, the power spectral density function $P(K_m)$ is computed by dividing T_E by the change in wavenumber $\Delta K_m (= 2\pi 2^{-m} dy^{-1} ln(2))$ so that

$$P(K_m) = \frac{<< (WC^{(m)}[i])^2 >> dy}{2\pi 2^{-m} ln(2)}$$

where $<< . >>$ is averaging over all values of the position index $[i]$ at scale index (m) (see Szilagyi et al., 1996 for discussion on alternative definition of ΔK_m). Notice that the wavelet power spectrum at wavenumber K_m is directly proportional to the average of the squared wavelet coefficients at that scale. Because the power at K_m is determined by averaging many squared wavelet coefficients, the wavelet power spectrum is generally smoother than its Fourier counterpart (see Katul and Parlange, 1994, Katul and Chu, 1998) and does not require the usual windowing and tapering.

The Haar wavelet co-spectrum of two turbulent flow variables (e.g. w and c) can be calculated from their respective wavelet coefficient vectors ($WC_u^{(m)}$ and $WC_c^{(m)}$) at wavenumber K_m using

$$P_{wc}(K_m) = \frac{<< WC_w^{(m)}[i] WC_c^{(m)}[i] >>}{2\pi 2^{-m} ln(2)}$$

where,

$$< wc > = \frac{1}{N} \sum_{m=1}^{m=M} \sum_{i=0}^{i=2^{M-m}-1} (WC_w^{(m)}[i] WC_c^{(m)}[i])$$

as discussed by Hayashi (1994), Katul and Parlange (1994, 1995), Katul and Vidakovic (1996), and Yee et al. (1996). The above wavelet expansion does preserve variances and co-variances; hence, it differs from the continuous wavelet analysis used by Brunet and Collineau (1994), Collineau and Brunet (1993), Gao and Li (1993), Lu and Fitzjarrald (1994), Qiu et al. (1995), Turner and Leclerc (1994), and Turner et al. (1994) in which wavelet analysis was used to identify specific structural elements in the turbulence measurements without regards to variance and covariance preservation.

22.2.3 Wavelet Thresholding

In Vidakovic (1995) and Katul and Vidakovic (1996, 1998), it is demonstrated that discrete orthonormal wavelet transformation disbalances energy of turbulent flow variables but preserves their variances. That is, energy of the transformed time series is concentrated in only few wavelet coefficients (usually less than 10% for many turbulent flow variables). Such a disbalance between number of wavelet coefficients containing energy and dimension (=N) offers a practical method to extract "active" turbulence from vertical velocity measurements. The energy extraction is based on the premise that much of the energy in vertical velocity time series is due to active turbulence (Raupach et al., 1996). Extracting energy containing events from their respective wavelet coefficients can be readily achieved via Lorentz wavelet filtering. Prior to discussing Lorentz wavelet filtering, the distinction between active and inactive eddy motion is first considered.

Active/Inactive Eddy Motion

The decomposition of eddy motion into active and inactive was first proposed by Townsend (1961, 1976) for a region in the atmosphere in which turbulent fluxes do not vary appreciably with height (see Katul et al., 1996a; Katul and Vidakovic, 1996, 1998 for review). Commonly, this region is referred to as equilibrium or constant flux region. Townsend's (1961) hypothesis decomposes eddy motion into an active component, which is a function of ground shear stress per unit density (u_*) and height (z), and an inactive component, which is produced by turbulence in the outer region (i.e. a region well above the constant flux layer). The inactive component is partly produced by the irrotational motion due to pressure fluctuations and by the large-scale vorticity field in the outer layer, which the equilibrium region senses as rapid variability in the mean flow (Katul et al., 1996a; and Katul and Chu, 1998 for review). Raupach et al. (1991) suggested that inactive eddy motion scales with u_* and boundary layer depth (see Katul et al., 1996a for details). In the context of canopy turbulence, eddy-motion responsible for much of the vertical transport is termed as "active" while eddy motion with scales much larger than the canopy height are termed inactive. The characteristics of active turbulence can be well described from a mixing layer analogy considered next.

The Mixing Layer Analogy

Raupach et al. (1996) proposed a mixing layer (ML) analogy to arrive at the universal characteristics of active turbulence close to the canopy atmosphere interface in uniform and extensive canopies. Their analogy is derived from solutions to the linearized perturbed two- dimensional inviscid momentum equations (Rayleigh's equation) using hydrodynamic stability theory. HST predicts unstable mode generation of two-dimensional transverse Kelvin-Helmholz (KH) waves with streamwise wavelength (Λ_x) if the longitudinal velocity profile has an inflection point (a necessary but not sufficient condition). Raupach et al. (1996) argued that such instabilities are the origins of organized eddy motion in plane mixing layers; however, a KH eddy motion cannot be produced or sustained in boundary layers due to the absence of such an inflection point in the velocity profile. A plane mixing layer is a "boundary-free shear flow" formed in a region between two co-flowing fluid streams of different velocity but same density (Tennekes and Lumley, 1972 p. 104; Raupach et al., 1996). Raupach et al. (1996) suggested that a strong inflection point in the mean velocity profile at the canopy-atmosphere interface results in a flow regime resembling a mixing layer rather than a boundary layer neighboring this interface. Raupach et al.'s (1996) mixing layer analogy is considered the first theoretical advancement to analyzing the structure of turbulence close to the canopy-atmosphere interface of a horizontally extensive uniform forest. We use the mixing layer analogy to verify that the wavelet thresholded time series resembles active turbulence.

Lorentz Thresholding

A measure of energy disbalance for a given measurement vector ($f(i)$, $i=1,2,..,N$) in a particular domain whether be it time, wavelet, or Fourier is the Lorentz curve $L(p)$. The $L(p)$ curve is a convex curve describing the cumulative energy that is contained in p smallest energy components. The $L(p)$ curve is computed by ranking, in ascending order, the energy (squared amplitude) in the wavelet coefficients and plotting the cumulative loss in energy (ordinate axis) incurred upon removing sequentially one coefficient until all N coefficients are removed (plotted on abscissa). A sample Lorentz curve is shown in Figure 1 for which the abscissa p is the cumulative number of coefficients thresholded (in fraction) and the ordinate $L(p)$ is the cumulative loss in energy (in fraction). That is, when all the coefficients are thresholded (abscissa is unity), the cumulative loss in energy is unity.

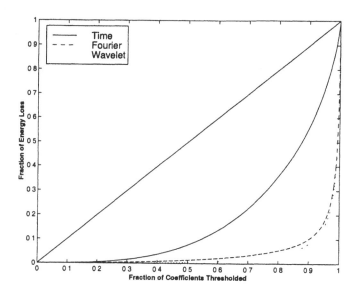

FIGURE 1. Sample Lorentz Curve for the vertical velocity time series in the time, Fourier and Haar wavelet domains. The diagonal corresponds to perfectly balanced (zero variance) time series

In Figure 1, the convexity in $L(p)$ relative to the diagonal line (ideally balanced energy) is directly proportional to energy disbalance. Commonly, energy disbalance is quantified by the area between $L(p)$ and the diagonal line. The more the signal is disbalanced (i.e. energy concentrated in fewer coefficients), the larger the area between $L(p)$ and the diagonal. Notice in Figure 1 that energy disbalance in the wavelet domain is larger than its time or Fourier counterpart. Typically, high energy disbalance in the time domain is associated with strong intermittency so that it suffices to retain only a small fraction of time intervals within a time series, while high energy disbalance in the Fourier domain is associated with concentration of energy at low wavenumbers so that it suffices to retain only few Fourier modes. Since canopy turbulent flows are both intermittent in time and exhibit energy concentration at low wavenumbers, it is natural that energy be concentrated in even fewer wavelet modes given that wavelet decompositions are sensitive to the signal time-frequency properties.

The well defined and unique reference point for energy balance in the wavelet Lorentz curve $L(p)$ suggest an objective method to threshold wavelet coefficients. This point can be identified as the point for which the $L(p)$ curve slope is unity. Hence, in Lorentz thresholding, wavelet coefficients are set to zero if

$$WC_k^2 < \beta; \beta = \frac{1}{N} \sum_{k=1}^{k=N} WC_k^2$$

where WC_k $(k=1,...,N)$ are the N wavelet coefficients of some turbulent flow variable. This thresholding criterion is based on the premise that gain in parsimony by thresholding an additional coefficient is exactly balanced by loss in energy (Vidakovic, 1995; Katul and Vidakovic, 1996). Since "active" motion is the main energy contributor to w (σ_w^2) and since Lorentz thresholding retains only large amplitude wavelet coefficients, then the non-zero wavelet coefficients are associated with "active turbulence". In analogy to conditional sampling, it is convenient to define the relative frequency D_w as the percent ratio of non-zero wavelet coefficients to total wavelet coefficients. In wavelet thresholding, D_w is considered as a measure of compression efficiency (Farge et al., 1992).

To graphically demonstrate the effects of Lorentz wavelet thresholding on a given vertical velocity time series, Figure 2 displays measured (top left) and wavelet thresholded (top right) vertical velocity a 27 minute run from a hardwood forest at the canopy-atmosphere interface collected on July 8, 1996 around noon. The wavelet thresholded time series is computed by wavelet transforming the original time series, applying Lorentz threshold, and finally inverse transforming the thresholded series back to the time domain. For this run, 91% of the wavelet coefficients were set to zero (i.e. $D_w=0.09$). The remaining 9%, which contain 91% of the variance, were used to reconstruct the time series from non-zero wavelet coefficients. The reconstruction is also shown for two sub-sampling intervals to better illustrate reconstructed rapid changes in vertical velocity. We associate the wavelet thresholded time series with "active turbulence". We note that an earlier study by Howell and Mahrt (1994) established the filtering potentials of multiresolution wavelet analysis to extract energy containing events. The above approach differs from their adaptive multiresolution filtering scheme since neither a window width (i.e. constant cutoff scale) nor a small-scale cutoff variance are needed.

22.3   Experimental Setup

Eddy correlation measurements from two forest experiments at the Blackwood Division of the Duke Forest ($35°58'N$, $79°08'W$, Elevation $= 163m$) near Durham, North Carolina were collected and analyzed for this study.

22.3.1   Pine Forest Experiment

These measurements were collected in 1995 and 1996 for few days but at different seasons. The site is a uniform-even aged managed Loblolly pine forest that extends 1000 m in the north-south direction and at least 300 m in the east-west direction. The stand, originally grown from seedlings planted at $2.4m$ by $2.4m$ spacing in 1983, is $14m$ $(=h)$ tall (Ellsworth et

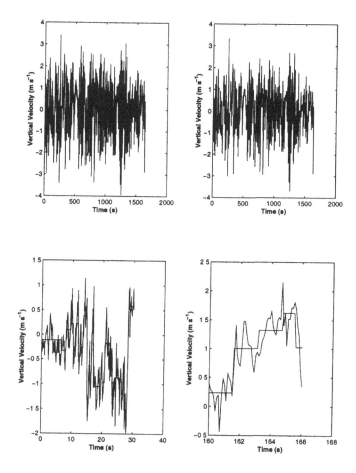

FIGURE 2. The effect of Lorentz thresholding on vertical velocity time series. Top Left panel is the original time series, Top Right panel is the Lorentz thresholded (in the wavelet domain) time series, Lower Left and Lower Right panels demonstrate the effect of thresholding on rapid transients of the signal, with the step functions representing the thresholded and reconstructed time series.

al., 1995; Katul et al., 1997a). Eddy correlation instruments, consisting of a Campbell Scientific Krypton ($KH2O$) hygrometer co-located with a triaxis Gill sonic anemometer having a path length d_{sl}=0.149m, were positioned on a $22m$ aluminum walkup tower at z/h=1. The sonic anemometer was placed 150cm away from surrounding leaves to avoid potential sonic wave reflection by pine needles and minimize the "wake turbulence" effects. The U, V, W, T, Q time series were sampled at $10Hz$ and segmented into runs for which the sampling duration per run (T_p) was 27.3 minutes. The run fragmentation resulted in $16,384$ ($=2^{14}$) measurements per flow variable per run (M=14). The raw measurements for each run were then transformed so that the mean longitudinal velocity $< U >$ was aligned along the mean horizontal wind direction and $< V >$=0. In general, data collection started around 0900 DST and was terminated at about 1800 DST unless storms or rain events were forecasted the previous day. The amount of runs collected each day varied from 8 to 19 runs (see Katul et al., 1997b, 1998b for further details).

22.3.2   Hardwood Forest Experiment

This stand is a $33m$ tall (standard deviation in height is $8m$) uneven-aged mature second growth deciduous hardwood mix with the oldest individual exceeding $180years$ (Katul et al., 1997b). At this stand, a $40m$ tall aluminum walkup tower was used to mount eddy correlation instruments. The tower is in a flat area known as the Meadow Flats, which describes the topography for $500m$ in all directions (Conklin, 1994; Katul et al., 1997b). In this data set, $10Hz$ velocity, temperature, and water vapor measurements were collected using the same instruments from the pine experiment at $z/h = 1.0$. These instruments were moved from the pine forest, positioned at the hardwood forest on June 16, 1996, and operated from June 17 to July 11.

The basic distinction between these two stands is best illustrated by the vertical variation in leaf area density ($a(z)$) normalized by the total leaf area index (LAI) and is shown in Figure 3. Notice that for the hardwood forest (LAI=6.0), much of the foliage is concentrated in the upper third of the canopy, while for the pine forest (LAI=3.1), much of the foliage is concentrated in the middle layers of the canopy. The combined total number of runs from both experiments is 121 with 69 runs collected above the pine forest.

22.4   Results and Discussion

The results and discussion section is divided into two parts. We first present measured wavelet spectra and co-spectra to identify energetic scales and

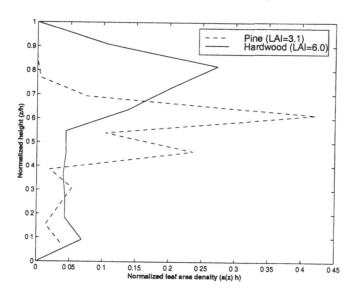

FIGURE 3. Normalized leaf area density variations for pine and hardwood forests as a function of normalized height. The canopy heights and leaf area index are 14 meters and 3.1 for the pine forest and 33 meters and 6 for the hardwood forest.

investigate their consistency with HST predictions. These calculations can (i) indirectly confirm signatures of active turbulence, (ii) assess the scales influenced by fine scale turbulence (which are high dimensional), and (iii) investigate similarity in eddy sizes responsible for momentum and scalar transport. The latter is necessary since much of the derivations by Raupach et al. (1996) considered only momentum transport. In the second part, the active turbulence time series is constructed via Lorentz wavelet thresholding. Using this constructed time series, its energy and flux transporting properties are discussed.

22.4.1 Wavelet Energy, Power Spectra and Cospectra

In Figure 4, the dimensionless wavelet power spectra of u ($=P_u$), w ($=P_w$), c ($=P_c$, $c=q$ as an example) are shown along with the -1 and $-5/3$ power-laws for both forest stands, respectively. The wavelet spectra are normalized with characteristic velocity, scalar, and length scales corresponding to u_*, c_* ($=< wc > /u_*$), and h. The wavenumber, computed from Taylor's (1938) frozen turbulence hypothesis and adjusted by 1.8 ($=U_c/< U >$) as discussed by Shaw et al. (1995) and Raupach et al. (1996), is also normalized by h. Notice from Figure 4 that for $Kh < 1$, a -1 power-law is evident in P_u but not P_w. As discussed in Katul et al. (1996a) and Katul and Chu (1998), a -1 power-law in the u spectrum at low wavenumber is a signature of inactive eddy motion. Therefore, the spectra in Figure 4 suggest that inactive eddy motion contributes to σ_u^2 but not σ_w^2 at $z = h$ and for both stands. While the -1 power-law reflects inactive eddy motion (which scales with boundary-layer height), the -1 power-law in Figure 4 does not extend beyond $Kh < 0.003$ because of buoyancy effects. Evidence of -1 power laws in the water vapor concentration power spectra also suggest that inactive eddy motion may contribute to scalar variances (see Kader and Yaglom, 1991 for analogous argument in the atmospheric surface layer). At $Kh > 2$, all power spectra exhibit a $-5/3$ power-law in accordance with Kolmogorov's scaling for the inertial subrange (Kaimal and Finnigan, 1994). The wavelet power cospectra of the normalized uw ($=P_{uw}$), wT ($=P_{wT}$), and wq ($=P_{wq}$) are shown for all runs. Notice that for $Kh > 2$, an approximate $-7/3$ power-law (regression slope$=-2.08$), typically observed in the inertial subrange (fine scales) of surface layer turbulence is also present for nearly 3 decades. Hence, much of the wavelet coefficients are associated with fine scale turbulence (i.e. high dimensional) at $z/h=1$ and contribute little to the overall mass and momentum exchange.

22.4.2   Energetic Scales of Active Turbulence

In Figure 5, define E_i ($= K_m P_i$, $i = u, w, q, wu, wt, wq$) for vertical velocity (top left), heat (bottom left), momentum (top right), and water vapor (bottom right) to identify energetic scales characterizing vertical velocity and scalar/momentum flux transporting eddies. From Figure 5, all spectral and cospectral peaks occur at about $Kh = 1.5$ in agreement with HST predictions for both canopy stands. As discussed in Raupach et al. (1996), the peak frequency (f_{wp}) in the w spectrum occurs at $f_{wp}=0.45 < U > /h$ (for a generic canopy). For $Kh=1.5$, the peak frequency in the flux wavelet cospectra of Figure 5 is $1.5[1.8/(2\pi)] < U > /h$ (i.e. $=0.43 < U > /h$), and is in excellent agreement with Raupach et al. (1996) and Irvine and Brunet's (1996) data.

22.4.3   Wavelet thresholding and Active Turbulence

Having demonstrated the strong signature of active turbulence in the w time series, we consider whether Lorentz wavelet thresholding is capable of extracting the low dimensional component from the w time series. An independent test that the thresholded w time series reconstructs active turbulence for each run must be devised. Since active turbulence is responsible for much of σ_w^2 and $< wc >$ ($c = T, q, u$), then the reconstructed vertical velocity time series (w^f) variance and its covariance with a scalar (i.e. $< w^f c >$) must be preserved. We tested this hypothesis by:

- Wavelet transforming the original vertical velocity time series for each run ($=WC_w$).

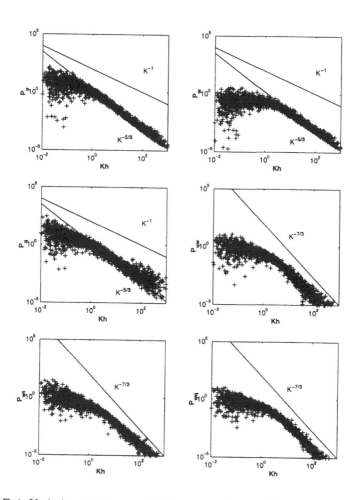

FIGURE 4. Variations with normalized wave numbers of Haar wavelet spectra and co-spectra for all measurement runs at both stands. Top panels are for the longitudinal and vertical velocities, Middle Left is for water vapor, Middle Right and Bottom panels are for longitudinal velocity, temperature, and water vapor cospectra with vertical velocity, respectively.

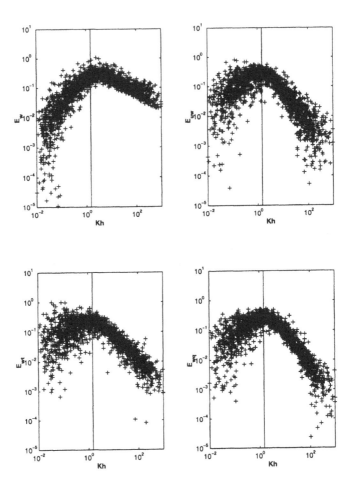

FIGURE 5. Variations with normalized wave number of the product of energy spectra and cospectra for w, w (Top Left), u, w (Top Right), w, T (Bottom Left), and w, q (Bottom Right). The vertical line corresponds to $Kh = 1.5$, estimated from hydrodynamic stability theory.

- Lorentz thresholding wavelet coefficients $(=WC_w^{(f)})$.

- Reconstructing the vertical velocity time series from non-zero wavelet coefficients for each run $(=w^f)$.

- Comparing variances of thresholded (w^f) and original (i.e. non-thresholded) vertical velocity time series for each run.

- Comparing covariances between original $< wc >$ and thresholded $< w^f c >$.

The variance and scalar covariance (with water vapor) comparisons of the above steps are shown in Figure 6. From Figure 6, it is evident that few wavelet coefficients reproduce much of the vertical velocity variances and scalar fluxes demonstrating that less than 9% of the vertical velocity wavelet coefficients are associated with active turbulence (D_w for all runs was 0.084 ± 0.02). We found that the mean D_w for all runs is about 0.08 for heat and momentum as well.

22.5   Conclusions

Turbulent velocity and scalar fluctuation measurements at the canopy-atmosphere interface of two morphologically distinct forests were collected and analyzed. We demonstrated that orthonormal wavelet transformation tend to concentrate energy in view wavelet coefficients. These "energetic" wavelet coefficients describe the primary momentum and scalar flux-transporting events at the canopy-atmosphere interface. Wavelet energy cospectra revealed similarity in eddy sizes responsible for much of the scalar and momentum covariances at dimensionless wavenumber $Kh=1.5$ for both stands. We showed that these co-spectral wavenumbers are associated with energy containing eddies of vertical velocity perturbations in agreement with the mixing layer analogy.

Acknowledgments: The authors would like to thank Joselyn Fenstermacher and Debbie Engel for their help in setting up the instruments at the hardwood site in 1996, Judd Edeburn and the Duke Forest staff for their general assistance at the Duke Forest, David Ellsworth for providing us with the leaf area index measurements, and George Hendrey for his support during the 1995-1996 Pine Forest Experiment. This project was funded, in part, by the U.S. Department of Energy (DOE) through the Southeast Regional Center at the University of Alabama, Tuscaloosa (DOE Cooperative agreement No. DE-FC03-90ER61010) and through the FACE-FACTS project (DE- FG05-95ER62083), the National Science Foundation (NSF Grants No. BIR-12333 and DMS- 9626159), and the U.S. Environmental Protection Agency (EPA Grant No. 91-0074-94).

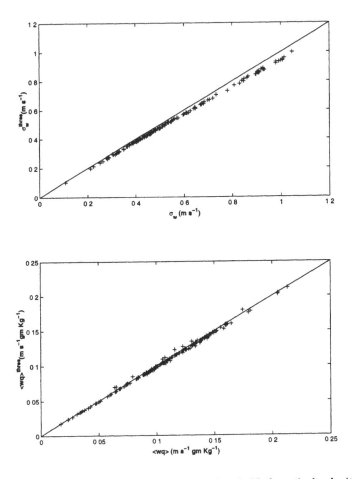

FIGURE 6. Comparisons between original and thresholded vertical velocity signals. Top panel is for vertical velocity standard deviation, and Bottom panel is for water vapor flux (water vapor signal not thresholded).

References

Amiro, B.D., 1990: Comparison of turbulence statistics within three Boreal forest canopies, *Bound. Layer Meteorol.*,**51**, 99-121.

Amiro, B.D., and P.A. Davis, 1988: Statistics of atmospheric turbulence within a natural black spruce forest canopy, *Bound. Layer Meteorol.*, **44**, 267-283.

Baldocchi, D., and T.P. Meyers, 1989: The effects of extreme turbulent events of the estimation of aerodynamic variables in a deciduous forest canopy, *Agric. Forest Meteorol.*, **48**, 117-134.

Baldocchi, D., and T.D. Meyers, 1988: Turbulence structure in a deciduous forest, *Bound. Layer Meteorol.*, **43**, 345-364.

Bergstrom, H., and U. Hogstrom, 1989: Turbulent exchange above a pine forest II: organized structures, Part II., *Bound. Layer Meteorol.*, **49**, 231-263.

Brunet, Y., and S. Collineau, 1994: Wavelet analysis of diurnal and nocturnal turbulence above a maize crop, in *Wavelets in Geophysics*, E. Foufoula-Georgiou and P. Kumar, Eds., Academic Press, 129-150.

Collineau, S., and Y. Brunet, 1993: Detection of turbulent coherent motion in a forest canopy: wavelet analysis, *Bound. Layer Meteorol.*, **65**, 357-379.

Conklin, P., 1994: *Turbulent wind, temperature, and pressure in a mature hardwood canopy*, Ph. D Dissertation, School of the Environment, Duke University, 105 pp.

Ellsworth, D.S., R. Oren, C. Huang, N. Phillips, and G. Hendrey, 1995: Leaf and canopy responses to elevated CO_2 in a pine forest under free-air CO_2 enrichment, *Oecologia*, **104**, 139- 146.

Farge, M., E. Goirand, Y. Meyer, F. Pascal, and M.V. Wickerhauser, 1992: Improved predictability of two-dimensional turbulent flows using wavelet packet compression, *Fluid Dyn. Res.*, **10**, 229-250.

Gao, W., amd B.L. Li, 1993: Wavelet analysis of coherent structures at the atmosphere-forest interface, *J. Appl. Meteorol.*, **32**, 1717-1725.

Gao, W., R.H. Shaw, and K.T. Paw U, 1992: Conditional analysis of temperature and humidity microfronts and ejection/sweep motions within and above deciduous forest, *Bound. Layer Meteorol.*,**59**, 35-57.

Hayashi, T., 1994: An analysis of wind velocity fluctuations in the atmospheric surface layer using an orthonormal wavelet transform, *Bound. Layer Meteorol.*, **70**, 307-326.

Howell, J.F. and L. Mahrt, 1994: An adaptive multiresolution data filter: Applications to turbulence and climatic time series, *J. Atmos. Sci.*, **51**, 2165-2178.

Irvine, M.R., and Y. Brunet, 1996: Wavelet analysis of coherent eddies in the vicinity of several vegetation canopies, *Phys. Chem. Earth*, **21**, 161-165.

Kader, B.A., and A.M. Yaglom, 1991: Spectra and correlation functions of surface layer atmospheric turbulence in unstable thermal stratification, in *Turbulence and Coherent Structures*, ed. O. Metais and M. Lesieur, Kluwer Academic Press, pp. 388-412.

Kaimal, J.C., and J.J. Finnigan, 1994: *Atmospheric Boundary Layer Flows: Their Structure and Measurements*, Oxford University Press, 289 pp.

Katul, G.G., C.D. Geron, C.I. Hsieh, B. Vidakovic, and A. Guenther, 1998a: Active turbulence and scalar transport near the forest-atmosphere interface", *J. Appl. Meteorol.*, to appear in December.

Katul, G.G., J. Schieldge, C.I. Hsieh, and B. Vidakovic, 1998b: Skin temperature perturbations induced by surface layer turbulence above a grass surface, *Water Resour. Res*, **34**, 1265-1274.

Katul, G.G., and B. Vidakovic, 1998: Identification of low-dimensional energy containing/flux transporting eddy motion in the atmospheric surface layer using wavelet thresholding methods, *J. Atmos. Sci*, **55**, 377-389.

Katul, G.G., and C.R. Chu, 1998: A theoretical and experimental investigation of the energy-containing scales in the dynamic sublayer of boundary-layer flows, *Bound. Layer Meteorol.*, **86**, 279-312.

Katul, G.G., R. Oren, D. Ellsworth, C.I. Hsieh, and N. Phillips, 1997a: A Lagrangian dispersion model for predicting CO_2 sources, sinks, and fluxes in a uniform loblolly pine stand, *J. Geophys. Res.*, **102**, 9309-9321.

Katul, G.G., C.I. Hsieh, G. Kuhn, D. Ellsworth, and D. Nie, 1997b: Turbulent eddy motion at the forest-atmosphere interface, *J. Geophys. Res.*, **102**, 13409-13421.

Katul, G.G., J.D. Albertson, C.I. Hsieh, P.S. Conklin, J.T. Sigmon, M.B. Parlange, and K.K. Knoerr, 1996a: The inactive eddy motion and the large-scale turbulent pressure fluctuations in the dynamic sublayer, *J. Atmos. Sci.*, **153**, 2512-2524.

Katul, G.G., C.I. Hsieh, R. Oren, D. Ellsworth, and N. Phillips, 1996b: Latent and sensible heat flux predictions from a uniform pine forest using surface renewal and flux variance methods, *Bound. Layer Meteorol.*, **80**, 249-282.

Katul, G.G., and B. Vidakovic, 1996: The partitioning of the attached and detached eddy motion in the atmospheric surface layer using Lorentz wavelet filtering, *Bound. Layer Meteorol.*, **77**, 153- 172.

Katul, G.G., and M.B. Parlange, 1995: The spatial structure of turbulence at production wavenumbers using orthonormal wavelets, *Bound. Layer Meteorol.*, **77**, 153-172.

Katul, G.G., and M.B. Parlange, 1994: On the active role of temperature in surface layer turbulence, *J. Atmos. Sci.*, **51**, 2181-2195.

Katul, G.G., M.B. Parlange, and C.R. Chu, 1994: Intermittency, local isotropy, and non-Gaussian statistics in atmospheric surface layer turbulence, *Phys. Fluids*, **6**, 2480-2492.

Lu, C.H., and D.R. Fitzjarald, 1994: Seasonal and diurnal variations of coherent structures over a deciduous forest, *Bound. Layer Meteorol.*, **69**, 43-69.

Mahrt, L., and J. Howell, 1994: The influence of coherent structures and microfronts on scaling laws using global and local transforms, *J. Fluid Mech.*, **260**, 247-270.

Mahrt, L., 1991: Eddy asymmetry in the sheared heated boundary layer, *J. Atmos. Sci.*, **48**, 472- 492.

Meneveau, C., 1991: Analysis of turbulence in the orthonormal representation, *J. Fluid Mech.*, **232**, 469-520.

Moritz, E., 1989: Heat and momentum transport in an oak forest canopy, *Bound. Layer Meteorol.*, **49**, 317-329.

Mulhearn, P.J., and J.J. Finnigan, 1978: Turbulent flow over a very rough, random surface, *Bound. Layer Meteorol.*, **15**, 109-132.

Paw U, K.T., Qiu, J., Su, H.B., Watanabe, T., and Y. Brunet, 1995: Surface renewal analysis: A new method to obtain scalar fluxes, *Agric. Forest Meteorol.*, **74**, 119-137.

Paw U, K.T., Y. Brunet, S. Collineau, R.H. Shaw, T. Maitani, J. Qiu, and L. Hipps, 1992: On coherent structures in turbulence above and within agricultural plant canopies, *Agric. Forest. Meteorol.*, **61**, 55-68.

Qiu, J., K.T. Paw U., and R.H. Shaw, 1995: Pseudo-wavelet analysis of turbulent patterns in three vegetation layers, *Bound. Layer Meteorol.*, **72**, 177-204.

Raupach, M.R., R.A. Antonia, and S. Rajagopalan, 1991: Rough-wall turbulent boundary layers, *Appl. Mech. Rev.*, **44**, 1-25.

Raupach, M.R., J.J. Finnigan, and Y. Brunet, 1996: Coherent eddies and turbulence in vegetation canopies: the mixing layer analogy, *Bound. Layer Meteorol.*, **78**, 351-382.

Shaw, R.H., K.T. Paw U, and W. Gao, 1989: Detection of temperature ramps and flow structures at a deciduous forest site, it Agric. For. Meteorol., bf 47, 123-138.

Shaw, R.H., Y. Brunet, J.J. Finnigan, and M.R. Raupach, 1995: A wind tunnel study of air flow in waving wheat: two-point velocity statistics, *Bound. Layer Meteorol.*, **76**, 349-376.

Szilagyi, J., G.G. Katul, M.B. Parlange, J.D. Albertson, and A.T. Cahill, 1996: The local effect of intermittency on the inertial subrange energy spectrum of atmospheric surface layer, *Bound. Layer Meteorol.*, **79**, 35-50.

Taylor, G.I., 1938: The spectrum of turbulence. *Proc. Roy. Soc. London*,bf A, 164, 476-490.

Tennekes, H., and J.L. Lumley, 1972: *A First Course in Turbulence*, MIT Press, 300 pp.

Thompson, N., 1979: Turbulence measurements above a pine forest, *Bound. Layer Meteorol.*, **16**, 293-310.

Townsend, A.A., 1976: *The Structure of Turbulent Shear Flow*, Cambridge University Press, 428 pp.

Townsend, A.A., 1961: Equilibrium layers and wall turbulence, *J. Fluid Mech.*, **11**, 97-120.

Turner, B.J., and M.Y. Leclerc, 1994: Conditional sampling method of coherent structures in atmospheric turbulence using wavelet transform, *J. Atmos. Ocean. Tech.*, **11**, 205-209.

Turner, B.J., M.Y. Leclerc, M. Gauthier, K.E. Moore, and D.R. Fitzjarald, 1994: Identification of turbulence structures above a forest canopy using a wavelet transform, *J. Geophys. Res.*, **99**, 1919-1926.

Vidakovic, B., 1995: Unbalancing data with wavelet transformations, *Wavelet Applications in Signal and Image Processing III*, San Diego, *CA SPIE 2569* vol 2, A. Laine and M. Unser, Eds. 845-857.

Wilson, J.D., 1989: Turbulent transport within the plant canopy, *Estimation of Areal Evapotranspiration*, ed. T.A. Black, D.L. Spittlehouse, M.D. Novak, and D.T. Price, *IAHS Publication* No. 177, pp.43-80.

Yee, E., R. Chan, P.R. Kosteniuk, C.A. Biltoft, and J.F. Bowers, 1996: Multiscaling properties of concentration fluctuations in dispersing plumes revealed using an orthonormal wavelet decomposition, *Bound. Layer Meteorol.*,**77**, 173-207.

23

Latent Structure Analyses of Turbulence Data Using Wavelets and Time Series Decompositions

Omar Aguilar

ABSTRACT In some applications, a decomposition of a time series into a low frequency (signal) and a high frequency (noise) components is of key interest. One illustrative example is Turbulence data, Katul and Vidakovic (1996), where the structure of turbulent eddy motion in the atmospheric surface layer plays a central role in the transport mechanism of heat, mass, and momentum from ground into the atmosphere. Organized and coherent eddies, usually called attached eddies, are responsible for much of the heat and momentum transfer in boundary layer flows. These eddies are surrounded by a fluid that contains fine-scale eddies, called detached eddies, that follow Kolmogorov's theory (1941) **K41**. We explore the decomposition into attached and detached eddy motion of turbulence data using Lorentz thresholding criterion for wavelet coefficients dealing with the problem of locality and non-periodicity of the organized events. As an alternative approach, a time series decomposition is developed to find latent cyclical components in the series and decompose it after fitting higher order auto-regressions, West (1995, 1997). The time series is decomposed into the sum of time-varying components corresponding to the auto-regressive roots. In this way it is possible to isolate sub-components of the series and their contribution to the decomposition. Validations of the results are presented for both methods in order to test the consistency with Townsend's (1976) attached eddy hypothesis.

23.1   Introduction

In some applications, it is important to decompose a time series into a low frequency (signal) and a high frequency (noise) components. One illustrative example is Turbulence data, Katul and Vidakovic (1996), where the decomposition is of key interest. The structure of turbulent eddy motion in the atmospheric surface layer plays a central role in the transport mechanics of heat, mass, and momentum from ground into the atmosphere.

Organized and coherent events, usually called **attached eddies**[1], are responsible for much of the heat and momentum transfer in boundary layer flows. These eddies are surrounded by a fluid that contains fine-scale eddies, called **detached eddies**, which do not contribute significantly to the production of turbulent fluxes and kinetic energy. These less-organized eddies are known to be statistically isotropic and follow Kolmogorov's theory (1941) **K41** hereafter. The quantification of the large-scale eddy motions from time series measurements of turbulent flow variables is of key interest to understand land-atmosphere interactions and thus a decomposition of turbulence data into attached and detached eddy motion becomes relevant.

The major problem in separating the attached eddy motion from time series measurements is the locality and non-periodicity of the organized events. The problem is addressed using two different approaches. First, the locality of wavelets in time is used to isolate the scale contributions of events in space. Lorentz thresholding criterion, (Goel and Vidakovic, 1995; Vidakovic, 1995), is used to eliminate wavelet coefficients with small contributions to the total turbulent energy. Second a time series decomposition is developed to explore latent cyclical components in the time series after fitting higher order auto-regressions, West (1995, 1997). The time series is decomposed into the sum of time-varying components corresponding to the auto-regressive roots. In this way it is possible to isolate sub-components of the series and their contribution to the decomposition. Scientific theory in this area, suggests that the results for both methods may be validated if they are consistent with the Townsend (1976) attached eddy hypothesis.

- The wavelet filtered series and the low frequency components in the time series decomposition correspond to the organized and attached eddies and should explain the majority of the variance of heat and momentum from the observed data.

- The difference between the original signal and the attached eddy motion should follow Kolmogorov's **K41** theory. That is, the detached and less organized eddies should be close to the "-5/3 Power Law", $E_x(K) \propto K^{-5/3}$ where $E_x(K)$ is the Fourier power spectral density.

23.2   The Experiment

In 1993, a total of 50,000 measurements of three velocity components and air temperature were taken over a uniform dry lake bed in Owens valley, California at an elevation of $1,100m$. The momentum roughness length for

---

[1]The name comes from the theory that the mean-flow vorticity and the energy containing turbulent motions are caused by anisotropic coherent eddies attached to the wall, Townsend (1976).

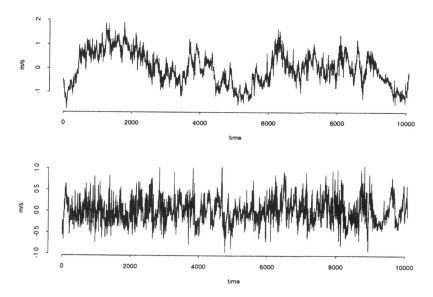

FIGURE 1. Turbulence Data: Velocity components measured over longitudinal direction U (top) and over vertical direction W (bottom).

this sandy surface was $0.13mm$; see Katul (1994) for details. The velocity components were measured in a range of $2m - 3.5m$ above the surface using a 56 Hz triaxal ultrasonic anemometer with a sampling period of 9.75 minutes. The velocity components were rotated to obtain measurements along longitudinal, lateral and vertical velocities. For the purpose of illustration in this paper, an equally spaced sample of 10,000 was taken from the 50,000 observations on two of the velocity components. These samples are presented in Figure 1 for the longitudinal and vertical velocities denoted by U and W respectively.

23.3   Wavelet Decomposition for Time Series.

The discrete wavelet transformation has the property of "disbalancing" the signal by concentrating and preserving the energy of the data in a small number of wavelets coefficients. Therefore, wavelets give parsimonious transformations ensuring that the high frequency features of a signal are described by a relative small number of wavelet coefficients. Based on this principle, Donoho and Johnston (1994) developed the wavelet shrinkage technique with the idea that some detail coefficients might be omitted without affecting the important features of a signal. For instance, additional noise in the signal can be removed by shrinking some low frequency wavelets coefficients towards zero, despite the fact that this noise will indeed affect all wavelet coefficients. The procedure can be summarized in three steps:

1. Transform the observed signal **y** with the Discrete Wavelet Transformation and obtain a set of wavelet coefficients **d**,

2. shrink some or all the wavelet coefficients with a shrinkage function, say $\delta_\lambda(x)$, to obtain a new vector of coefficients $\hat{\mathbf{d}}$ and

3. reconstruct the signal by applying the Inverse Wavelet Transformation to the shrunken coefficients.

There are different shrinkage functions $\delta_\lambda(x)$, of which the simplest case is to remove some of the wavelet coefficients (i.e. thresholding). The most common thresholding rules are **hard** and **soft** shrinkage techniques which replace the coefficients in **d** that are smaller in absolute value than a fixed threshold λ,

$$\delta_\lambda(x) = \left\{ \begin{array}{ll} x & \mathbf{1}\{|x| > \lambda\} \qquad \text{Hard} \\ \text{sign}(x)(|x| - \lambda) & \mathbf{1}\{|x| > \lambda\} \qquad \text{Soft} \end{array} \right.$$

where $\mathbf{1}\{A\}$ is the indicator function on the set A.

A crucial point here is the choice of the threshold parameter λ. Donoho and Johnstone (1994) propose the universal threshold method to set $\lambda = \sqrt{2log(n)}\hat{\sigma}$, where $n = 2^J$ is the number of observations as before and $\hat{\sigma}$ is an estimate of the scale of the noise, traditionally computed as the sample standard deviation of the coefficients at the finest level of detail. The idea is to remove all the wavelet coefficients that are smaller than the expected maximum of an assumed uncorrelated Gaussian noise.

The use of wavelet techniques for stationary processes is an emerging research area that is already impacting theoretical and applied time series analyses. Some key references are Chiann and Morettin (1994), Dijkerman and Mazumdar (1994) and McCoy and Walden (1996). The use of wavelet shrinkage as a de-noising tool to get better estimates of the time series signal and the parameters involved in the model is very appealing. In addition, the data can be decomposed into low frequency and high frequency components via multiresolution analysis resulting in a better understanding of the phenomena. However, the selection of the shrinkage function and the threshold parameter could add misleading information about the correlation structure of the actual signal and hence bad estimates of the parameters could be obtained. This is due to the fact that the correlation structure of the signal interacts with the correlations inherent in the wavelet transformation. Moreover, in almost all classical and non-classical thresholding rules applied to time series data the variances and covariances are not conserved after the reconstruction. The main problems are that, first, in time series data the high frequencies and the low frequencies are interacting by nature; second, a traditional assumption, for example in universal thresholding, is that the fine scales are modeled by uncorrelated Gaussian noise, which is obviously not satisfied in time series.

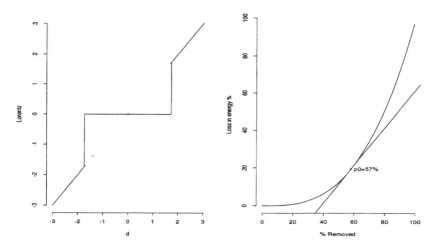

FIGURE 2. Lorentz shrinkage function.

23.3.1   Lorentz curve thresholding.

An alternative distribution-free shrinkage method that takes into account
the autocorrelation structure of the data is required, as stated before. In
time series data, the distribution of energy in the wavelet domain is more
disbalanced than the distribution of energy in the signal. Vidakovic (1995)
proposed a thresholding method based on the Lorentz curve, which is a
graphical representation of distribution inequality and a general measure
of disbalance of energy. This method replaces the low energy coefficients
by zero under the assumption that low energies should come from the
noise component of the data. The idea is to replace the $100 \times p_0\%$ of the
coefficients with the smallest energy with zero, where

$$p_0 = \frac{1}{n} \sum_i \mathbf{1}(d_i^2 \leq \bar{d}^2),$$

with d_i denoting the wavelet coefficients and $\bar{d}^2$ is the mean of the energies
$(d_1^2, d_2^2, \ldots, d_n^2)$.

The value p_0 represents the proportion at which the gain by thresholding
an additional element will be smaller than the loss in the energy, both losses
are measured on a 0-1 scale and are equally weighted. This is illustrated
in Figure 2. The left frame shows the Lorentz thresholding rule applied
to a linear function (dotted line) and the right frame shows the loss in
energy (y-axis) by removing different percentages of wavelet coefficients (x-
axis) with a tangent marked at $p_0 = 57\%$. As can seen in the graphs, the
described thresholding procedure is equivalent to the hard thresholding for
the particular threshold level of $\lambda = \sqrt{\sum_i d_i^2/n}$. This choice of the threshold
parameter will be more adequate for data where the low frequency terms
and the high frequency terms interact a lot.

23.4   Time Series Decomposition.

An AR(p) model $x_t = \sum_{i=1}^{p} \phi_i x_{t-i} + \epsilon_t$ is capable of exhibiting quasi-cyclical behavior at various distinct frequencies as described in West (1995, 1997). This is useful in isolating subcomponents of series with state space representations of certain classes of dynamic linear models, particularly state-space autoregressions.

An important decomposition arises from a state-space representation of the AR(p) model, West and Harrison (1997), namely

$$x_t = \mathbf{F}'\mathbf{z}_t + \nu_t$$
$$\mathbf{z}_t = \mathbf{G}\mathbf{z}_{t-1} + \omega_t$$

with $\mathbf{F} = (1, 0, 0, \ldots, 0)'$ and the $p \times p$ evolution matrix

$$\mathbf{G} = \begin{pmatrix} \phi_1 & \phi_2 & \cdots & \phi_{p-1} & \phi_p \\ 1 & 0 & \cdots & 0 & 0 \\ 0 & 1 & \cdots & 0 & 0 \\ \vdots & & \ddots & & \vdots \\ 0 & 0 & \cdots & 1 & 0 \end{pmatrix},$$

where $\mathbf{z}_t = (x_t, x_{t-1}, \ldots, x_{t-p+1})'$ is the state space vector at time t, ν_t the observation noise following Normal distribution with zero mean and known variance υ_t and $\omega_t = (\epsilon_t, 0, \ldots, 0)'$ the state evolution noise following a Normal distribution with zero mean and known variance $\mathbf{W}_t$. Assuming $\nu_t = 0$ and $\mathbf{W}_t = \mathbf{W}$ $\forall t$, write $\mathbf{G} = \mathbf{E}\mathbf{\Lambda}\mathbf{E}^{-1}$ and note that the eigenvalues are the reciprocal roots α_j, $j = 1, 2, \ldots, p$ of the characteristic polynomial $\phi(B)$. Assume there are q pairs of complex conjugates and $s = p - 2q$ real and distinct eigenvalues. The conditional predictive mean is $E(x_{t+k}|\mathbf{z}_t) = \mathbf{F}'\mathbf{G}^k\mathbf{z}_t$ and evaluating it at $k = 0$ we obtain the direct decomposition,

$$x_t = \sum_{j=1}^{r} y_{tj} + \sum_{j=1}^{c} z_{tj}$$

The c complex components are related to **quasi-periodic ARMA(2,1)** processes z_{tj} with **stochastically varying amplitudes and phases** but **fixed wavelength and moduli**. The r real components are related to **AR(1) processes** y_{tj} with **fixed moduli**. We can order the components according to the estimated periods or moduli of the corresponding roots. Measurement error can be incorporated by fitting higher order autoregressions in higher frequency terms, see West(1995, 1997) and West and Harrison (1997) for further details. More formal extensions for general models to include non-Gaussian observation errors, outliers and possibly non-stationary trend terms can be found in Aguilar (1998, chapter 5).

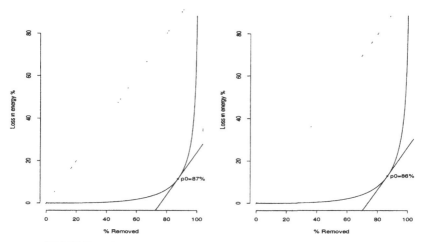

FIGURE 3. Lorentz wavelet thresholding of velocity components.

23.5 Results and Validations.

The two methodologies described above were applied to the two flow variables in the context of Townsend's attached and detached eddy hypothesis. First, a wavelet decomposition of both signals was performed using the "symmlets8" wavelet basis. Thresholding criterion was then used to remove the wavelet coefficients with smallest contributions to the total energy. Lorentz curve criterion was established when the gain in increasing the number of wavelet coefficients and the loss in energy were in balance. Approximately 87.42% and 86.39% wavelet coefficients below this threshold were removed, representing loss of energy of about 11.43% and 12.82% for U and W respectively. Figure 3 shows the Lorentz wavelet thresholding curves for U and W where the diagonal line represents a well balanced signal.

The point, p_0 is over the tangent line parallel to the diagonal representing clearly the break-even point between percentage of coefficients rejected and loss in energy. The reconstruction was performed applying the inverse wavelet transformation to the new set of coefficients. The resulting series represents the attached eddies in each one of the velocity components. Furthermore, the difference between the original signal and the filtered series will be the detached eddy motion part of the series.

The second analysis of the series was done using the time series method. For this kind of data a traditional way to start the analysis is by fitting higher order AR models to approximate what may be lower order ARMA models or non-linear features in the series. In this case, two constant AR(10) models were fitted to each one of the velocity components with traditional reference priors. Figure 4 presents 95% posterior intervals for each one of

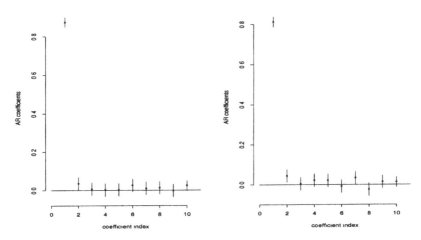

FIGURE 4. 95% posterior intervals for the AR(10) coefficients of the U (left) and W (right) velocity components

the AR coefficients ϕ_j for both flow variables (U left frame and W right frame). As can be seen from the picture, both series show similar patterns having the first coefficient around 0.9 and the rest of the coefficients close to zero. Monte Carlo techniques were used to sample the posterior distribution of the reciprocal roots of the characteristic polynomial. That is, a random sample was drawn from the posterior distribution of the AR coefficients ϕ for each one of the velocity components and the reciprocal roots were computed for each sample. In both cases, two real and four pairs of complex roots were observed. For the longitudinal velocity component U, posterior distributions of the wavelengths/periods λ_j and moduli a_j for the two dominant complex roots are displayed in Figure 5. The posterior means for the two largest periods were 12.06 and 5.57 with corresponding moduli of 0.73 and 0.65. The results were almost identical for the vertical component W and are not displayed here. Both series were then decomposed into the sum of time-varying components corresponding to the autoregressive roots and both decompositions are displayed in Figure 6. The roots were ordered by wavelength (alternatively they could have been ordered by moduli or amplitude). The decomposition was just the sum of six real components, two corresponding to the real roots following AR(1) processes and four corresponding to the sum of the complex conjugates following quasi-cyclical ARMA(2,1) processes. Most of the implied latent components with lower moduli or very high frequencies are introduced to adequately capture the correlation structure in the series but do not represent physical meaningful components. For instance, in this particular application the interest lies only on two components corresponding to the attached and detached eddy motion. In each one of the decompositions of Figure 6, the original series,

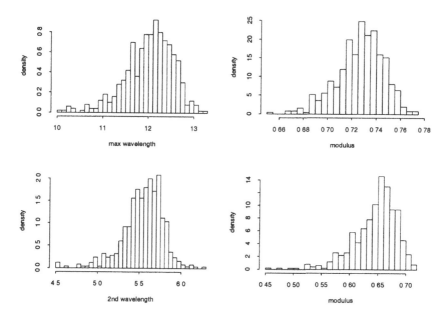

FIGURE 5. Posterior distributions of the wavelength and modulus of the two dominant complex roots for the longitudinal velocity component U.

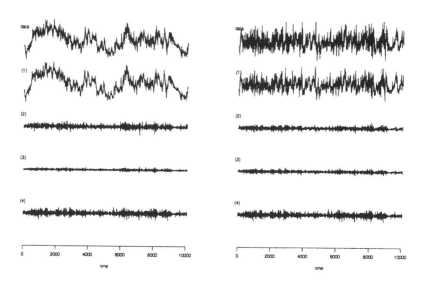

FIGURE 6. Time series decompositions of velocity components, U (left) and W (right).

top row, is decomposed into three components. The dominant component, plotted in the second row, is related to a real root and can be interpreted as the trend of the series. The next component, in the third row, is related to the pair of complex roots with largest wavelength and the last row is the sum of the rest four components. In both cases the most dominant component was that first real root and the rest of the components have smaller periods, smaller moduli and are negligible in amplitude as expected. In the context of turbulence data, the first real component represents the attached eddy motion and the sum of rest of the components the detached eddy motion.

23.5.1   Results and Validations

The top two frames of Figure 7 display the first 250 points of the original signal (dotted line) for the longitudinal velocity component U on the left and the vertical velocity component W on the right. The estimated attached eddies for each flow variable series calculated with the time series decomposition is overlaid correspondingly. The bottom two frames show the same comparison using the estimated attached eddies computed using the Lorentz curve thresholding criterion. It is clear that large scale-eddies are well captured in both flow variables and using two different approaches. Note however, that the wavelet thresholding method results are smoother than those from the time series decomposition maybe because of the choice of the wavelet basis. The decompositions found so far need to be validated with Townsend's theory. This theory claims that velocity components can be decomposed into

$$u_i = u_i^a + u_i^d \quad \text{and} \quad w_i = w_i^a + w_i^d,$$

where a and d represent attached and detached eddy motion respectively. Theory also suggests that for the attached eddy motion the following relations should hold:

- $E(U^a) = E(W^a) = 0,$
- $\sigma_U^2 = E((U^a)^2), \quad \sigma_W^2 = E((W^a)^2),$ variance conservation and
- $E(UW) = E(U^aW^a)$ covariance conservation.

Validations for Lorentz thresholding criterion and time series decomposition are presented in Table 1 for variance conservation and Table 2 for covariance conservation.

As can be seen from Tables 1 and 2, in both flow variables about 99% for U and 95% for W of the turbulence variance and fluxes are retained by the estimated attached eddy motion with Lorentz Thresholding criterion. On the other hand, with the proposed time series decomposition about 97% and 93% of the variance is conserved for U and W respectively. The

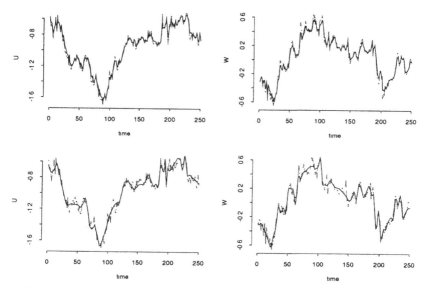

FIGURE 7. Original signal(dashed line) vs estimated attached eddy motion (solid line) calculated with time series decomposition (top frames) and with wavelet thresholding (bottom frames).

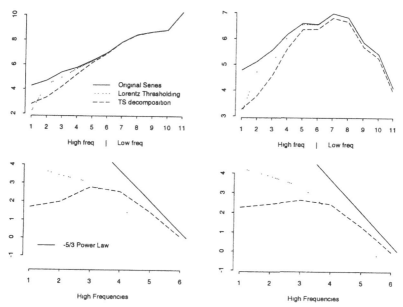

FIGURE 8. Energy at different frequencies (attached eddy motion top frames) and (dettached eddy motion bottom frames)

	U Comp.		W Comp.	
Original	0.4420		0.0741	
Lorentz	0.4382	99.14 %	0.0692	93.39 %
TS Decomp.	0.4284	96.92 %	0.0689	92.98 %

TABLE 1. Variance Conservation

covariance between U and W is preserved in more than 90% with both methodologies.

	Covariance	% Conserved
Original Series (UW)	-0.0230	
Lorentz Thresholding	-0.0224	97.39 %
Time Series Decomp.	-0.0210	91.31 %

TABLE 2. Covariance Conservation (UW)

A different validation procedure is presented in Figure 8. Each frame shows on the horizontal axis the resolution level number in a wavelet decomposition, 1 for the finest level and 11 for the coarsest level. On the vertical axis, $\log_2(\sum d_i^2)$, the logarithm of the energy is plotted for each level. This plot is useful to validate how the original series was reproduced by the estimated attached eddies at different frequencies. The top frames display the frequency composition of the attached eddies compared with the original series for U and W on the left and right frames respectively and for both methodologies. In any case, at low frequencies the differences between the original series and the estimations are negligible in both variables.

The detached eddies were validated according to Townsend's hypothesis as well. Specifically, these fine-scale eddies should follow the frequently called "-5/3 Power Law" suggested in Kolmogorov's **K41** scaling theory:

$$E_x(K) \propto K^{-5/3},$$

where $E_x(K)$ is the Fourier power spectral density and K is the wavenumber. The detached eddies were calculated taking the difference between the original signal and the filtered series for the Lorentz thresholding criterion. In the case of the time series decomposition, the fine-scale eddies were represented by the sum of all but the dominant real component. The bottom frames of Figure 8 illustrate a comparison of the energies of the estimated detached eddies with the $-5/3$ line at high frequencies (wavelet levels 1 through 6). It seems that the fine-scale eddies for both variables follow the $-5/3$ power law consistent with Kolmogorov's **K41** theory.

23.5.2 Conclusions

Two different methodologies have been used to make inferences on latent components that are present in time series data. In the wavelet decomposition, the reconstructed signal after thresholding is the estimation of a latent process and usually conserves most of the energy of the signal. Consequently, the wavelet based-solution is a good candidate for the attached eddy motion part of the velocity components in the turbulence example described above. In addition, the attached eddies are consistent with Townsends's theory and the detached eddies, calculated as the difference between the original signal and the attached eddies, follow Kolmogorov's **K41** theory. Nevertheless, the wavelet solution usually observes smooth features in the reconstruction of the latent process possibly due to the choice of the wavelet basis. Related to this, Katul and Vidakovic (1996) developed an algorithm to find the optimal wavelet basis for turbulence data based on minimizing a relative entropy measure, which allows to maximize the discrimination procedure between organized and less organized eddies.

In a second analysis, a time series decomposition is used to explore and isolate quasi-cyclical components of AR models at different frequencies. The time series is decomposed into time-varying components related to the roots of the characteristic polynomial of the model. The results in the turbulence example are also consistent with Townsend's attached eddy motion theory and the detached eddies follow Kolmogorov's theory as well. The two velocity components analyzed evidence a dominant real root interpreted as the attached eddies having a value very close to one in absolute value. This suggests the possibility of non-stationary models for the data and hence the need of decomposition results of models that allow for non-stationary components as in the case of time-varying autoregressions.

Acknowledgments. I would like to thank Mike West and Brani Vidakovic for their helpful comments, suggestions and support. I also would like to thank Gabriel Katul who kindly provided the data and for helpful discussions.

References.

Aguilar, O. (1998), *Latent structure in Bayesian multivariate time series models.* PhD thesis, Duke University.

Chiann, C. and Morettin, P. (1994), *Wavelet analysis for stationary processes.* Unpublished manuscript, University of Sao Paulo, Brazil.

Dijkerman, R.W. and Mazumdar, R.R. (1994), *Wavelet representations of stochastic processes and multiresolution of stochastic models.* IEEE Transaction and signal processing, 42, 1640-1652.

Donoho, D. and Johnstone, T., (1994), *Bank of wavelet-related reports.* Technical reports, Stanford University.

Goel P. and Vidakovic, B. (1995), *Wavelet transformations as diversity enhancers*. Discussion Paper 95-08, ISDS, Duke University.

Katul, G. (1994), *A model for sensible heat flux probability density function for near-neutral and slightly stable atmospheric flows*. Boundary-Layer Meteorology, 71, 1-20.

Katul, G. and Vidakovic, B. (1996), *The partitioning of attached and detached eddy motion in the atmospheric surface layer using Lorentz wavelet filtering*. Boundary-Layer Meteorology, 77, 153-172.

Kolmogorov, A. N. (1941), *The local structure of turbulence in incompressible viscous fluid for very large Reynolds numbers*. Dokl. Akad. Nauk. SSSR, 30, 301-305.

McCoy, E.J. and Walden, A.T. (1996), *Wavelet analysis and synthesis of stationary long-memory processes*. Journal of Computational and Graphical Statistics, 5, 26-56.

Towsend, A. (1976), *The Structure of Turbulent Shear Flow*. Cambridge University Press, 429 pp.

Vidakovic, B. (1995) *Unbalancing data with wavelet transformations*. SPIE, 2569, 845-857.

West, M. (1995), *Bayesian inference in cyclical components of dynamic linear models*. Journal of the American Statistical Association, 90, 845-857.

West, M. (1997), *Time Series Decomposition*. Biometrika, 84, 489-494.

West, M., & Harrison, P. J., (1997), *Bayesian Forecasting and Dynamic Models*. Second Edition, Springer-Verlag: New York.

Lecture Notes in Statistics

For information about Volumes 1 to 67
please contact Springer-Verlag

Vol 68: M. Taniguchi, Higher Order Asymptotic Theory for Time Series Analysis. viii, 160 pages, 1991.

Vol 69. N.J.D. Nagelkerke, Maximum Likelihood Estimation of Functional Relationships. V, 110 pages, 1992.

Vol. 70. K Iida, Studies on the Optimal Search Plan viii, 130 pages, 1992

Vol. 71 E.M.R.A. Engel, A Road to Randomness in Physical Systems ix, 155 pages, 1992.

Vol 72· J.K Lindsey, The Analysis of Stochastic Processes using GLIM vi, 294 pages, 1992.

Vol. 73· B C. Arnold, E. Castillo, J.-M. Sarabia, Conditionally Specified Distributions. xiii, 151 pages, 1992.

Vol 74. P. Barone, A. Frigessi, M. Piccioni, Stochastic Models, Statistical Methods, and Algorithms in Image Analysis. vi, 258 pages, 1992

Vol 75 P.K Goel, N.S Iyengar (Eds.), Bayesian Analysis in Statistics and Econometrics. xi, 410 pages, 1992.

Vol 76· L. Bondesson, Generalized Gamma Convolutions and Related Classes of Distributions and Densities. viii, 173 pages, 1992

Vol 77· E Mammen, When Does Bootstrap Work? Asymptotic Results and Simulations. vi, 196 pages, 1992

Vol. 78 L. Fahrmeir, B Francis, R Gilchrist, G. Tutz (Eds.), Advances in GLIM and Statistical Modelling Proceedings of the GLIM92 Conference and the 7th International Workshop on Statistical Modelling, Munich, 13-17 July 1992 ix, 225 pages, 1992.

Vol 79. N Schmitz, Optimal Sequentially Planned Decision Procedures xii, 209 pages, 1992

Vol 80· M. Fligner, J Verducci (Eds.), Probability Models and Statistical Analyses for Ranking Data xxii, 306 pages, 1992

Vol. 81 P Spirtes, C Glymour, R. Scheines, Causation, Prediction, and Search. xxiii, 526 pages, 1993

Vol. 82. A. Korostelev and A Tsybakov, Minimax Theory of Image Reconstruction. xii, 268 pages, 1993.

Vol 83· C Gatsonis, J Hodges, R. Kass, N. Singpurwalla (Editors), Case Studies in Bayesian Statistics. xii, 437 pages, 1993

Vol. 84. S Yamada, Pivotal Measures in Statistical Experiments and Sufficiency. vii, 129 pages, 1994.

Vol. 85: P. Doukhan, Mixing: Properties and Examples. xi, 142 pages, 1994

Vol. 86: W. Vach, Logistic Regression with Missing Values in the Covariates. xi, 139 pages, 1994.

Vol. 87: J. Müller, Lectures on Random Voronoi Tessellations.vii, 134 pages, 1994.

Vol. 88: J. E. Kolassa, Series Approximation Methods in Statistics. Second Edition, ix, 183 pages, 1997.

Vol. 89: P. Cheeseman, R.W Oldford (Editors), Selecting Models From Data: AI and Statistics IV xii, 487 pages, 1994.

Vol. 90· A Csenki, Dependability for Systems with a Partitioned State Space: Markov and Semi-Markov Theory and Computational Implementation. x, 241 pages, 1994.

Vol 91: J D. Malley, Statistical Applications of Jordan Algebras. viii, 101 pages, 1994

Vol. 92 M. Eerola, Probabilistic Causality in Longitudinal Studies. vii, 133 pages, 1994

Vol. 93. Bernard Van Cutsem (Editor), Classification and Dissimilarity Analysis xiv, 238 pages, 1994.

Vol. 94. Jane F. Gentleman and G.A. Whitmore (Editors), Case Studies in Data Analysis. viii, 262 pages, 1994

Vol. 95: Shelemyahu Zacks, Stochastic Visibility in Random Fields. x, 175 pages, 1994

Vol 96: Ibrahim Rahimov, Random Sums and Branching Stochastic Processes. viii, 195 pages, 1995

Vol. 97· R Szekli, Stochastic Ordering and Dependence in Applied Probability. viii, 194 pages, 1995

Vol. 98· Philippe Barbe and Patrice Bertail, The Weighted Bootstrap. viii, 230 pages, 1995

Vol 99· C.C. Heyde (Editor), Branching Processes Proceedings of the First World Congress viii, 185 pages, 1995.

Vol. 100 Wlodzimierz Bryc, The Normal Distribution Characterizations with Applications. viii, 139 pages, 1995

Vol 101. H.H. Andersen, M.Højbjerre, D Sørensen, P.S.Eriksen, Linear and Graphical Models· for the Multivariate Complex Normal Distribution. x, 184 pages, 1995

Vol 102: A.M. Mathai, Serge B. Provost, Takesi Hayakawa, Bilinear Forms and Zonal Polynomials x, 378 pages, 1995

Vol. 103. Anestis Antoniadis and Georges Oppenheim (Editors), Wavelets and Statistics. vi, 411 pages, 1995

Vol 104: Gilg U.H. Seeber, Brian J. Francis, Reinhold Hatzinger, Gabriele Steckel-Berger (Editors), Statistical Modelling: 10th International Workshop, Innsbruck, July 10-14th 1995 x, 327 pages, 1995.

Vol. 105. Constantine Gatsonis, James S. Hodges, Robert E Kass, Nozer D. Singpurwalla(Editors), Case Studies in Bayesian Statistics, Volume II x, 354 pages, 1995